AutoCAD®

19th Edition

by Ralph Grabowski

AutoCAD® For Dummies®, 19th Edition

Published by: **John Wiley & Sons, Inc.,** 111 River Street, Hoboken, NJ 07030-5774, www.wiley.com

For general information on our other products and services, please contact our Customer Care Department within the U.S. at 877-762-2974, outside the U.S. at 317-572-3993, or fax 317-572-4002. For technical support, please visit https://hub.wiley.com/community/support/dummies.

Wiley publishes in a variety of print and electronic formats and by print-on-demand. Some material included with standard print versions of this book may not be included in e-books or in print-on-demand. If this book refers to media such as a CD or DVD that is not included in the version you purchased, you may download this material at http://booksupport.wiley.com. For more information about Wiley products, visit www.wiley.com.

Library of Congress Control Number: 2022930466

ISBN 978-1-119-86876-7 (pbk); ISBN 978-1-119-86877-4 (ebk); ISBN 978-1-119-86878-1 (ebk)

SKY10032973_020122

Table of Contents

Introduction

Welcome to the wonderful world of AutoCAD and to the fame and fortune that awaits you as an AutoCAD user. (Would I lie to you?)

Believe it or not, AutoCAD is around 40 years old, having been born in December 1982, when most people thought that personal computers weren't capable of industrial-strength tasks like CAD. The acronym stands for Computer-Aided Drafting, Computer-Aided Design, or both, depending on who you talk to. What's equally surprising is that many of today's hotshot AutoCAD users, and most of the readers of this book, weren't even born when the program first hit the street and when the grizzled old-timer writing these words began using it.

AutoCAD remains the king of the PC computer CAD hill by a tall margin, and, on top of that, is one of the longest-lived computer programs ever. It's conceivable that the long-term future of CAD may belong to special-purpose, 3D, web-connected software such as the Autodesk Fusion and Forge programs. Until then, AutoCAD's DWG file format is the de facto standard, and a lot of design software works with that file format. For the foreseeable future, AutoCAD is where the action in CAD will be.

You may have heard that AutoCAD is complex and therefore difficult to learn and use. Yes, the user interface includes about 1,300 icons. But it has been my observation that the easier any software is to learn and use, the sooner you bump up against its limitations. A simple car with no accelerator, one forward gear, no steering, and no brakes would be easy to use until you reach a hill, a curve, or a stop sign, or you need to back out of a parking space.

Yes, AutoCAD is complex, but that's the secret to its success. Some claim that few people use more than 10 percent of AutoCAD's capabilities. Closer analysis reveals that most people use the same basic 5 percent and everyone else uses a different 5 percent after that. The trick is to find *your* 5 percent, the sweet spot that suits your particular discipline. If you follow my advice, I think you'll find that using AutoCAD is as simple and intuitive as driving a car.

It should be perfectly clear that if your career path has put you in a position where you need to know how to use AutoCAD, you're no dummy!

About This Book

Unlike many other *For Dummies* books, this one often tells you to consult the official software documentation. AutoCAD is just too big and powerful for a single book to attempt to describe it completely. The book that ultimately covers every AutoCAD topic would need a forklift to move it. Literally. Autodesk stopped shipping paper instruction manuals with the software somewhere around 1995, when the full documentation package grew to about a dozen volumes and more than 30 pounds.

In *AutoCAD For Dummies,* I occasionally mention differences from previous releases so that everyone gains some context and so that upgraders can more readily know what has changed; plus, you're bound to encounter a few of the billions and billions of drawings created using methods that predominated in older releases of the software. I mention the important differences between AutoCAD and AutoCAD LT. In particular, AutoCAD LT has no programming language and has extremely limited support for parametrics (see Chapter 19) and 3D (see Chapter 21).

This book is not *Mechanical Drafting For Dummies,* or *Architectural Drafting For Dummies,* or even *Crash Testing For Dummies.* It doesn't cover traditional drafting principles and procedures, but it does cover the AutoCAD commands necessary to create and edit drawings. Remember, though, that knowing AutoCAD's commands won't make you a great designer, just as knowing how to touch-type with a word processor doesn't necessarily make you a great author. The job title *CAD operator* doesn't exist, but almost all drafters and designers use CAD.

In addition, the book does *not* cover the discipline-specific features found in vertical market products that run on top of AutoCAD, such as AutoCAD Electrical or AutoCAD Mechanical, although most of the information in this book applies to the general-purpose features of those programs as well.

This book covers AutoCAD 2015 through 2023. The obvious major differences between these versions and 2014 and earlier are the initial startup screen and the format of the Ribbon. The underlying principles remain the same. I draw your attention to differences where appropriate.

Late in 2010, Autodesk released the first non–Microsoft Windows version of AutoCAD in 20 years. Although AutoCAD for Mac is available, *AutoCAD For Dummies* covers only the Windows version. The two versions are file-compatible but differ in how they look and what they can do. If you have AutoCAD for Mac, you can get a grasp of basic concepts from this book, but you might be better off with a Mac-specific book such as *AutoCAD 2022 For Beginners (For Mac Users),* by CADfolks.

Foolish Assumptions

AutoCAD has a large, loyal, dedicated group of longtime users. If you've been using AutoCAD for a decade or more, if you plan your vacation time around Autodesk University, if you used AutoCAD to create your wedding invitations, if you tell police officers that you can walk a straight line if they will press F8 first, or if you read 1,200-page technical tomes about AutoCAD for pleasure, *AutoCAD For Dummies* is not for you. This book *is* for you if you want to get going quickly with AutoCAD and understand the importance of developing proper CAD techniques right from the beginning.

However, you do need to have an idea of how to use your computer before tackling AutoCAD and this book. And you must have a computer system running AutoCAD or AutoCAD LT (preferably the 2023 version but at least 2015 or newer). A printer or plotter and a connection to the internet are helpful, too.

You also need to know how to use your version of Windows to copy and delete files, create folders, and find files. You need to know how to use a mouse to select (highlight) or to choose (activate) commands, how to close a window, and how to minimize and maximize windows. You should be familiar with these basics of your operating system before you start using AutoCAD.

Conventions Used in This Book

Here are some conventions that you'll run across in this book.

Using the command line

The command line is that gray rectangle near the bottom of the AutoCAD screen that says *Type a command*. One way of using AutoCAD is to type the names of commands in this area. In addition, this is where AutoCAD talks back to you when it needs more information. Examples of AutoCAD prompts appear in this book with a `special typeface`, as does any other text in the book that replicates a message, a word, or text that appears on the screen. Sequences of prompts that appear at the AutoCAD command line have a shaded background in this book, like this:

```
Specify lower left corner or [ON/OFF] <0.0000,0.0000>:
```

When there is a specific action that I want you to take at one of these prompts, look for the italic passage at the end of the line, such as when I want you to press the Enter key on the keyboard:

```
Specify ending width <5.0000>: Press Enter
```

Text that I want you to type into the program at the command line, in a dialog box, in a text box, or elsewhere appears in **boldface** type, like the 3 at the end of the following line.

```
Specify starting width <0.0000>: 3
```

Many figures in this book also show AutoCAD command-line sequences that demonstrate AutoCAD's prompts and sample responses.

Using aliases

Many AutoCAD commands have *aliases* — shortcut versions of commands that have fewer letters than the full command names, in case you like typing commands at the AutoCAD command line. In this book, I show aliases in uppercase as part of the command names. To start a command with an alias, you have to type only that uppercase letters that I show you. For example, to draw a line, type either **Line** (the official command) or **L** (its alias) and then press Enter to execute the command. When I tell you to start a command, I spell it out in full (such as Line, Circle, or COpy), but you need to enter only the letters I show in uppercase (**L**, **C**, or **CO**, respectively). Note also that the uppercase letters aren't always the initial letters nor are they always adjacent. For example, the eXit command can be entered as the full word or as just the letter **X** and DimANgular can be entered as **DAN**.

As you begin to type a command name at the keyboard, the program will try to guess the ones you might want by displaying a list of suggestions. You can click the name you want or keep typing until your choice rises to the top, at which point you simply press Enter or the space bar.

Icons Used in This Book

Throughout this book, I point out certain morsels of particularly important and useful information by placing handy icons in the margin. Naturally, different icons indicate different types of information.

TIP

The Tip icon points to insights that can save you time and trouble as you use AutoCAD. In many cases, Tip information acts as a funnel on AutoCAD's impressive but sometimes overwhelming flexibility: After telling you many of the ways that you *can* do something, I tell you the way that you *should* do it, in most cases.

TECHNICAL STUFF

The Technical Stuff icon points out places where I delve a little more deeply into AutoCAD's inner workings or point out information that most people don't need to know most of the time. These paragraphs definitely are not required reading, so if you see one at a point when you've reached your techie-detail threshold, feel free to skip it.

WARNING

The Warning icon tells you how to stay out of trouble when working close to the edge. Failure to heed its messages may have unpleasant consequences for you or your drawing — or both.

REMEMBER

The Remember icon knows that you have a lot to remember when you're using AutoCAD, so I've remembered to remind you not to forget about some of those things that you should remember.

Beyond the Book

I have written a lot of extra content that you won't find in this book. Go online to find

- **AutoCAD drawings:** Drawings that you can use with this book are at `www.dummies.com/go/autocadfd19`. The drawings, which are on the Downloads tab, are in Zip format; download and unzip them to a folder, and they'll be ready to open in AutoCAD. The Zip files are named according to chapter and contain one or more drawing files. For example, `afd03.zip` contains the drawings for Chapter 3. Note that not all chapters have drawing files associated with them.
- **Cheat sheet:** The cheat sheet for this book has a roadmap for setting up new drawings, as well as a list of keyboard shortcuts. To get to the cheat sheet, go to `www.dummies.com` and type *AutoCAD For Dummies* in the Search box.
- **Updates:** If I have any updates to the book, you can find them at `www.dummies.com/extras/autocad19`.

Where to Go from Here

Because you're reading this Introduction, you are like me — you like to read. (The cut-to-the-chase people tend to flip to the index right away and look up what they need to know at that instant.) If you're a total AutoCAD newbie, you might want to read this book in order, from front to back; it follows a straightforward route from setting up the drawing environment to outputting your masterworks on paper to sharing your work with others.

If you're an experienced user, you'll probably be an index-flipper who looks for the missing information needed to complete a specific task. You can probably find the index on your own, but I encourage you to browse through this book anyway, with a highlighter or sticky notes in hand, so that you can find those particularly important places when you need them again.

Whichever route you choose, I hope that you enjoy your time with *AutoCAD For Dummies*. A-a-and, you're off!

1 Getting Started with AutoCAD

IN THIS PART . . .

Find your way around the AutoCAD screen.

Discover the half-dozen commands that will account for much of your AutoCAD activities.

See the big picture: from how to start a drawing to printing it.

Correctly set up AutoCAD and create templates for drawing success.

Navigate through your drawing by panning and zooming.

» Creating your first drawing in AutoCAD

» Seeing the complete picture

» Understanding the difference between pixels and vectors

» Comprehending the Cartesian coordinate system

Chapter 1
Introducing AutoCAD and AutoCAD LT

This chapter helps ease you into using AutoCAD to create engineering drawings. Although it's not uncommon to feel overwhelmed the first time you see AutoCAD, rest assured that you don't need to learn all the controls that you see in the default environment to be an efficient user of the program.

After a brief introduction to the program, I take you through an exercise to show you just how easy it can be to use AutoCAD. The exercise is followed up with some key concepts that you should understand when using AutoCAD, including how it differs from most other computer applications.

REMEMBER

When you're starting out with AutoCAD, heed this quote from *The Hitchhiker's Guide to the Galaxy:*

Don't panic!

Launching AutoCAD

The first thing you need to do to start using AutoCAD is to launch the AutoCAD program (well, duh!) and, if necessary, maximize its screen display. AutoCAD has so many tools and palettes that you'll almost always want to use it in full-screen mode. Follow these steps:

1. **Launch AutoCAD.**

   As indicated in the Introduction, I assume that you have a working knowledge of how to use your version of Windows, including how to launch applications. Depending on your version of Windows and how it is set up, you might have to double-click a desktop icon or find a suitable entry in the Start→[All] Programs menu or Start→All Apps menu on the start screen. The wording of the selections varies depending on the version of AutoCAD and Windows.

2. **Start a new drawing.**

   Click the rectangular New button towards the upper-left corner of the screen.

3. **If necessary, expand AutoCAD to full-screen mode.**

   Click the middle Windows button in the upper-right corner of the application window.

4. **If necessary, expand the graphic area (the big, gray area in the middle) to full-screen size.**

   Click the middle button in the upper-right corner, near the compass rosette.

5. **Place the cursor in the gray graphics area (midscreen), and then press the Esc key twice to make sure that no commands are active.**

   Now you're ready to start drawing in AutoCAD, as shown in Figure 1-1.

Your screen may look a little different from Figure 1-1 depending on your version of AutoCAD and Windows and your screen resolution. Note too that although you'll draw using white on dark gray (refer to Figure 1-1), I drew using black on white (see Figure 1-2), and my menu icons have a white background compared to your gray background. I made the color change so that the figures would be clearer on the printed page.

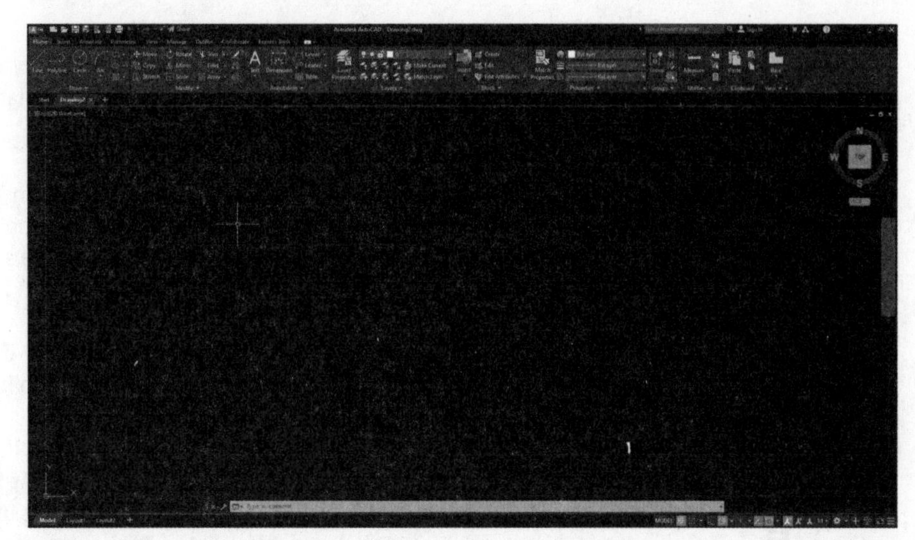

FIGURE 1-1: Your AutoCAD, ready to draw!

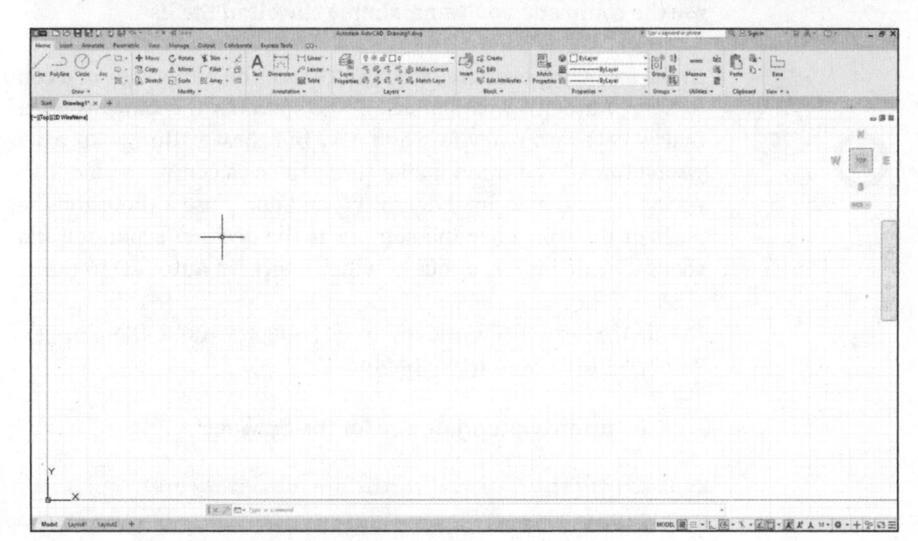

FIGURE 1-2: My AutoCAD, ready to draw!

Drawing in AutoCAD

AutoCAD offers a wide range of commands to create, modify, and annotate 2D and 3D designs. Don't feel as though you need to learn and master every one of the approximately 1,300 (and counting) commands and options that AutoCAD offers to be a proficient drafter; most veteran drafters probably use only 20 or so commands for most basic drafting tasks.

The following simple exercise introduces a few of the commonly used commands to establish the size of your drawing area as well as the commands for creating straight line segments and circles.

You can start a command by clicking its button on the Ribbon menu, which is across the top of the screen, or by entering the command's name in the *command line*, which is the light-gray text-entry area at the bottom of the screen that reads *Type a command*.

In this exercise and others in this book, AutoCAD's command line entries `look like this`, and you type the commands and responses shown in **bold**. Press Enter or the spacebar after each command or response that you type.

TIP

You don't even have to move the cursor to the command line. As you type, AutoCAD tries to guess which command you want and displays a list of possibilities at the command line, even when the cursor is on the Ribbon menu area. When you see the command you want, simply click it in the list.

WARNING

In the following exercise, don't add spaces on either side of a comma! In most situations, AutoCAD treats pressing the spacebar the same as pressing Enter, which makes keyboard entry fast and easy but messes things up when you do it at the wrong time. In addition, make sure you use a comma as the X,Y separator and the period (.) as the decimal delimiter, and don't use a thousands separator (,). Some parts of the world use the comma as the decimal separator and the space as the thousands delimiter, either of which confuses AutoCAD to no end.

In this first exercise I ask you to do things without explaining why. Trust me; all will become clear in later chapters:

1. **Set up an appropriate size for the drawing:**

```
LIMITS
Reset Model space limits:
Specify lower left corner or [ON/OFF] <0.0000,0.0000>: 0,0
Specify upper right corner <12.0000,9.0000>: 60,40
```

   Now, to be able to see the entire drawing area, type the letters **Z A** and press Enter. Note that there must be a space between the Z and the A.

2. **Disable Dynamic Input mode to work with the command line:**

```
DYNMODE
Enter new value for DYNMODE <3>: -3
```

3. **Draw the frame:**

```
Line
Specify first point: 26,12
Specify next point or [Undo]: 13,12
Specify next point or [Undo]: 22,24
Specify next point or [Close Undo]: 40.5,24
Specify next point or [Close Undo]: 41,22
Specify next point or [Close Undo]: 26,12
Specify next point or [Close Undo]: 20.6667,28
Specify next point or [Close Undo]: 25,28
Specify next point or [Close Undo]: Enter
```

4. **Draw a bit more:**

```
Line
Specify first point: 45,12
Specify next point or [Undo]: 42.87,14.53
Specify next point or [Undo]: 39.38,28.5
Specify next point or [Close Undo]: 35.3,30
Specify next point or [Close Undo]: Enter
```

5. **Draw a round thing:**

```
Circle
Specify center point for circle or [3P 2P Ttr (tan tan
  radius)]: 13,12
Specify radius of circle or [Diameter]: 8
```

6. **Draw another round thing:**

```
Circle
Specify center point for circle or [3P 2P Ttr (tan tan
  radius)]: 45,12
Specify radius of circle or [Diameter]: 8
```

Figure 1-3 shows the bicycle you've drawn, and you didn't even need training wheels!

It has been claimed that Line and Circle are the second- and third-most-used commands after UNDO. You should now SAVE your drawing as an historic artifact. That was easy, wasn't it?

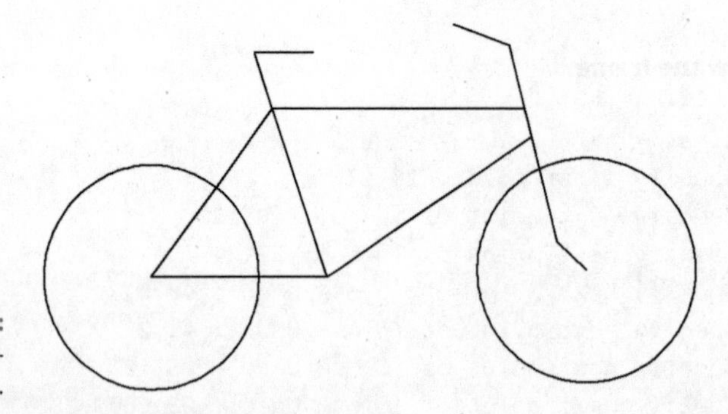

FIGURE 1-3: Your first AutoCAD drawing.

Understanding Pixels and Vectors

To use AutoCAD effectively (or even at all) you need to understand how an image is displayed on your computer screen, and how the image is stored when it is not being displayed.

- **An image on a computer screen is made up of pixels.** When you look closely at the screen with a strong magnifying glass, you'll see that the image is formed from a large number of small dots of light, as shown in Figure 1-4, called *pixels.* This has nothing to do with Tinker Bell, except that an onscreen image of her would indeed be made up of pixels.

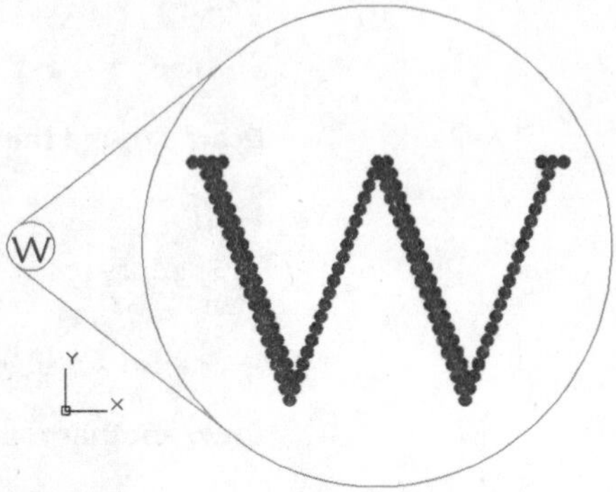

FIGURE 1-4: Pixels.

- **All programs that display a graphic image simply turn on or off suitable spots to build the picture.** This is a *raster* image. A straight line in a raster image is just a fortuitous alignment of appropriate dots, and after it's been created, it can't be edited as a single object.
- **A major difference between CAD programs and computer graphics programs (such as Microsoft Paint) lies in how they save the image to disk.** When the image from a Paint-type program is saved to disk, it's stored as a bitmap that simply lists the color of each pixel. It's simply a snapshot of what you see onscreen.
- **All CAD programs work with and store on a vector file on disk.** A *vector file* is a big collection of numbers and words that list the type, size, and location of every entity in the drawing. When a CAD program displays your drawing

onscreen, it analyzes the vector data and calculates which pixels to turn on or off, depending on which portion of the drawing you're viewing. CAD programs understand that a circle is a closed curve with a center point and a constant radius. If you change its radius, the CAD program redraws the image onscreen to show the new size.

- **AutoCAD doesn't limit you to working only with what you can see onscreen.** You can include as much detail in a drawing as needed. You can zoom in to see more detail and zoom out to see the big picture. At any time, the screen shows only those entities and their detail that the screen is capable of showing.

  Some screens can show more pixels than others can. The number ranges from the 320 per row by 200 rows (320 x 200) of the very old Color Graphics Adapter (CGA) of the 1980s to 3840 x 2160 and beyond for today's 4K monitors. However, the drawing file always contains the same information. If it were moved to a computer with a higher resolution graphics adapter and monitor, then greater detail would show without you having to zoom in as far. Conversely, a drawing file moved to a computer with a lower screen resolution does not lose any detail, but you'll need to zoom in more closely to see details clearly.

  How big is "the big picture"? AutoCAD can draw a circle with a radius of 10^{99} (a 1 followed by 99 zeros) units, but the observable part of the universe is "only" about 5 x 10^{23} miles in diameter, depending on how you measure and whose numbers you use (subject to change without notice). Check out `en.wikipedia.org/wiki/Observable_universe` for the latest number.

  Conversely, AutoCAD can draw a circle with a with a radius as small as 10^{-99} (which equals 0.00000[plus 90 more zeros plus]0001) units in diameter, as opposed to the classical radius of an electron, which is positively huge at 2.8179403267 x 10^{-13} cm.

- **It's possible for a drawing file to contain much more than you can see at any one time.** The computer screen is not really the drawing; it is just a viewer that lets you look at all or part of the drawing file.

The Cartesian Coordinate System

AutoCAD uses the Cartesian coordinate system to define all locations in the drawing. This includes things such as the starting and ending points of lines, the centers of circles, the locations of text notes, and so on. Cartesian coordinates are named for French philosopher René Descartes, who is famous for statement "I think, therefore I am," although today he might say, "I tweet, therefore I am" — although tweeting doesn't always involve thinking.

In his *Discourse on Method,* Descartes, wearing his mathematician's hat, came up with the idea of locating any point on a planar surface by measuring its distance from the intersection of a pair of axes (called, by convention, the X-axis and the Y-axis). (That's *axes* as in more than one axis, not several tools for chopping wood.) By convention, the intersection of these axes are perpendicular to one another, and their intersection point is identified as 0,0 — or the origin.

For example, if your address is 625 East 18th Street in a typical town, you live 6¼ blocks east of First Avenue and 18 blocks north of Main Street.

AutoCAD also uses the notation that the origin is at point 0,0. Positive values are to the right of and above this point, and negative values are to the left of and below it. You can identify any location on a drawing by its horizontal distance from the origin, followed by its vertical distance from the same starting point.

AutoCAD shows Cartesian coordinates as a pair of numbers separated by a comma. The number to the left of the comma is the X (horizontal) coordinate, and the value to the right is the Y (vertical) coordinate. You used this convention when creating your bicycle drawing. When working in three dimensions (see Chapter 21), AutoCAD adds a third coordinate: Z.

It's worth repeating my earlier warning: Make sure you use a comma as the X,Y separator and the period (.) as the decimal delimiter, and don't use a thousands separator.

» Going bar-hopping: Title bars and the status bar

» Unraveling the Ribbon

» Practicing with palettes

» Discovering the drawing area

» Using online help

Chapter 2
The Grand Tour of AutoCAD

Over the years, AutoCAD's interface has undergone many changes, starting with a simple text menu down the right side (still the second-fastest way of using AutoCAD) and then progressing to drop-down menus, toolbars, a Dashboard (which only survived two releases — 2007 and 2008), tool palettes, and, for now, the Ribbon menu.

Like the rest of this book, this chapter is written for someone who has used other Windows programs but has little or no experience with AutoCAD. Here and throughout the rest of the book, I show you how to do things by using AutoCAD's implementation of Microsoft's Fluent User Interface (or FUI; pronounced "foo-ey"). AutoCAD has always been big on backward compatibility, and this includes the interface. You can always shift between older and newer versions of the user interface.

Looking at AutoCAD's Drawing Screen

When you first open AutoCAD, you encounter the Start window. I can already hear your plaintive cry: "Where do I draw? The screen is full!" No problem. The Start screen (shown in Figure 2-1) is just a menu of available actions. It should be quite obvious as to what each item does. For now, just click the rectangular New button on the left side.

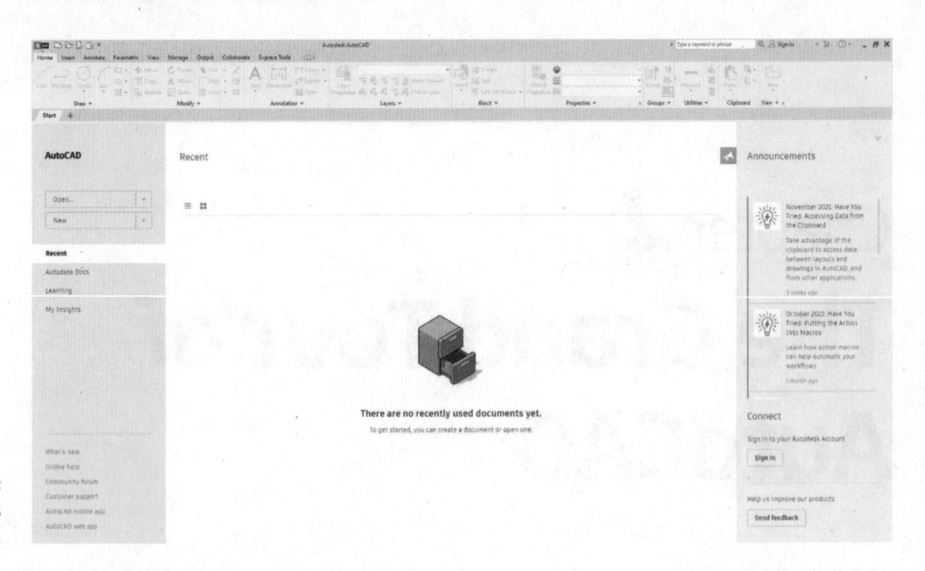

FIGURE 2-1: AutoCAD's Start screen.

TIP

You can always get the Start screen back by clicking its tab along the document tabs row, at the top of the large drawing area.

If the screen is partially obscured by any palettes, close them by clicking the X in their upper-left corners. AutoCAD remembers that you closed them, so the program doesn't open them next time.

Figure 2-2 shows the screen you then see, AutoCAD's initial drawing window.

REMEMBER

Your screen will have a very dark gray background. I'm using a white background because it is clearer on the printed page.

Now you're ready to get to work. Starting from the top down, AutoCAD's interface has eight main sections:

- **Application menu:** Click the Application button (known informally as "the red *A*") at the top-left corner of the AutoCAD window to open the AutoCAD Application menu. It presents mostly file-related commands; from this menu

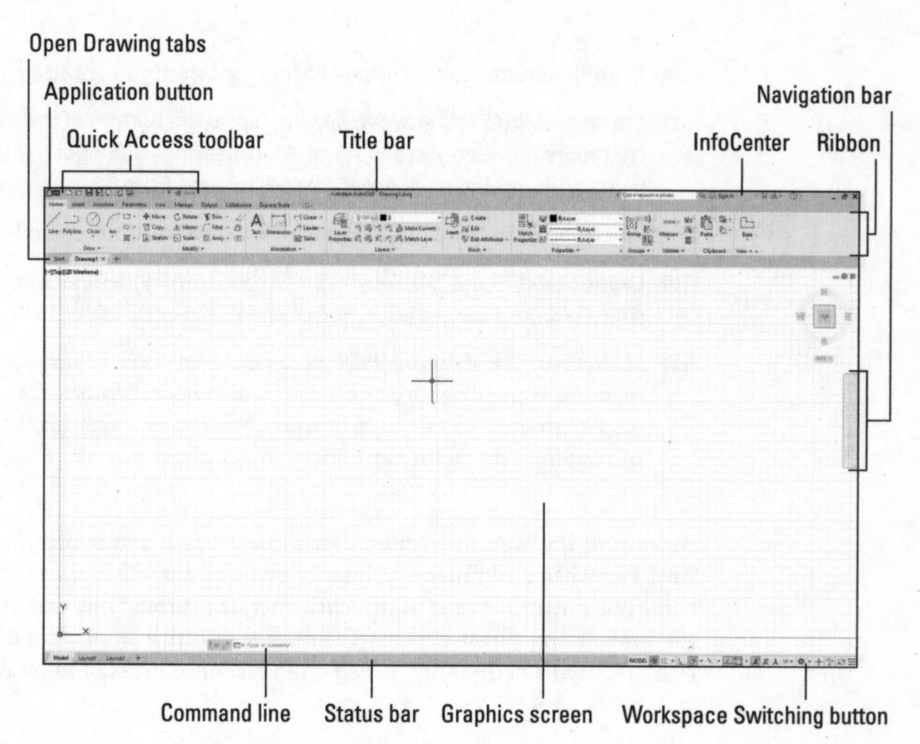

FIGURE 2-2: AutoCAD's initial drawing window.

you can create new drawings, open existing drawings, save files, or print masterpieces. It also gives access to the important Options command.

- **Quick Access toolbar:** This toolbar, in the top-left corner of the screen, includes buttons for some of the most commonly used functions, such as Save and Undo. You can add functions that you use all the time and delete unneeded buttons by clicking the down-facing triangle near the right-hand end of the toolbar.

TIP

If you're floundering, looking for the weird icon representing the command you want to use, click the down arrow at the right end of the Quick Access toolbar and select Show Menu Bar. The classic text-based menu appears above the Ribbon.

- **Ribbon:** Whereas the Application menu focuses on file management, the Ribbon holds commands to create and modify drawing objects.
- **Document tabs:** The series of tabs across the top of the graphic screen, one for each open drawing, makes it easy to jump from drawing to drawing, compared to earlier releases. New tabs appear as you open or start additional drawings. New in AutoCAD 2023, you can drag ("tear off") file tabs to place drawings on another monitor. To return them, drag the drawing back into AutoCAD.

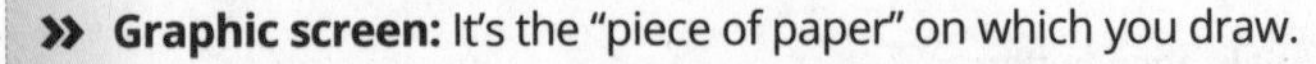

» **Graphic screen:** It's the "piece of paper" on which you draw.

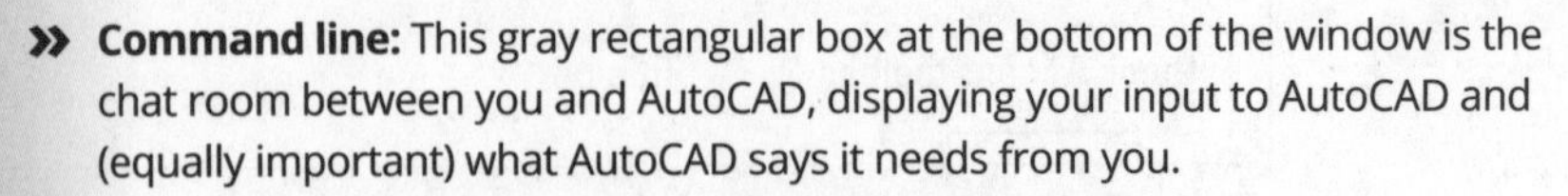

» **Command line:** This gray rectangular box at the bottom of the window is the chat room between you and AutoCAD, displaying your input to AutoCAD and (equally important) what AutoCAD says it needs from you.

REMEMBER

When you're having a problem and all else fails, read the command line.

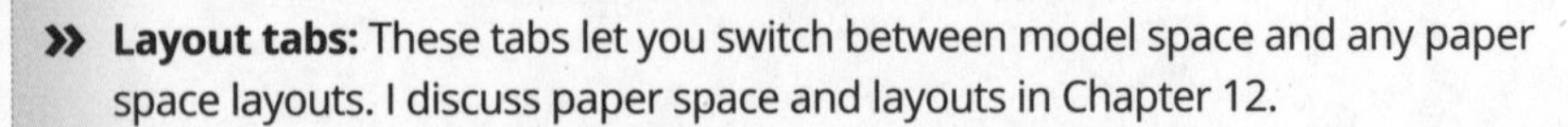

» **Layout tabs:** These tabs let you switch between model space and any paper space layouts. I discuss paper space and layouts in Chapter 12.

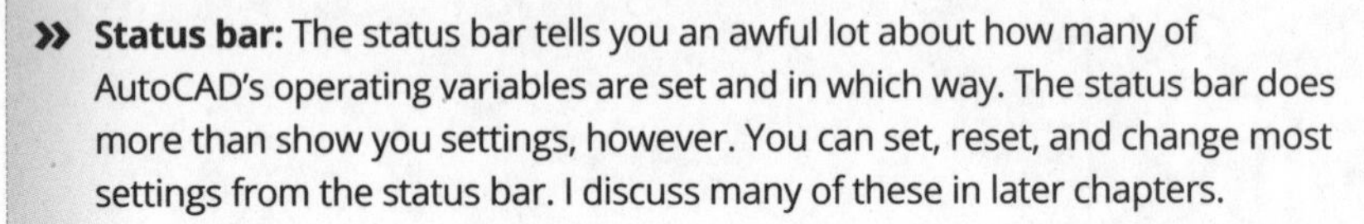

» **Status bar:** The status bar tells you an awful lot about how many of AutoCAD's operating variables are set and in which way. The status bar does more than show you settings, however. You can set, reset, and change most settings from the status bar. I discuss many of these in later chapters.

Because of the way that AutoCAD was developed, you usually have four or five ways to invoke a command, primarily by using the Ribbon, keyboard entry, toolbars, the menu bar, and right-click menus. Throughout this book, I focus on the Ribbon and direct keyboard entry because the other methods, from earlier releases, aren't necessarily turned on in recent releases of AutoCAD.

PROFILING YOUR DISPLAY

The illustrations and descriptions in this chapter and throughout the book show the *default* configuration of the AutoCAD 2023 Drafting & Annotation workspace. That is, I show the way the screen looks when you use the standard version of AutoCAD and not a flavored version, such as AutoCAD Architecture or AutoCAD Mechanical. The main change I make in this book from AutoCAD's default settings is to configure the drawing area background to be white because the figures show up better in print. The default background color in both AutoCAD and AutoCAD LT is dark gray, but many longtime users prefer a pure black background for less glare but this is much less of an issue with today's LCD flat-panel displays. You may want to set a black or a white background on your own system or stay with the default gray background. It's your choice, and there's no right or wrong way. Some AutoCAD object colors show up better on a light background, and some are better on a black one.

When you want to print in color, you may use a white screen so that WYSIWYG (What You See Is What You Get) on your monitor matches what you see on paper. This is discussed in Chapter 16.

You may also notice subtle differences in AutoCAD's appearance, depending on which version of Windows you're using and if you have used the Windows Control Panel to set it up for best performance, instead of the default setting of best appearance.

TIP

As slick as they are, navigating Ribbon panels and browsing the Application menu aren't always the most efficient ways of doing things. When you want to do real work, you need to combine the Ribbon panels with other methods, especially entering commands and options at the keyboard or choosing them from right-click menus.

A *workspace* defines the AutoCAD environment, including such things as which version of the Ribbon menu to display and whether or not toolbars are used. In addition to the default Drafting & Annotation workspace, a few additional pre-configured workspaces are available from the Workspace Switching button. You can customize workspaces. I stick with the out-of-the-box Drafting & Annotation workspace, except in Part 5, where I use the one for 3D modeling.

For your information

Located at the right side of the program title bar, InfoCenter serves as Information Central in AutoCAD. You can

- **Search for information.** Type a keyword and then click the magnifying glass for more information.
- **Sign in to your Autodesk account.** Click the Sign In link and log in with your username and password. Your Autodesk account may give you access to additional services, such as web-based file storage and collaboration service in the cloud.
- **Download free or inexpensive add-ons.** Click the button that looks like a shopping cart to open the Autodesk App Store website and download apps for designing staircases, creating rebar, and much more.
- **Connect to Autodesk via social media.** Click the triangle button to see a list of links for product updates and to connect to Autodesk via social networks such as YouTube, Twitter, and Facebook. (This button appears after you open a drawing.)
- **Find help.** Click the question mark button in the InfoCenter area to open the online Help site.

TIP

You may already be familiar with the Quick Access toolbar in Microsoft Office applications. Other features AutoCAD has in common with Office (and most Windows applications) are the capability to have more than one file open at a time, to cut or copy and paste between files, to tile or cascade multiple open files (see Figure 2-3), and to minimize, restore, and maximize individual drawings and the application itself.

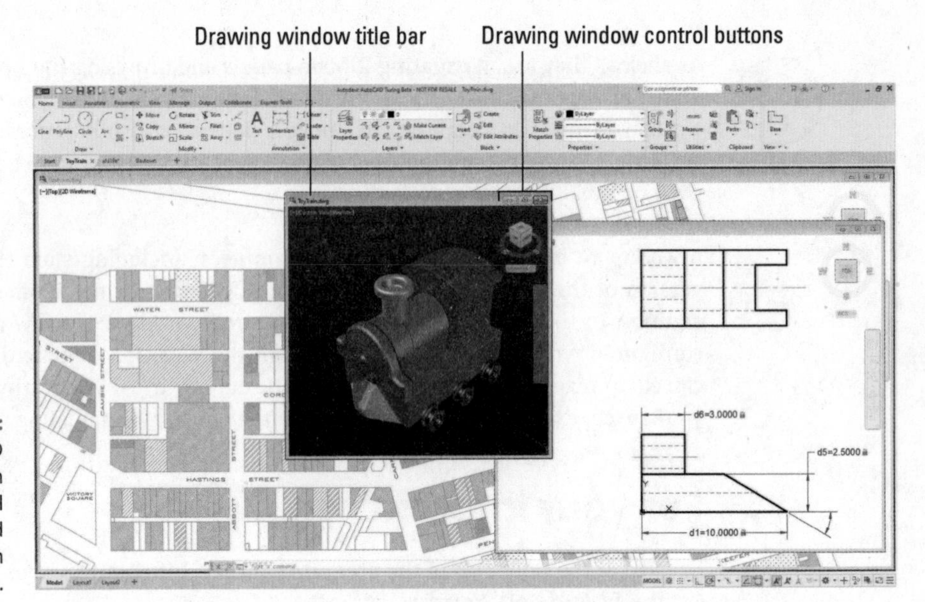

FIGURE 2-3: The AutoCAD screen with several windowed drawings in view.

Making choices from the Application menu

The Application menu is accessible from all workspaces. The AutoCAD Application menu has the following commands on the Application menu:

- **New:** Create a new drawing or a new sheet set. I discuss sheet sets in Chapter 25.
- **Open:** Open an existing drawing or sheet set.
- **Save:** Save the current drawing in the current location. If the current drawing hasn't been saved, you're prompted for a filename and a location.
- **Save As:** Save the current drawing with a new filename or in a different location or in another format, and make the newly named or located file the current drawing.
- **Import:** Open drawing files stored in formats other than DWG. I discuss this feature in Chapter 24.
- **Export:** Save the current drawing to one of a large variety of file formats, including Design Web Format (DWF), PDF, and several other CAD file formats.
- **Publish:** Send a 3D model to an outside 3D printing service, or create an archived sheet set. (AutoCAD LT doesn't support 3D.) Use eTransmit to create a package that includes all files referenced by the selected drawings, or share the current drawing by using your configured email client or Autodesk's Share service.

» **Print:** Print a single drawing or batch-plot multiple drawings, create or modify named page setups, and manage plotters and plot styles. I cover most of these operations in Chapter 16.

» **Drawing Utilities:** Set file properties or drawing units; compare the differences between two drawings; purge unused blocks, layers, and styles from the current drawing; and audit or recover damaged drawings.

» **Close:** Close the current drawing or close all drawings. If any drawings have changed, you're prompted to save them before AutoCAD closes the file.

The **CLOSEALLOTHER** command closes all open drawings except for the active one. (Unfortunately, the command doesn't appear on the Ribbon menu, so you have to type it at the command prompt.) You'll appreciate the significance of this feature after you open 20 or so drawings looking for a particular one and now want to close the rest.

In addition to the Application menu's file menu items, it has a few other features worth a mention:

» **Recent Documents:** When you choose this option, the right pane displays a list of recently edited drawings that aren't open. You can show them in a simple list or as thumbnail images. You can also pin them to stop them from scrolling off the list. Naturally enough, clicking a filename opens the drawing.

The Recent Documents feature is hardly necessary, however. Instead, click the Start tab near the upper-left corner of the screen to bring back the Start screen, which displays a scrollable series of thumbnail views of the last few open drawings, as shown in Figure 2-1 earlier in the chapter.

» **Open Documents:** Choose this option to see which documents are already open, and click an item to switch to it.

This feature, like Recent Documents, isn't necessary because a series of tabs appears across the top of the graphic screen, one for each open drawing. Hovering the cursor over a tab produces a quick preview of its drawing, which makes jumping from drawing to drawing much easier compared to earlier releases.

» **Options:** Click this button to open the Options dialog box, where you can adjust hundreds of system settings. You can also open Options by typing **OP** (the alias for the OPtions command).

» **Search:** When you're unsure of a command name or you want help on a topic, just start typing in the Search bar, at the top of the Application menu. AutoCAD quickly displays a categorized list, complete with links to start commands or to access the online Help system. See Figure 2-4.

You can also search directly from the command line. Just start typing the name of a command. As you type, AutoCAD displays a list of what it thinks you are searching for. When you see the one you want, click the question mark beside its name to get help about the selected item.

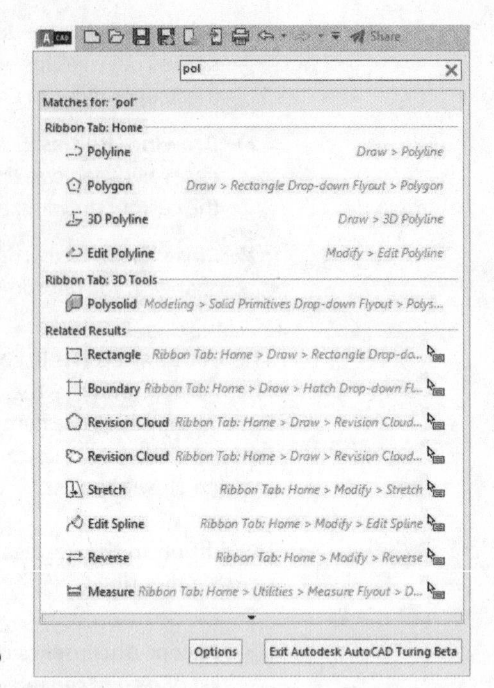

FIGURE 2-4:
When you can't find it on the Ribbon or in the tool buttons, just start typing!

TIP

A quick way to close AutoCAD is to double-click the red *A*. AutoCAD asks whether you want to save unsaved drawings before it shuts down.

Unraveling the Ribbon

The primary interface element in the Drafting & Annotation, 3D Basics, and 3D Modeling workspaces is the *Ribbon*, a customizable area that contains a series of tabbed, task-oriented collections of panels. Those panels marked with a little down-facing triangle on the panel label have more tools concealed on a slideout (see Figure 2-5). Click the panel label to open the slideout. You can click the push-pin icon to pin open the slideout if you don't want it to slide home when you've finished using it.

If you find yourself using a particular Ribbon panel often, click and drag it into the drawing area. If, for example, you're doing a lot of dimensioning, you can drag the Dimensions panel into the drawing or even to another monitor, and it stays put, even as you switch to other panels or tabs.

REMEMBER

You can fully customize the Ribbon, but I don't get into customizing AutoCAD in this book. If you want to find out more, enter **Customization Guide** in the AutoCAD search window.

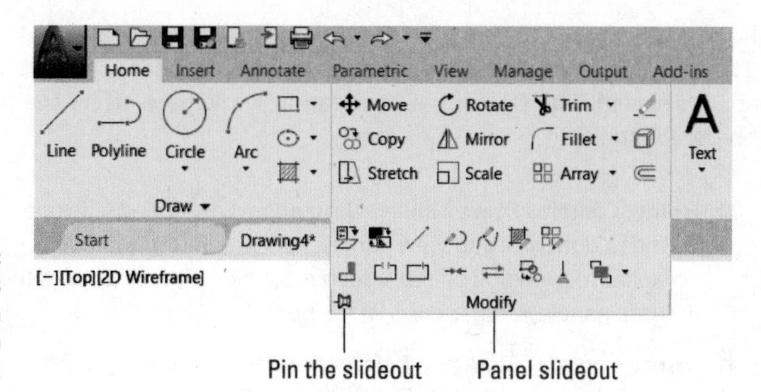

FIGURE 2-5: More tools than you can wave a Ribbon at.

To gain some screen space, you can click the little white button to the right of the name of the last tab on the Ribbon to reduce the amount of space the Ribbon occupies, or use the drop-down menu next to the little white button to tailor the Ribbon's display, or drag individual tabs over to a second monitor. By default, the Ribbon is docked at the top of the screen, but you can dock it against any edge or float it on another monitor.

REMEMBER

Depending on your screen's resolution, the Ribbon may not show all the command buttons in a panel, all the available panels in a tab, or even all available tabs. If a panel looks empty, click the down arrow at its lower edge to see all the buttons. If entire tabs or panels seem missing, simply right-click anywhere on the Ribbon and then click Show Tabs to see a list of all the tabs, or click Show Panels to see all available panels in the current tab.

EXPRESS SERVICE?

AutoCAD includes the Express Tools tab (at the far right end of the Ribbon — refer to Figure 2-2); AutoCAD LT does not include or support the Express Tools.

So what are Express Tools? They are what programmers play with in their spare time. They can be extremely useful but aren't fully supported by Autodesk. When users show enough interest in particular tools, they often get rolled into future releases. Examples include a tool to arrange text around a circular arc, or to draw a jagged break symbol when the real line is longer than line in the drawing such as a long shaft.

The tabs on the Ribbon are organized by task and differ according to the workspace selected. The Drafting & Annotation workspace offers the following panels on the Ribbon:

- **Home:** Contains Draw, Modify, Annotation, Layers, Block, Properties, Groups, Utilities, Clipboard, and View panels. Some panels may be displayed as collapsed, depending on the screen resolution. I cover most of the commands in these panels in other parts of the book.
- **Insert:** Contains Block and Reference panels as well as Import commands, and a series of commands for working with nongraphical information, such as attributes, fields, data links, and geographic data. I cover blocks (see Chapter 17) and external references (see Chapter 18), but a description of geographic and data tools is beyond the scope of this book.
- **Annotate:** Expands on the minimalist Annotation panel on the Home tab, with many more options for placing text, dimensions, leaders, and tables, as well as markup functions and a few annotation-scaling tools.
- **Parametric:** Serves as the home base of one of AutoCAD's most powerful features. You can apply geometric or dimensional parameters or constraints (rules of behavior) to drawing objects so that, say, two circles always are the same distance apart or the length of a rectangle is always twice its width. I introduce parametric drawing in Chapter 19.

  AutoCAD LT is limited when it comes to parametrics. You can modify or delete existing constraints, but you need the full version of AutoCAD to create them.
- **View:** Contains tools and panels for controlling drawing display, working with viewports, loading various palettes, and organizing Windows functions, such as cascading open files or displaying different parts of the application window. I explain most of the features on this tab later in the book.
- **Manage:** Contains panels that access the Action Recorder and CAD Standards, neither of which is in AutoCAD LT, and a set of drawing management and customization tools. I don't cover anything on the Manage tab in this book.
- **Output:** Has panels that allow you to get those drawings off your hands by printing (also known as *plotting*) or publishing them, exporting them to PDF or DWF files, or simply sending them electronically to others. I cover some of these functions in Chapter 16.
- **Collaborate:** Most of the buttons on this tab require you to pay a subscription before you can access the services. You can upload drawings to a secure website and then access them on an iPad (you *do* have an iPad, right?) or Android tablet or on any computer with a web browser. You can also invite non-AutoCAD users to view and mark up your drawings or to join you for a discussion, using a standard web browser.

» **Express Tools:** Holds an invaluable set of custom commands that streamline your work procedures in pretty well every aspect of AutoCAD. The Express Tools are created when a programmer says, "Hey, how about if we try . . ." The tools often serve as the final beta-test version of new features that sometimes are incorporated directly into later releases. They're officially unsupported, but they are installed with AutoCAD 2023 automatically and most work well. Express Tools are not available in AutoCAD LT. In earlier versions of AutoCAD you may need to install them separately with the Setup program.

REMEMBER

Some Ribbon buttons may be hidden under other, similar ones. For example, POLygon may be hidden under RECtangle or vice versa, depending on which you used last.

REMEMBER

The Ribbon displays somewhat different sets of tabs with other workspaces, such as 3D Basics and 3D Modeling. Other Ribbon tabs may exist if you purchased AutoCAD as part of a *collection* — a group of related Autodesk products that are sold in one package.

Getting with the Program

In most of this book, I focus on 2D drafting, which is by far the easiest way to get your feet wet with AutoCAD. Just don't drip water on your computer. If you're not already in the Drafting & Annotation workspace (the default workspace), you can return to it by clicking the Workspace Switching button, towards the right end of the Status bar (and shown in the margin). I cover workspace switching in more detail in Chapter 21, where I discuss 3D modeling that has a different workspace. When you are in the Drafting & Annotation workspace, AutoCAD displays the interface shown earlier, in Figure 2-2.

Like all good Windows programs, AutoCAD has *tooltips,* those short descriptions that appear in little text boxes whenever you hover the mouse pointer over a button. In AutoCAD, tooltips display two levels of information. When you first hover the mouse pointer over a tool button, you see a short identification of the command. When you continue hovering, a longer description of the icon's function, often with a descriptive image, appears in an extended tooltip. As helpful as they are when you're starting with AutoCAD, you'll probably want to remove these training wheels sooner or later because they cover up some of your drawing. You can do so in the Options dialog box. See the online Help system for more information.

I'm intrigued by the computer industry's fascination with icons. Traditional Chinese uses about 450,000 icons, but in the 1950s Simplified Chinese came into being with only 4,500 icons. On the other hand, Korean is generally considered to be the best written language in the world. It has 24 "letters," no punctuation marks, does not distinguish between uppercase and lowercase, and is perfectly phonetic. On the other, other hand, AutoCAD uses about 1,300 icons.

Looking for Mr. Status Bar

The *status bar* (see Figure 2-6), which appears at the bottom of the AutoCAD screen, displays — and allows you to change — several important drawing modes, aids, and settings that affect how you draw and edit in the current drawing.

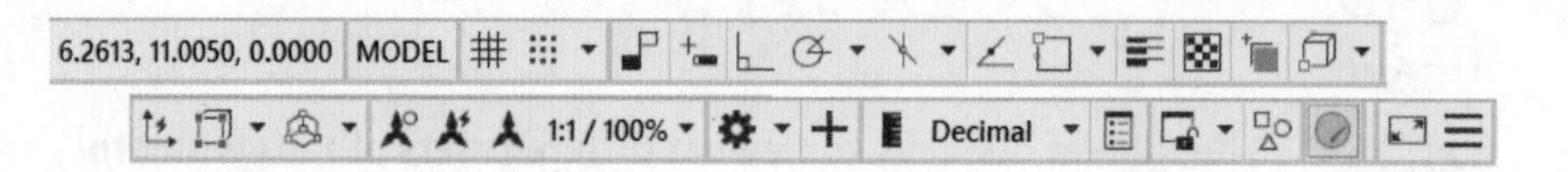

FIGURE 2-6: Status (bars) check.

Earlier AutoCAD releases allowed you to change status bar buttons to display icons or traditional text labels. Now, unfortunately, we're stuck with icons only. The good news is that the F keys (F3, F5, and so on) can also be used to turn on and off the most commonly used drawing modes.

REMEMBER

In Figure 2-6, I present the status bar in two rows so you can see the icons at a reasonable size. On the AutoCAD screen, the status bar is one long line of icons (unless your computer has a lower screen resolution, which forces the bar onto two lines).

I cover each status bar setting as appropriate when I discuss its relevant commands. See the online extras for a description of each of the status bar buttons.

Using Dynamic Input

The F12 key turns Dynamic Input on and off. When it's active, most command input and responses from the command line are repeated in the graphic area of the screen, close to the current cursor location. Dynamic Input can increase efficiency because you don't have to shift your focus between the command line and the current cursor location. On the other hand, some users find it distracting to have so much information dancing and flashing around the cursor. When you find it irritating, just turn it off by pressing the F12 key.

WARNING

When Dynamic Input is on, coordinate values you enter from the keyboard are relative to the *current location* in the drawing. When Dynamic Input is off, values are absolute relative to the *drawing origin* of 0,0 (also known as absolute coordinates).

REMEMBER

Don't get in the habit of relying on Dynamic Input. Sometimes there simply isn't room in the Dynamic Input tooltip to show as much information as you get at the command line.

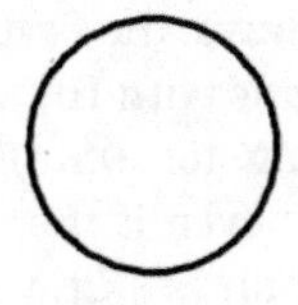

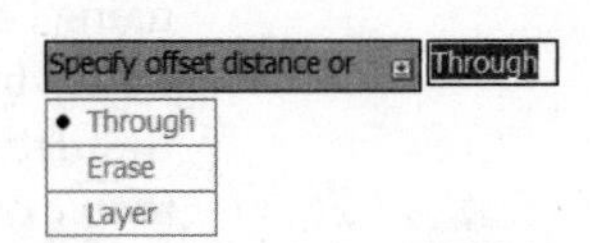

FIGURE 2-7: Choosing command options from the Dynamic Input menu.

The Dynamic Input tooltip doesn't display options unless you press the down-arrow key, as shown in Figure 2-7. Pressing the up-arrow key displays previous input.

WARNING

I like Dynamic Input. Really, I do. But sometimes it fights with normal command input, and that can make things really confusing.

Let your fingers do the talking: The command line

The command line (or command window, or command prompt, or command area, or whatever else you want to call it), as shown in Figure 2-8, is a throwback to the dark ages of AutoCAD. It puzzles newcomers and delights AutoCAD aficionados. Despite the razzle-dazzle of Dynamic Input, the command line is still the most efficient way to perform many AutoCAD operations.

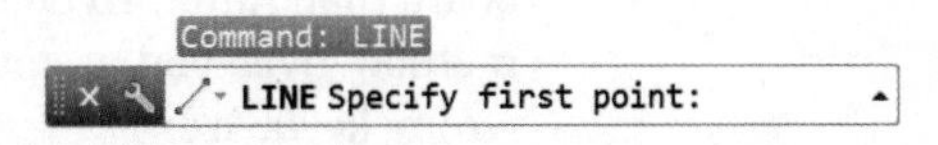

FIGURE 2-8: Obey the command line; that is an order.

Cozy up to the command line, because it's AutoCAD's primary communications conduit to you. AutoCAD displays prompts, warnings, and error messages in the command line that Dynamic Input may not show. Even when Dynamic Input gives you the same info, glancing at the command line is more efficient.

REMEMBER

If you take away only one tip from this book, it should be this: "When all else fails, read the command line!"

The *command line* is a semitransparent toolbar that can float anywhere on the screen, allowing drawings to show through it. This recovers some of the vertical window area that was lost to the Ribbon. Hey, Microsoft, computer screens are getting wider, not taller! By default, the command line displays one command input line plus the last three input prompts, but you can change the number of prompts that are displayed.

As a command is running, the command line often displays several options that can be chosen during the command. You choose an option by entering the uppercase letter or letters shown in each option, such as Close or LWeight. It's usually, but not always, the first letter. Some commands have more than one option whose name begins with the same letter, such as Entity versus Exit, so you enter E for Entity but X for eXit. For consistency, some options use a letter other than the first letter even if the current command doesn't have conflicts. The most common example is eXit, which always uses X even if no other options in the current prompt start with E. Entering command option choices is not case-dependent. X or x will both work. However, ON and OFF must be entered in full because they are shown in all uppercase.

You don't even need to type the option letter. You can select an option by using the mouse and clicking it in the command line.

The key(board) to AutoCAD success

Despite (or because of) AutoCAD's long heritage as the most successful CAD software for personal computers, newcomers are still astonished at the amount of typing they have to do. Modern programs have much less dependency on the keyboard than AutoCAD does, but as you get used to it, you'll find that no other input method gives you as much flexibility or speed as pounding the ivories — oops, wrong keyboard!

Typing at your computer's keyboard is an efficient way to run certain commands, and it's the *only* way to run others. Instead of hunting for a button or scrolling through a menu, you can start a command by typing the command name, seeing which ones AutoCAD presents to you, and pressing Enter when you find the right one. Even better, for many common commands, you can type the short form for a command name (known as an *alias)* and then press Enter. For example, you can simply type L for the Line command, C for Circle, and CO for the COpy command. Hands up, everyone who'd rather type **APPLY** instead of **APPLYGLOBALCAPACITIES**!

Aliases are not case-sensitive, so entering, for example, **C** or **c** will start the Circle command.

Print a list of aliases and highlight the ones you find yourself using most often. To see a complete list of command aliases, look in the AutoCAD (or AutoCAD LT) Program Parameters (PGP) file by going to the Manage tab and clicking Edit Aliases on the Customization panel. When Windows Notepad opens with the `acad.pgp` (or `acadlt.pgp`) file loaded, scroll down to the Sample Aliases for AutoCAD Commands section.

Get comfortable using the keyboard and command line. Several everyday commands are nowhere to be found on the Ribbon. If you want to run those commands, you have to type them!

AutoCAD's AutoComplete feature can help you become a keyboard jockey. Start typing a command name, and a list appears at the cursor showing all command names that start with the letters you've typed. You can continue until only the command you want appears, or you can scroll down the list and select a command with your mouse.

The command line features more functions beyond simply entering commands and options. It includes a spel checquer for command names so that close enough is good enough, the same as hand grenades and dancing. For example, it recognizes the misspelled LABLE as the correct LABEL. It also has a synonym list so that entering ROUND (an otherwise non-existent command) starts the FILLET command, and it's adaptive so that commands you use more frequently rise to the top of the suggestion list.

The following steps demonstrate how to use the keyboard to run commands and view and select options. If Dynamic Input is toggled on, press F12 to turn it off — temporarily, at least. Follow these steps to work with the command line:

1. **Type L and then press Enter.**

   AutoCAD starts the Line command and displays the following prompt in the command line:

   ```
   LINE Specify first point:
   ```

2. **Click a point anywhere in the drawing area.**

   The command line prompt changes to

   ```
   LINE Specify next point or [Undo]:
   ```

3. **Click another point anywhere in the drawing area.**

   AutoCAD draws the first line segment.

4. **Click a third point anywhere in the drawing area.**

   AutoCAD draws the second line segment and prompts you:

   ```
   LINE Specify next point or [Close Undo]:
   ```

5. **To activate the Undo option, type U and press Enter.**

   You can type the option letter in lowercase or uppercase.

   AutoCAD undoes the second line segment.

   If you type an option that the command line doesn't recognize (for example, X isn't a valid option for the Line command), the command line displays an error message and prompts you again for another point — in this case, a point for the Line command:

```
Point or option keyword required.
Specify next point or [Undo]:
```

Option keyword is programmer jargon for letters, shown in uppercase, that activate a command option. This error message is AutoCAD's way of saying, "Huh? I don't understand what you mean by typing X. Either specify a point or type a letter that I do understand."

6. **Type 3,2 (with no spaces) and press Enter.**

   AutoCAD draws a new line segment to the point whose X coordinate is 3 and Y coordinate is 2.

7. **Click several more points anywhere in the drawing area.**

   AutoCAD draws additional line segments.

8. **Type C and then press Enter.**

   The C instructs AutoCAD to draw a final line segment, which creates a closed figure and ends the Line command. A blank command line returns, indicating that AutoCAD is ready for the next command:

```
Type a command
```

Here are a few other tips and tricks for effective keyboarding:

» **Display the much larger text window.** The normal one-line command line usually shows you what you need to see, but occasionally you want to review a larger chunk of command-line history. Press F2 to see the AutoCAD *text window*, which is simply an enlarged, scrollable version of the command line, as shown in Figure 2-9, left.

  Press Ctrl+F2 to see the editable version, as shown in Figure 2-9, right. In this mode, you can copy and paste command history. Question: How do you think I make sure that all my examples and exercises work properly? Answer: I have an excellent technical editor who checks everything.

» **Press Esc to bail out of the current operation.** Sometimes, you might get confused about what you're doing in AutoCAD or what you're seeing in the command line. If you need to bail out of the current operation, press Esc one or more times until you see a command line that has only "Type a command." This indicates that AutoCAD is resting, waiting for your next command.

» **Press Enter to accept the default action.** Some command prompts include a default action in angled brackets. For example, the first prompt of the POLygon command is

```
Enter number of sides <4>:
```

The default is four sides, and you can accept it by simply pressing Enter. (That is, you don't have to type **4** first.)

REMEMBER

» **AutoCAD uses two kinds of brackets when it prompts.**

- *Command options* appear in regular square brackets: `[Close Undo]`. To activate a command option, type the letter(s) that appear in uppercase and then press Enter, or click the option directly in the command line.
- A *default value or option* appears in angled brackets: `<4>`. To choose the default value or option, press Enter.

TIP

You can also right-click and choose Enter as input to a command. Easier yet, you can use the spacebar instead of Enter, as long as you're not entering text.

» **Watch the command line.** You can discover a lot about how to use the command line by simply watching it after every action you take. When you click a toolbar button or menu choice, AutoCAD displays the name of the command in the command line. If you're watching the command line, you absorb the command names more or less naturally.

» **Leave the command line in the default configuration.** The command line, like most other parts of the AutoCAD screen, is resizable and movable. The default location (at the bottom of the AutoCAD screen) and size (one line in the command line and three fading semitransparent lines extending into the drawing area) work well for many people. Resist the temptation to mess with the command line's appearance, at least until you're comfortable with using it.

» **Right-click in the command line for options.** When you right-click in the command line, you see a menu with some useful choices. For example, Recent Commands shows the last six commands you ran.

» **Press the up- and down-arrow keys to cycle through the stack of commands you've used recently.** This is another handy way to recall and rerun a command. Press the left- and right-arrow keys to edit the command-line text you've typed or recalled.

TIP

Here's the easiest way to run a command again. Suppose you want to draw several circles. Start the Circle command and draw the first circle. To draw the next one, simply press the spacebar and the Circle command repeats. If you press the spacebar whenever AutoCAD is waiting for you to start a new command, it repeats the last command you used.

FIGURE 2-9: My, how you've grown: Pressing F2 (on the left side) or Ctrl+F2 (right side) expands the command line to a command text window.

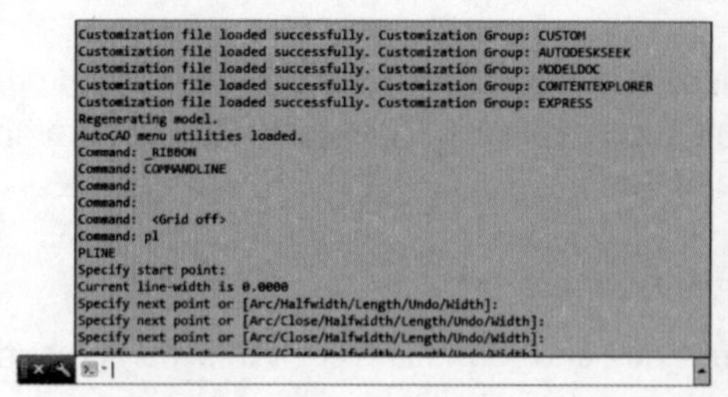

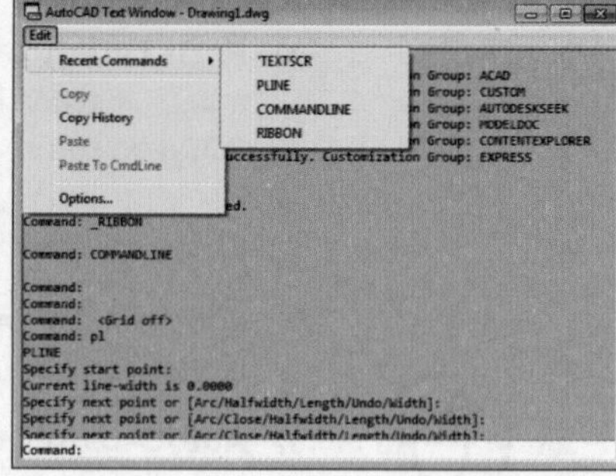

Keeping tabs on palettes

Palettes are refined (well-mannered) versions of dialog boxes. Unlike regular dialog boxes, which insist on your undivided attention while they're open, palettes stay discreetly on the screen as you carry on with other tasks. AutoCAD still has many dialog boxes, but over the past several releases, palettes have replaced quite a few former dialog boxes.

AutoCAD contains more than a dozen palettes (fewer than a dozen in AutoCAD LT). Unless noted otherwise, you can open any of these palettes from the Palettes panel of the View tab. I describe just a few of them, in the appropriate chapters.

Down the main stretch: The drawing area

After you've taken some warm-up laps, you're probably itching for the main event. The AutoCAD *drawing area* is where you do your drawing. Gee, what a surprise! In the course of creating drawings, you click points to specify locations and distances, click objects to select them for editing, and zoom and pan to get a better view of what you're working on.

Most of this book shows you how to interact with the drawing area, but you should know a few things upfront.

Model space and paper space layouts

AutoCAD operates in two parallel universes, called *spaces*, which AutoCAD indicates with a status bar button and two or more tabs in the lower-left section of the drawing area:

» **Model:** Where you create and modify the objects that represent things in the real world, such as wheels, wires, walls, widgets, waterways, or whatever.

- **Paper:** Where you create particular views of these model-space objects in preparation for printing, often with a title block around them. Paper space comprises one or more *layouts*, each of which can contain a different arrangement of model space views and different title block information. You can create many layouts of a single drawing. See Chapter 12 for information about creating paper space layouts, and see Chapter 16 for the lowdown on plotting them.

Drawing on the drawing area

Here are a few things you should know about the AutoCAD drawing area:

- **Get in the habit of looking at the command line after every action you take.** Efficient, confident use of AutoCAD requires that you continually glance from the drawing area to the command line (to see those all-important prompts) and then back up to the drawing area. This sequence isn't a natural reflex for most people, and that's why the Dynamic Input tooltip at the cursor was introduced. But you still get information from the command line that you don't get anywhere else.
- **When you click in the AutoCAD drawing area, you're almost always performing an action.** Clicking at random in the drawing area isn't quite as harmless in AutoCAD as it is in many other Windows programs. AutoCAD interprets clicks as specifying a point or selecting objects for editing. If you get confused, press Esc a couple of times to clear the current operation and return to the waiting command line.
- **You can still right-click.** In most cases, you can right-click in the drawing area to display a menu with some options for the current situation.

Fun with F1

Unfortunately, in AutoCAD, F1 doesn't stand for Formula One. Pressing F1 at any time opens the online Help window, shown in Figure 2-10, as does clicking the question mark.

TIP

Click the down arrow beside the question mark to open the Help menu with additional help-related options.

As is the case with most Windows programs, AutoCAD Help is context-sensitive. For example, if you start the Line command and just don't know what to do next, Help will, er, help. You can browse the online Product Documentation from the AutoCAD Help page or type words in the Search box to look for specific topics. In this book, I sometimes direct you to the AutoCAD online Help system for information about advanced topics.

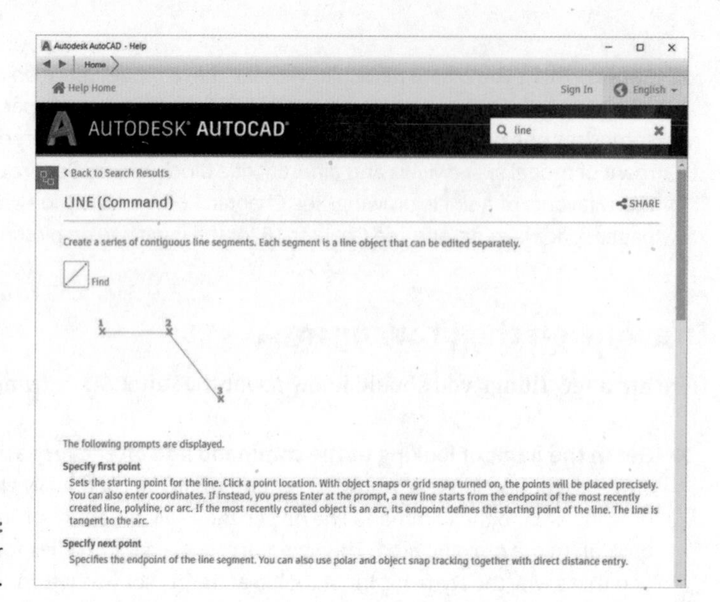

FIGURE 2-10: Help is at your F1 fingertip.

TIP

As mentioned, when you enter a command name in the command prompt, AutoCAD displays a list of commands with similar names. When you move the cursor to a listed command name, a question mark appears next to the name. Click the question mark to display the Help screen. To locate where the command is positioned on the Ribbon, click the Find button near the top of the Help window, and AutoCAD switches the Ribbon menu to the tab that contains the command in question and highlights its specific panel and icon.

REMEMBER

You really do need to take advantage of the online Help resources because AutoCAD contains so many commands (nearly a thousand) and options. Everyone from the greenest beginner to the most seasoned expert can find out something by using online Help. Take a moment to peruse the home page of the main Help system so that you know what's available. Throughout this book, I direct you to pages in the Help system that I think are particularly useful, but don't be afraid to explore on your own when you get stuck or feel curious.

TIP

The good news is that the online Help system is always being monitored and updated to reflect how people are using it and to update any errors or omissions. The bad news is that if you don't have a current internet connection, all you get is the local stripped-down version. The more good news is that you can click the down arrow next to the question mark icon and choose Download Offline Help to download and install the latest version of the full-meal deal. After you do so, AutoCAD always looks for an internet connection first so that you get the latest help — but if it can't connect, it falls back to the downloaded version you installed. The install program offers to download Help, which should be pretty well mandatory for your laptop.

IN THIS CHAPTER

» **Setting up a simple drawing**

» **Drawing some objects**

» **Zooming and panning in a drawing**

» **Editing some objects**

» **Plotting a drawing**

Chapter 3 A Lap around the CAD Track

Chapters 1 and 2 introduce you to the AutoCAD world and to the AutoCAD interface. Other chapters in this book present the techniques that underlie good drafting practice. By now, you're probably eager to start moving the cursor around and *drawing* something. This chapter leads you on a gentle tour of the most common CAD drafting functions, including setting up a new drawing, drawing and editing objects, zooming and panning the view, and printing (or plotting) a drawing. I don't go into full detail about every option of every command, but I give you a feel for what it can do. Go ahead and slam the tires, and don't worry about putting a dent in the doors!

In this chapter, you create the drawing of an architectural detail of a base plate and column. Even if you don't work in architecture or building construction, this exercise gives you some simple shapes to work with and demonstrates commands you can use in most drafting disciplines.

Throughout this book, I show AutoCAD running in the Ribbon-based Drafting & Annotation workspace that is present in both AutoCAD and AutoCAD LT. Likewise, I tell you where to find commands and what to select by using the Ribbon.

Although the drafting example in this chapter is simple, the procedures that it demonstrates are real, honest-to-CAD-ness, proper drafting practices. I emphasize from the beginning the importance of proper drawing setup, putting objects on appropriate layers, and drawing and editing with due concern for precision. Some of the steps in this chapter may seem a bit strange at first, but they reflect the way that experienced AutoCAD users work. My goal is to help you develop good CAD habits and do things the right way from the start.

The steps in this chapter, unlike the steps in most chapters in this book, form a sequence. You must complete the steps in order. Figuring out how to use AutoCAD is a little like figuring out how to drive, in which parallel parking comes before Indy car racing, except that with AutoCAD you're free to stop in the middle of the street and take a break. If things get away from you, press Esc two or three times to terminate any command that's in progress, and type the letter **U** and then press the Enter key to undo the last thing you did. Incidentally, U and UNDO are two similar commands and between them are probably the most-often-used commands in AutoCAD.

If you find that selecting and editing objects work differently from the way I describe in this chapter, you have (or someone else has) probably changed the configuration settings on the Selection tab in the Options dialog box. Chapter 24 describes these settings and how to restore the AutoCAD defaults.

A Simple Setup

In this chapter, I walk you through the steps to create, edit, view, and plot a new drawing. See Figure 3-1 to get an idea of what the finished product looks like. I use *Imperial units* in this chapter (you know — inches and feet, for example). I describe how to set imperial units versus metric units in Chapter 4.

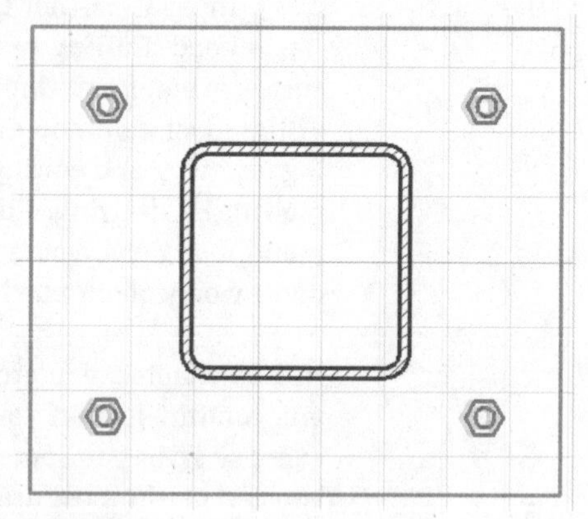

FIGURE 3-1: How base is my plate.

In the following exercise, you use several of the status bar buttons at the bottom of the screen. I don't like the icons that AutoCAD displays, and I grew bored from hovering the mouse cursor over each button until the tooltip appeared. Earlier versions of AutoCAD allowed us to switch from icons to text mode, wherein each button had a text label instead of an icon. Unfortunately, AutoCAD no longer supports this feature, so I have to refer to each button by its tooltip description along with the relevant icon in the margin. The good news is that you can toggle ten of those settings by using the F keys (such as F3), which I reference in the text in parentheses, as in (F3).

A gray icon indicates that the button's function is Off, while a light blue icon indicates On. (Your screen might show icons against a dark gray background, but I used a white background for a clearer image on the printed page.)

You can find the files I use in this sequence of steps at this book's companion website: Go to `www.dummies.com/go/autocadfd19` and download `afd03.zip`. The Zip file contains imperial and metric versions of the base plate exercise at various stages, and the Read Me file on the Downloads tab describes the files in details.

REMEMBER

Pay attention to feedback from AutoCAD. Glance at the messages the program sends you after each step via the command line at the bottom of the screen or the Dynamic Input tooltip near the cursor so that you become familiar with the names of commands and their options. If you want to see messages next to the cursor as you use the program, click the Dynamic Input (F12) button on the status bar so that its icon is light blue rather than gray.

In this first set of steps, you create a new drawing from a template, change a few settings to establish a 1:10 scale (that is, 1 inch on the drawing is equivalent to 10 inches on the real object), and save the drawing:

1. **Start AutoCAD.**

   As indicated in Chapter 1, the startup procedure can vary depending on your version of Windows and how it is set up.

2. **Start a new drawing that uses AutoCAD's Imperial template.**

   a. *Click the down arrow next to the New button on the Start tab.*

WARNING

   Don't click the big New button itself or the New button on the Quick Access toolbar. You must use the down arrow. I explain why in Chapter 4 — humor me for now. Have I lied to you recently?

   A list of drawing templates (DWT files) appears. Templates in AutoCAD are similar to templates in Microsoft Word and other programs. Chapter 4 describes how and why to create and use custom drawing templates.

   b. *If the acad.dwt file name is missing from the list, select Browse Templates. In the Select Template dialog box that appears, select `acad.dwt`, as shown in Figure 3-2. (In AutoCAD LT, select `acadlt.dwt`.)*

   AutoCAD creates a new, blank drawing that uses the settings in `acad.dwt`. The `acad.dwt` template (`acadlt.dwt` in AutoCAD LT) is AutoCAD's default, plain-Jane (or should that be plain-Pat to be gender-neutral?) template for creating drawings that use imperial units (units expressed in inches or feet or both). Chapter 4 contains additional information about these and other templates.

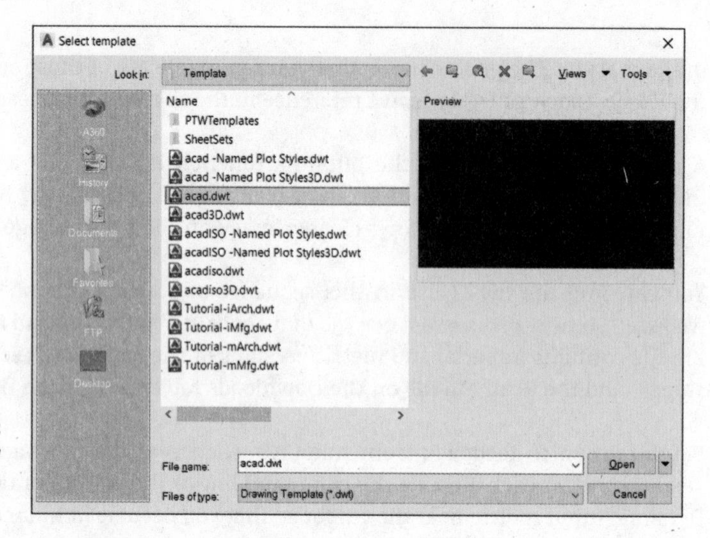

FIGURE 3-2: Starting a new drawing from a template.

3. **Check your workspace.**

The workspaces in AutoCAD look similar. To make sure that you're using the same workspace as I am in this chapter (and in most of this book), click the gear icon, near the right end of the status bar (and shown in the margin). If the check mark in the drop-down list *isn't* beside *Drafting & Annotation,* click *Drafting & Annotation*.

4. **Set up the initial working environment.**

Make the following settings:

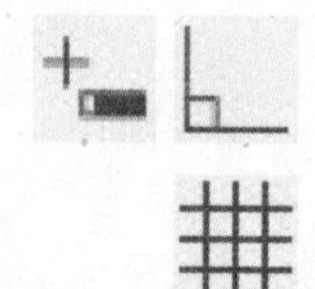

a. Click all the buttons on the left half of the status bar as necessary until they look gray, except Dynamic Input (F12) and Ortho (F8), which should be blue.

b. Turn the background grid (F7) on or off, your choice.

Most beginners seem to like it turned on, and most experienced users turn it off.

Some of the other settings can make point selection difficult. Start with them all turned off, and then toggle them on and off as needed. I tell you which ones to use in the remaining steps.

5. **Set the drawing limits that define the working area.**

a. Type **LIMITS** and press Enter.

(I explain drawing limits in more detail in Chapter 4.)

b. For now, ignore the Dynamic Input tooltip next to the cursor, and look at the command line. AutoCAD echoes the command name and replies to it. Set the limits of the drawing area by responding to the prompts:

```
Specify lower left corner or [ON/OFF] <0.0000,0.0000>:
   Press Enter
Specify upper right corner <12.0000,9.0000>: 100,50
```

REMEMBER

Don't add a space before or after the comma of a coordinate value. It's the final frontier. If you press the spacebar before you enter the entire coordinate value, AutoCAD will process only the text you've typed.

6. **Display the defined working area.**

   a. *Start the **Z**oom command.*

   The Zoom command is hidden away on the Ribbon, so it is usually far faster and easier to simply type **Z** and press Enter.

```
ZOOM Specify corner of window, enter a scale factor (nX or
   nXP), or [All/Center/Dynamic/Extents/Previous/Scale/
   Window/Object] <real time>:...
```

   b. *Type **A** and press Enter.*

   AutoCAD zooms so that the screen shows the limits (that is, *All*) of your drawing.

REMEMBER

I use a `special font` to show you what AutoCAD is saying. You either type the characters in **bold** and press Enter, or just press Enter (such as in Step 4a) to accept AutoCAD's default option or recent value. For simplicity, rather than show AutoCAD's complete response, I often show only what's necessary for you to understand what's going on.

7. **Establish the snap and grid drafting settings.**

   a. *Right-click the Display Drawing Grid button near the left end of the status bar and then choose Grid Settings.*

   The Snap and Grid tab in the Drafting Settings dialog box appears, as shown in Figure 3-3. Note that AutoCAD LT lacks some options here that are present in the full version of AutoCAD.

   b. *Change the values in the Drafting Settings dialog box to match those in Figure 3-3, and click OK when you're finished:*

   - *Snap On:* When selected, constrains the cursor to moving in an invisible grid of equally spaced points (0.5 units apart, in this case).
   - *Grid On:* When selected, displays a visible grid of little dots or grid lines on the screen (5 units apart, in this case), which you can use as reference points. The grid doesn't appear on printed drawings.
   - *Snap X Spacing:* 0.5
   - *Snap Y Spacing:* 0.5

- *Grid X Spacing:* 0.5
- *Grid Y Spacing:* 0.5
- *Major line every*: 5

You see a network of grid lines, 0.5 units apart, with brighter lines every 5 units, in the drawing area. If you move the mouse pointer around and watch the coordinate display area at the left side of the status bar, you notice that the values still change in tiny increments, just as they did before. I have more information on snap mode and grid display in Chapter 4.

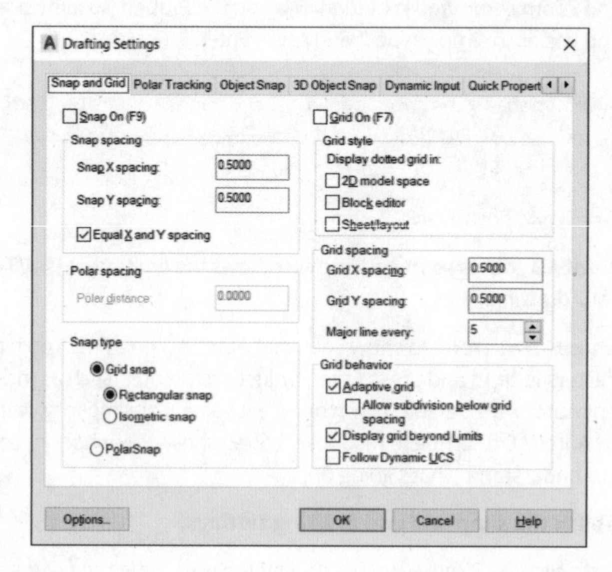

FIGURE 3-3: Snap and Grid settings.

8. Save your work:

a. Click the Save button on the Quick Access toolbar, or press Ctrl+S.

Because you haven't saved the drawing yet, AutoCAD opens the Save Drawing As dialog box. Navigate to a suitable folder by choosing from the Save In drop-down list or double-clicking folders in the list of folders below it.

I remembered to remind you to remember where you save the file so that you can find it later. (And remember the three signs of old age: The first is loss of memory . . . I forget the other two.)

b. Type a name in the File Name text box.

For example, type **My Plate Is Base**. Depending on your Windows Explorer or File Explorer settings, you may or may not see the `.dwg` extension in the File Name text box. In any case, you don't need to type it — AutoCAD adds it for you.

c. *Click Save.*

AutoCAD saves the new DWG file to the folder you specified.

If you forget the name of the file later, simply click the Start tab in the upper-left corner of the drawing or check the Recent Documents list of the Application menu.

The good news is that it isn't necessary to complete all these steps for every new drawing. Chapter 4 goes into more detail about drawing setup, describes why all these gyrations are necessary, and tells you how to avoid doing them more than once by defining your own template files.

Drawing a (Base) Plate

When your drawing is properly set up, you're ready to draw some objects. In this example, you use the Line command to draw a steel base plate and column, the Circle command to draw an anchor bolt, and the POLygon command to draw a hexagonal nut.

AutoCAD, like most CAD programs, uses layers as an organizing principle for all objects you draw. Chapter 9 describes layers and other object properties in detail. For now, think of layers as sheets of transparent plastic lying on top of the drawing. You can hide or display individual drawing elements by simply removing or replacing their particular sheet of plastic.

In this example, you create separate layers for the base plate, column, anchor bolts, nuts, and crosshatching. This might seem like layer madness, but when you're creating complex drawings, you need to use layers properly to keep everything organized.

The following steps demonstrate how to create and use layers, as well as how to draw lines, circles, rectangles, and polygons. You also see how to move, copy, mirror, offset, and apply fillets.

1. **Complete the drawing setup in the earlier section "A Simple Setup."**
2. **Prepare to create the five layers you will use to organize the objects that will make up the base plate example by invoking the LAyer command.**

   Do one of the following:

   - *On the Layers panel of the Home tab, click the Layer Properties button.*

     The Layer Properties button is at the left edge of the Layers panel.

   - *Enter **LA** from the keyboard.*

As you type, a list of commands that start with *LA* appears near the cursor. The full command name is LAYER, but as soon as *LA* is highlighted, press Enter.

The LAyer command starts and AutoCAD displays the Layer Properties Manager palette, as shown in Figure 3-4. If you want to move any palette to a different screen location, just move the cursor to the vertical bar on the palette's left edge and then drag the palette to a new location.

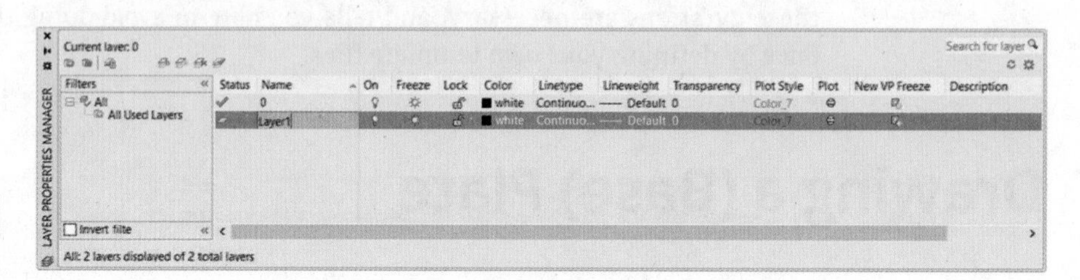

FIGURE 3-4: Creating a new layer.

3. Follow these steps to create the needed layers:

a. Click the New Layer button.

It's the leftmost of the group of four buttons that include the green check mark and the red *X*. AutoCAD adds a new layer to the list and gives it the default name Layer# where # is a number. For the first layer, the default name is Layer1 (see Figure 3-4).

b. Name That Layer.

Type **Plate** as the name of the first layer on which you'll draw the base plate, and press Enter.

c. Click the color swatch or name (white) in the Color column of the new Plate row.

The Select Color dialog box appears, as shown in Figure 3-4.

d. Click the fourth color square (cyan/light blue, index color 4), for the first layer in the single, separate row to the left of the ByLayer and ByBlock buttons, and then click OK.

The Select Color dialog box closes, and the color for the Plate layer becomes cyan.

e. With Layer Properties Manager still open, repeat Steps 3a through 3d to create four more layers with these names and colors:

Layer Name	Color
Column	5 (blue)
Anchor Bolts	3 (green)
Nuts	1 (red)
Hatch	6 (magenta)

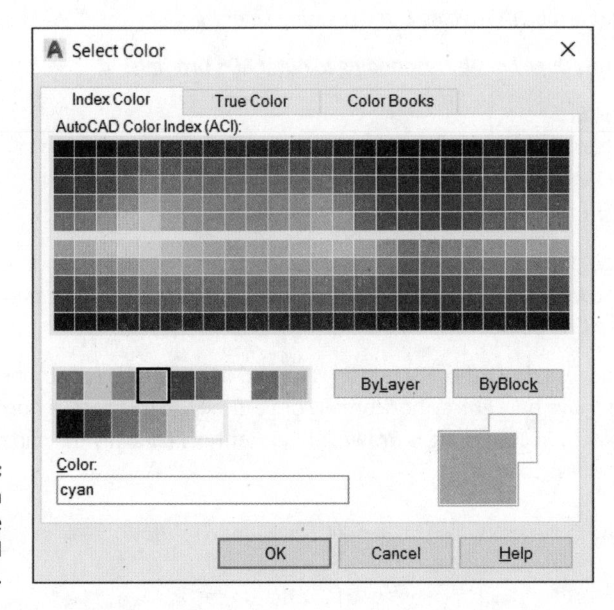

FIGURE 3-5: Select Cyan from the standard color tiles.

You don't have to create a separate layer for every type of object you draw. For example, you can draw both the anchor bolts and nuts on a Hardware layer. Layer names and usage depend on industry and office practices, in addition to a certain amount of individual judgment. Having too many layers is better than having too few, because lumping together two or more layers is much easier than dividing the objects on one layer into two or more layers. I cover working with layers in more detail in Chapter 9.

Start with the objects that will be drawn on the Plate layer by following these steps:

1. **Make the Plate layer current before drawing the objects that make up the plate.**

   a. Double-click the Plate layer name.

   A green check mark appears beside the layer name, indicating that it's current. Anything you draw now will be on the Plate layer and will inherit its color.

b. Click the Close button (the X at the upper-left corner of the palette shown in Figure 3-4) to close the Layer Properties Manager palette.

The Layer Properties Manager palette can be left open on the screen or reduced to a vertical bar, but you might as well close it because you won't need it again in this drawing session.

2. Draw the base plate.

*a. Start the Line command by clicking its button on the Draw panel of the Ribbon's Home tab or by typing **L** and pressing Enter.*

b. Draw three lines by responding to AutoCAD's prompts:

```
LINE
Specify first point: 68,7 and press Enter.
Specify next point or [Undo]: Move the cursor to the left
  and then enter 36
```

TIP

You don't need to enter a negative value for the second point. AutoCAD knows which way to draw the line because the cursor is located to the left of the first point.

TIP

Because *the* Ortho (F8) button is turned on, AutoCAD constrains the cursor to move only *exactly* horizontally or vertically. In this case, a horizontal line exactly 36 units long is drawn, no matter how far away the cursor is from the first point.

```
Specify next point or [Close Undo]: Move the cursor up and
  then enter 36
Specify next point or [Close Undo]: Move the cursor to the
  right and then enter 36
Specify next point or [Undo]: C
```

Entering **C** (uppercase or lowercase) closes the line back to the starting point.

If you didn't complete the previous steps, start from this point by using drawing `afd03a-i.dwg`, available in the `afd03.zip` download at `www.dummies.com/go/autocadfd19`. Download it to a suitable location, click the Application button, and then click Open. Browse to the file.

Time to add a bolt. You first need to change to the Anchor Bolts layer, and then draw the bolt:

1. **Go to the Layers panel of the Home tab.**
2. **Click the arrow at the right-hand end of the Layer drop-down list.**

   The current entry (from the running example) is Plate. Layer names are listed in alphabetical order, not in the order in which you created them.

3. **Choose the Anchor Bolts entry to set the Anchor Bolts layer as the current layer.**

TIP

Using the Layer drop-down list, in the upper-right corner of the Layers panel of the Home tab of the Ribbon, saves you from having to open Layer Properties Manager, select the layer, click the Set Current button, and close Layer Properties Manager when you want to change to a different layer. Becoming an AutoCAD master is all about efficiency. The real heart of all human progress is laziness — we will work hard at finding ways to avoid work.

4. **Draw the bolt:**

   a. *Start the Circle command, either from the Draw panel or by simply typing* **C** *and pressing Enter.*

   b. *Draw a circle by responding to AutoCAD's prompts this way:*

```
CIRCLE
Specify center point for circle or [3P/2P/Ttr (tan tan
  radius)]: int
Of: Move the cursor until it is close to the lower-left
  corner of the rectangle and then click when a green X
  appears at the intersection of the two lines.
Specify radius of circle or [Diameter]:
AutoCAD defaults to the radius of a circle, but you want to
  specify the diameter, so enter D.
Specify diameter of circle: 1.5
```

5. **Save your drawing.**

WARNING

AutoCAD automatically backs up your work, but you could still lose the last ten minutes or so. Now is a good time to press **Ctrl+S**.

Now add a hexagon nut for the bolt:

1. **Make the Nuts layer current by using the Layer drop-down list from the Layers Ribbon panel.**
2. **Start the POLygon command.**

   Its button is either in the upper-right corner of the Draw panel or on the drop-down list with the RECtangle button, depending on which command you last used. You can also type POL and press Enter.

3. **Draw a hexagon.**

   a. *Respond to these AutoCAD prompts:*

```
POLYGON Enter number of sides <4>: 6
Specify center of polygon or [Edge]: cen
Of: Click anywhere on the circle.
Enter an option [Inscribed in circle/Circumscribed about
  circle] <I>: C
```

REMEMBER

The Inscribed option draws a polygon whose corners touch the circumference of the imaginary circle. The Circumscribed option draws a polygon whose sides are tangent to the circumference of the circle and has nothing to do with a medical procedure. Bolt heads and nuts are normally specified by the width across the flats (unless you own a vintage British car such as a 1937 Rolls-Royce, where Whitworth wrenches are arbitrarily defined by the diameter of the thread), so choose the **C** option.

   b. *Move the cursor.*

   The hexagon rotates about its center. You should still have ORTHO turned on, so AutoCAD shows only two possible choices for the vertex of the polygon: horizontal or vertical from the center of the imaginary circle.

   c. *Move the mouse until a vertex is horizontal from the center of the circle.*

   d. *Respond to this AutoCAD prompt:*

```
Specify radius of circle: 1.5
```

   e. *Press Enter, and AutoCAD draws the hexagon.*

Okay, suppose that you draw the circle and the hexagon in the wrong location. Easy enough. Just move objects to another location as follows:

1. **Start the Move command, which is on the Modify panel.**

   When AutoCAD prompts you, respond this way:

```
MOVE
Select objects: Click the circle or the hexagon. 1 found
Select objects: Click the hexagon or the circle. 1 found, 2
  total
Select objects: Press Enter.
Specify base point or [Displacement]: 6,6
Specify second point or <use first point as displacement>:
  Press Enter.
```

   Like magic, AutoCAD moves the hexagon and the circle six units to the right and six units up.

2. **Take a break.**

   Press **Ctrl+S** to save the drawing. It may be time for a hydraulic break; remember that the mind can't hold more than the bladder.

There are several different ways to produce additional nuts and bolts. The next two procedures combine to demonstrate one method:

1. **Start the COpy command, which is on the Modify panel. Then follow these prompts.**

   ```
   COPY
   Select objects: p Selects the Previous selection set, being
     the objects you selected to move. 2 found
   Select objects: Press Enter
   Current settings: Copy mode = Multiple
   Specify base point or [Displacement/mOde]
     <Displacement>: 0,24
   Specify second point or [Array] <use first point as
     displacement>: Press Enter.
   ```

   Presto! AutoCAD creates a copy of the circle and the hexagon 24 units above the first set.

2. **Start the MIrror command, found on the Modify panel. Then follow the prompts.**

   ```
   MIRROR
   Select objects: p Selects the Previous selection set, being
     the first nut and bolt. 2 found
   Select objects: Click the upper circle or the upper
     hexagon. 1 found
   Select objects: Click the upper hexagon or the upper
     circle. 1 found, 4 total
   Select objects: Press Enter.
   Specify first point of mirror line: mid
   of: Select the upper horizontal line.
   Specify second point of mirror line: Make sure Ortho (F8)
     is still on and pick a point above or below the upper
     line.
   Erase source objects? [Yes/No] <N>: Press Enter.
   ```

   That was a little faster than drawing four circles and four hexagons by hand, wasn't it?

3. **Draw the column. Start by selecting Column to set it as the current layer.**

   a. *On the Layers panel of the Home tab, click the Layer drop-down list to display the list of layers. Select Column to set it as the current layer.*

      Now you're ready to create the hollow column, starting with its outside profile. Be sure to keep the cursor in the drawing area during this step and the next one.

   b. *Start the RECtang command, found on the Draw panel.*

```
RECTANG
Specify first corner point or [Chamfer/Elevation/Fillet/
  Thickness/Width]: 44,16
Specify other corner point or [Area/Dimensions/Rotation]:
  @12,18
```

      A rectangle is drawn in the middle of the square base plate.

   c. *Start the FILlet command, on the Modify panel, and watch the command line closely.*

      In this step, you round the corners of the column with the FILlet command and then use OFFset to give it a wall thickness.

```
FILLET
Current settings: Mode = TRIM, Radius = 0.0000
Select first object or [Undo/Polyline/Radius/Trim/
  Multiple]: r
Specify fillet radius <0.0000>: 2
Select first object or [Undo/Polyline/Radius/Trim/
  Multiple]: p
Select 2D polyline or [Radius]: Move your cursor over the
  new rectangle and observe how AutoCAD shows you a
  preview. Click the rectangle.
4 lines were filleted
```

      You can pick the lines at each corner that need to be filleted (that's eight picks), but because the column is a continuous polyline, a more efficient method in this case is to use the FILlet command's Polyline option to fillet all four corners in one fell swoop. I discuss polylines in Chapter 6 and fillets in Chapter 11.

   d. *Now give the column a ¾-inch wall thickness. Start the OFFset command. Once again, it's a Modify item.*

```
OFFSET
Current settings: Erase source=No Layer=Source
  OFFSETGAPTYPE=0
```

```
Specify offset distance or [Through/Erase/Layer]
   <Through>: .75
Select object to offset or [Exit/Undo] <Exit>: Click the
   filleted rectangle.
```

WARNING

Make sure that the Object Snap (F3) status bar button is toggled off for these next prompts, or AutoCAD may offset the object back on top of itself:

```
Specify point on side to offset or [Exit/Multiple/Undo]
   <Exit>: Move your cursor over the new rectangle and
   observe how AutoCAD shows you a preview. Click inside the
   rectangle.
Select object to offset or [Exit/Undo] <Exit>: Press Enter.
```

There you go! You created a hollow structural shape with wall thickness and radiused corners (see Figure 3-6) in just four simple steps.

4. **Press Ctrl+S to save the drawing.**

   AutoCAD saves the drawing and renames the previously saved version `drawingname.bak` — for example, `My Plate is Base.bak`. AutoCAD uses the filename extension `.bak` for a backup file. When (not *if*) things get out of hand, you can always rename the `.bak` file to `.dwg` to return to the most recent version.

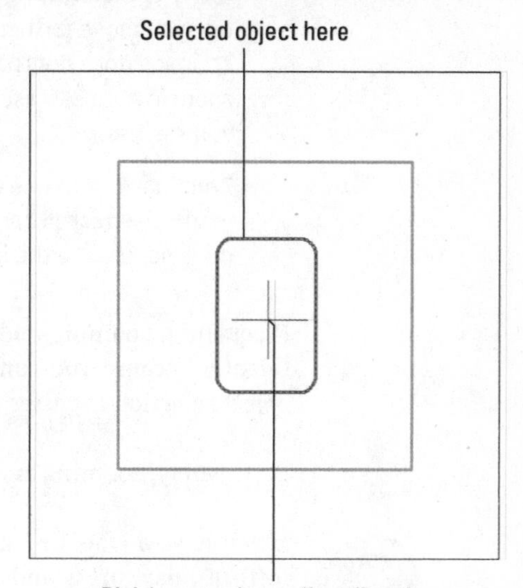

FIGURE 3-6: Give the column some thickness with OFFSET.

TIP

Now that you've given Dynamic Input a test drive, turn it off for the rest of this chapter by clicking its button on the status bar so that the button looks dimmed or by pressing F12. If you like using Dynamic Input, toggle it back on.

The drawing `afd03b-i.dwg` contained in the `afd03.zip` download includes the base plate and column and one anchor bolt.

Taking a Closer Look with Zoom and Pan

The drawing example in this chapter is uncluttered and manageable, but most real CAD drawings are neither — technical drawings are usually jam-packed with lines, text, and dimensions. CAD drawings are often plotted on sheets of paper that measure 2 to 3 feet on a side (in the hundreds of millimeters, if you're a metric maven). Anyone who owns a monitor of that size probably can afford to hire a room full of drafters and therefore isn't reading this book. You need to zoom and pan in drawings — a lot. I cover zooming and panning in detail in Chapter 5. These quick definitions should suffice for now:

- **Zoom:** Change the magnification of the display. When you zoom in, you move closer to the drawing objects so that you can see more detail; when you zoom out, you move farther away so that you can see more of the drawing area. Zooming does *not* change the size of the objects in the file. You can think of zoom in AutoCAD like the zoom function on your camera: You get closer, but you see less.
- **Pan:** Move from one area to another without changing the magnification. If you've used scroll bars in an application, you've panned the display. Panning does *not* change the location of objects in the file.

Frequently, zooming and panning let you see details better, and draw more confidently because you can see what you're doing, and edit more quickly because object selection is easier when a bazillion objects aren't on the screen.

Fortunately, zooming and panning in AutoCAD are as simple as they are necessary.

If you have a wheel mouse, you can zoom by simply rolling the wheel back and forth. To pan, press and hold the wheel (yes, the wheel is also a button) and drag the view around. These actions can be performed even when another command is active. If you don't have a wheel mouse, run out and buy one. The small amount you'll pay ($20 for cordless or $5 for a USB corded version) is easily recovered by your new operating efficiency. (*Disclaimer:* I have no financial interest in any mouse company or wheel company.)

Meanwhile, until you buy a wheel mouse, check out Chapter 5 to see how to use AutoCAD's Zoom and Pan Realtime features.

The fastest way to return to a full view of the entire drawing is to type, at the command line **Z A**, which is short for Zoom All. Note the space between the two letters, and remember to press Enter or the spacebar after you type each letter.

Modifying to Make It Merrier

When you have a better view of the base plate by zooming, which I talk about in the preceding section, you can edit the objects on it more easily. In the following few sections, you use the Hatch command to add crosshatching to the column and use the Stretch command to change the shape of the plate. As always, I cover these commands in detail later in this book.

The drawing `afd03c-i.dwg` contained in the `afd03.zip` download adds the remaining anchor bolts.

Crossing your hatches

The next editing task is to add crosshatching to the space between the inner and outer edges of the column. The hatch lines indicate that the drawing shows a cross section of the column. To do so, follow these steps:

1. **Turn off Object Snap (F3), Ortho (F8), and Snap (F9) modes by clicking their respective buttons on the status bar or by pressing the indicated function key until the icons look dimmed.**

2. **Select the Hatch layer from the Layer drop-down list at the top of the Layers panel on the Home tab.**

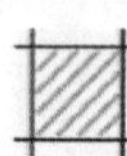

3. **On the Home tab's Draw panel, click the Hatch button (shown in the margin) or type Hatch at the keyboard.**

   The Hatch Creation tab appears on the Ribbon. For more information on this tab, and on hatching in general, see Chapter 15. Note that the Ribbon now has a light blue line around it and all the panels now show actions related to creating and formatting crosshatching.

4. **On the Hatch Creation tab's Pattern panel, select ANSI31.**

   Depending on your screen resolution, the panels may show more or less information. If you don't see an ANSI31 sample swatch, click the down arrow at the bottom of the lower-right corner of the Pattern panel. A scrollable list of all available patterns will appear.

TIP

The ANSI31 swatch is the American National Standards Institute (ANSI) standard pattern number 31 (ANSI 31) and not ANS 131.

a. Move the cursor over the drawing objects.

A live preview shows you the result if you click at the current crosshair position. AutoCAD prompts you:

```
Pick internal point or [Select objects/Undo/seTtings]: Move
  the cursor so it's between the inside and outside edges
  of the column. Zoom in if you need to get closer.
```

The live preview shows the ANSI31 hatch pattern filling the space between the two filleted rectangles. Live preview not only shows you the pattern but also lets you preview the hatch angle and scale. In this case, it looks like the hatch pattern may be too fine.

*b. In the Scale window in the lower-right corner of the Hatch Creation tab's Properties panel, change the value to **5** and press Tab to confirm it.*

c. Move the cursor back to the area between the two filleted rectangles to preview the hatch again. If it looks okay, click in the hatched area to confirm the hatch object, and then press Enter to finish the command.

The finished column and base plate should look much like the one shown in Figure 3-7.

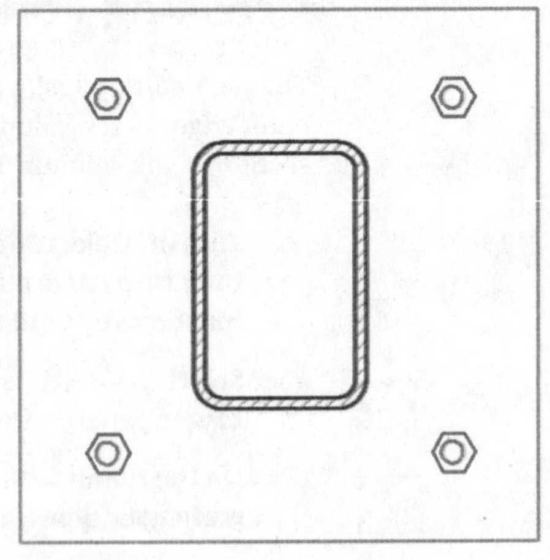

FIGURE 3-7: Button-downed base plate.

Perfect! But suppose that the nutty engineer (hey, I resemble that remark!) has decided that the column needs to measure 18 x 18 inches rather than 12 x 18 inches. Therefore, the base plate is too small, and the anchor bolts are in the wrong place. If you were drafting from paper on the drawing board, you'd pull out the eraser right about now to rub out half your effort — and if you were lucky, you'd have an electric eraser. At some point, you simply tore up the paper and started over, or labeled the appropriate dimensions as NTS (Not To Scale). It's AutoCAD to the rescue.

Now that's a stretch

The Stretch command is powerful: It can stretch or move objects, or stretch some and move others at the same time, depending on how you select them. The key to using Stretch is specifying a crossing selection box properly. Chapter 10 gives you more details about crossing selection boxes and how to use them with the Stretch and other editing commands.

REMEMBER

Begin with the drawing from the previous section open in AutoCAD, or open the file `afd03c-i.dwg` contained in the `afd03.zip` download.

1. **On the Modify panel of the Home tab click the Stretch button (shown in the margin).**

   The Stretch command starts, and AutoCAD prompts you to select objects. This is one of those times and one of those command that requires you to watch the command line.

2. **Follow the command line instructions to click points from right to left to define a crossing selection box.**

```
Select objects to stretch by crossing-window or
   crossing-polygon...
Select objects: Click a point above and to the right of the
   upper-right corner of the plate (Point 1 in Figure 3-8).
Specify opposite corner: Move the cursor down and to the
   left. Click a point below the plate, roughly under the
   center of the column (Point 2 in Figure 3-8).
```

   The pointer changes to a dashed rectangle enclosing a rectangular green area, which indicates that you're specifying a crossing selection box. The crossing selection box must cut through the plate and column for the Stretch command to work (see Figure 3-8). You see the following:

```
10 found
Select objects: Press
   Enter.
```

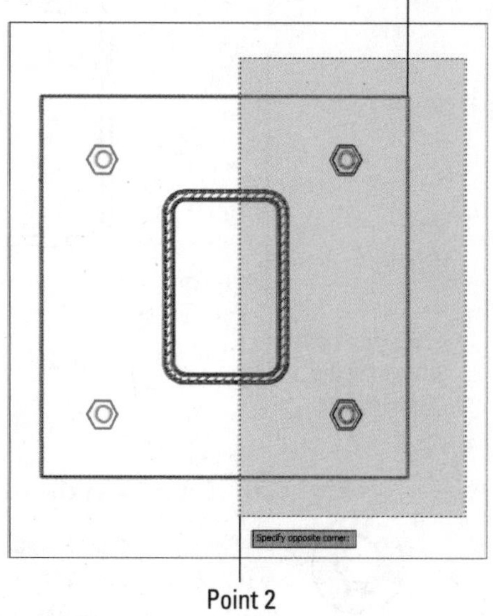

FIGURE 3-8: Specifying a crossing selection box for the Stretch command.

3. **If these drafting settings aren't already set this way, turn off Snap (F9) and turn on Ortho (F8) and Object Snap (F3). Then set a base point for the stretch operation.**

```
Specify base point or [Displacement] <Displacement>: Move
   the mouse pointer over the lower-right corner of the
   plate, and click when you see a square box with an
   Endpoint tooltip.
```

This point serves as the base point for the stretch operation. (Chapter 11 describes base points and displacements in greater detail.) As before, if you can't get the cursor to snap to the endpoint, right-click the **Object Snap** (F3) button and select Endpoint.

4. **Specify a displacement for the stretch operation.**

   AutoCAD prompts you at the command line:

   ```
   Specify second point or <use first point as displacement>:
   ```

   Move the cursor horizontally to the right, type **6**, and press Enter, as shown in Figure 3-9.

   AutoCAD stretches the plate and column with its hatching, and moves the anchor bolts by the distance you indicated (see Figure 3-9).

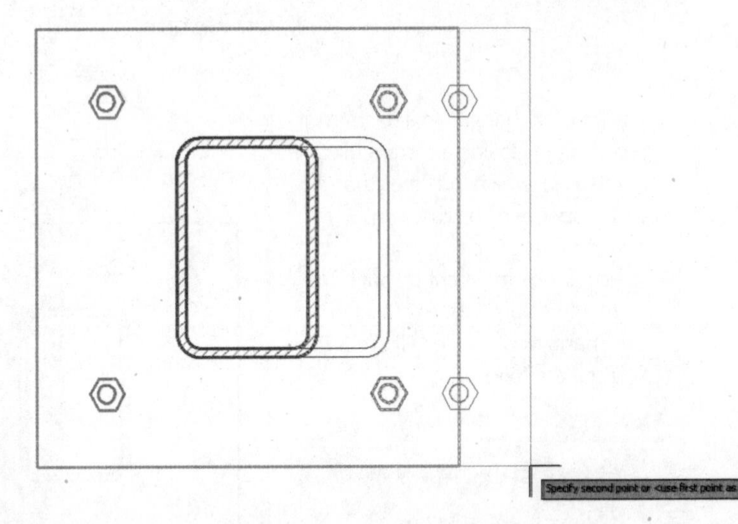

FIGURE 3-9: Stretching the base plate.

5. **Enter Z A at the command line to see the entire drawing.**

TIP

   If the first stretch didn't work properly, press Ctrl+Z to undo, and then try again. Stretch is an immensely useful command that makes you wonder how drafters used to work with only erasers and pencils.

6. **When the stretched drawing is as you want it, press Ctrl+S to save the drawing.**

The drawing `afd03d-i.dwg` contained in the `afd03.zip` download is the stretched version of the base plate. How does it compare with your version?

After some drawing and editing, you may wonder how you're supposed to know when to turn off or on the various status bar modes (Snap, Grid, Ortho, Object Snap, and others). Rest assured that you eventually begin to develop an instinctive sense of when they're useful and when they're in the way. If a mode is in your way or you realize that you need one, you can click the buttons or press the F keys at any time while using the editing and drawing commands. Later in this book, I give you more-specific guidelines.

Following the Plot

Looking at drawings on a computer screen and exchanging them with others via email or websites is all well and good, but sooner or later, someone — maybe you — will want to see the printed versions. *Plotting* drawings in AutoCAD (as CAD geeks refer to *printing* in AutoCAD) is a little more complicated, however, than printing a word processing document or a spreadsheet. You have to address issues such as drawing scales, lineweights, title blocks, and weird paper sizes. I delve deeper into plotting in Chapter 16, but this section describes an abbreviated procedure that can help you generate a recognizable printed drawing.

The steps in the following section show you how to plot the model space portion of the drawing. As Chapter 12 describes, AutoCAD includes a sophisticated feature — paper space layouts — for creating arrangements, usually including a title block, of the drawings you plot. Because I promised you a gentle tour of AutoCAD drafting functions, I save the discussion of paper space layouts and title blocks for later in this chapter. When you're ready for the whole plotting enchilada, turn to Chapter 12 for information about how to set up paper space layouts and see Chapter 16 for full plotting instructions.

Plotting the drawing

The following steps should produce a satisfactory plot in almost all circumstances, though specific brands and models of plotter or printer might sometimes vary the result. Writers call this introduction the CYA, or Cover Your Backside, statement.

Follow these steps to plot a drawing:

1. **Click the Plot button on the Quick Access toolbar or press Ctrl+P.**

   The Quick Access toolbar is at the left end of the program's title bar, just to the right of the Application button. The Plot icon looks like an ordinary desktop printer.

AutoCAD opens the Plot – Model dialog box, with the title bar showing what you're plotting (Model, in this case). See Figure 3-10.

If you don't see all the option windows shown down the right side of the dialog box in Figure 3-10, click the arrow in the lower-right corner.

2. **In the Printer/Plotter section, select a printer from the Name drop-down list.**

   When in doubt, the Default Windows System Printer usually works.

3. **In the Paper Size section, use the drop-down list to select a paper size that's loaded in the printer or plotter.**

   Anything Letter size (8½ x 11 inches) [A4 (210 x 297mm)] or larger works for this example.

4. **In the Plot Area section, select Limits from the What to Plot drop-down list.**

   This is the entire drawing area, which you specified when you set up the drawing in the section "A Simple Setup," earlier in this chapter.

5. **In the Plot Offset section, select the Center the Plot check box.**

   Alternatively, you can specify offsets of 0 or other amounts to position the plot at a specific location on the paper.

6. **In the Plot Scale section, deselect the Fit to Paper check box and choose 1:10 from the Scale drop-down list.**

   You receive no prize for guessing the metric equivalent of 1:10.

7. **In the Plot Style Table (Pen Assignments) section, click the drop-down list and choose `monochrome.ctb`.**

   The `monochrome.ctb` plot style table ensures that all lines appear in solid black rather than as different colors or weird shades of gray. See Chapter 16 for information about plot style tables and monochrome and color plotting.

8. **Click Yes when the question dialog box appears, asking Assign This Plot Style Table to All Layouts?**

   You can leave the remaining settings at their default values as shown in Figure 3-10.

9. **Click the Preview button.**

   If the plot scale you entered in the Plot dialog box is out of sync with the drawing's annotation scale, the Plot Scale Confirm dialog box appears, advising you that the annotation scale isn't equal to the plot scale. This drawing doesn't contain text or dimensions, and you didn't make the hatch annotative, so you can click Continue to generate the plot.

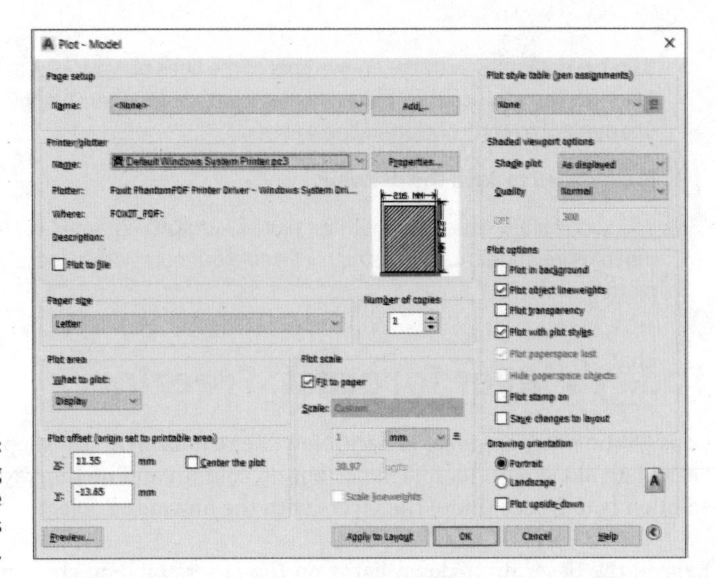

FIGURE 3-10: The Plot dialog box, with the More Options area visible.

Annotative scaling controls the printed size of text, dimensions, hatching, and other types of annotation objects at plot time — as long as the drawing's annotation scale matches the plot scale. I explain annotative objects in Chapters 13 and 14.

The Plot dialog box disappears temporarily, and AutoCAD shows how the plot will look on paper.

10. **Right-click in the preview area and choose Exit.**
11. **If the preview doesn't look the way you want, adjust the settings in the Plot dialog box and click the Preview button again. Continue to adjust the settings until the plot looks right.**
12. **Click OK.**

    The Plot Scale Confirm dialog box pops up again. You may be tempted to click Always Continue Under These Conditions, but I recommend against it until you're more familiar with annotative objects.

    The Plot dialog box closes. AutoCAD generates the plot and sends it to the printer. AutoCAD then displays a Plot and Publish Job Complete balloon notification from the right end of the status bar. A link labeled Click to View Plot and Publish Details displays more information about the plot job.

13. **Click the Close (X) button in the Plot and Publish Job Complete balloon notification.**

    The balloon notification disappears.

TIP

If you're not happy with the lineweights of the lines on your plot at this point, fear not: You can use the lineweights feature (see Chapter 9) or plot styles (see Chapter 16) to control plotted lineweights.

14. **Press Ctrl+S to save the drawing.**

    You successfully executed your first plot in AutoCAD. Chapter 16 tells you much more about AutoCAD's highly flexible but occasionally perplexing plotting system.

Today's layer forecast: Freezing

Oops — the nutty engineer is back. Now the requirement is a shop drawing of the base plate alone with its four holes but no column and no nuts. No problem. The solution is to freeze (hide) the layers with the unwanted objects.

Expand the Layer drop-down list from the Layers tab and click the 0 layer name to make it the current one, because the current layer (which had been Hatch) can never be frozen.

Expand the Layer drop-down again, but this time click the sunshine icons beside the Column, Hatch, and Nuts layers. As you do so, each sunshine turns to a snowflake and everything on the corresponding layer disappears. Frozen layers cannot be seen, edited, or plotted. Click a snowflake, and its layer thaws to sunshine. I discuss layers in Chapter 9.

Congratulations! You have now used the few commands that will account for the majority of your long and prosperous AutoCAD career!

» Starting new drawings

» Setting up model space

» Creating and using drawing templates

Chapter 4
Setup for Success

The good news is that AutoCAD is very powerful and versatile. And the not-quite-so-good news is that it's extremely powerful and versatile. This means that you can set it up to work in almost any segment of nearly any industry or application, just about anywhere in the world. AutoCAD is used for mechanical design, electrical and electronic circuit schematics, hydraulics, buildings, bridges, theatre stage layouts, cloth-cutting layouts in the clothing industry, designing big floppy clown shoes, keeping track of season's ticket holders' seats in a hockey arena, and so on. In short, everything from autos to zoos.

Because most companies and schools will already have set things up to suit their particular standards, all you need to do — and in fact, *must* do — is to use their template files. If you're working alone, however, you should read on to learn how to create template files of your own.

In the long (or even medium or short) term, AutoCAD is much easier to use when you start from a drawing that's already set up properly. There can be quite a few things to set up, but if you do things properly, you need to do it only once.

Sloppy setup really becomes apparent when you try to plot (note that CAD geeks say "Plot" whereas normal people say "Print") your drawing. Things that seemed more or less okay as you zoomed around on the screen are suddenly the wrong size or don't look right on paper. Chapter 16 covers plotting, but the information in this chapter is a necessary prerequisite to successful plotting and sheet setup.

If you don't get this stuff correct, there's a good chance you'll find that the plot sickens.

This chapter describes the decisions you need to make before you set up a new drawing, shows the steps for doing a complete and correct setup, and demonstrates how to save setup settings for reuse.

Don't assume that you can just create a new blank DWG file and start drawing things. *Do* read this chapter before you wander too far away from this book. Many AutoCAD drawing commands and concepts depend on proper drawing setup, so you'll have a much easier time drawing and editing things after you do your setup homework.

Then, after you digest the detailed drawing setup procedures described in this and the following chapters, use the "AutoCAD Drawing Setup Roadmap" in the book's cheat sheet as a quick reference. To get to the cheat sheet, go to `www.dummies.com` and type **AutoCAD For Dummies** in the Search box.

A Setup Roadmap

Before you start the drawing-setup process, you need to make only two initial decisions about your new drawing:

- What system of measure — metric or imperial — will you use?
- What drawing units will you use?

Choosing your units

AutoCAD is extremely flexible about drawing units; it lets you have them *your* way. Usually, you choose the type of units that you normally use to talk about whatever you're drawing: feet and inches for a building in the United States and most of Canada, millimeters for almost everything in almost all the rest of the world, and so on.

Speaking of millimeters, here's another choice you have to make even before you choose your *units* of measure — and that's your *system* of measure.

Most of the world abandoned local systems of measure generations ago. Even widely adopted ones, like the imperial system, have mostly fallen by the wayside — except, of course where feet, inches, pounds, gallons, and degrees Fahrenheit still rule.

Make sure everyone agrees on the units being used. NASA once crashed a very expensive space probe onto Mars because of a mix of imperial and metric systems.

During drawing setup, you choose settings for *length* units (for measuring linear objects and distances) and *angle* units (for measuring angles between nonparallel objects or points on arcs or circles) in the Drawing Units dialog box, as shown in Figure 4-1. (I show you how to specify these settings in the section "Setting your units," later in this chapter.) AutoCAD's length unit types are as follows:

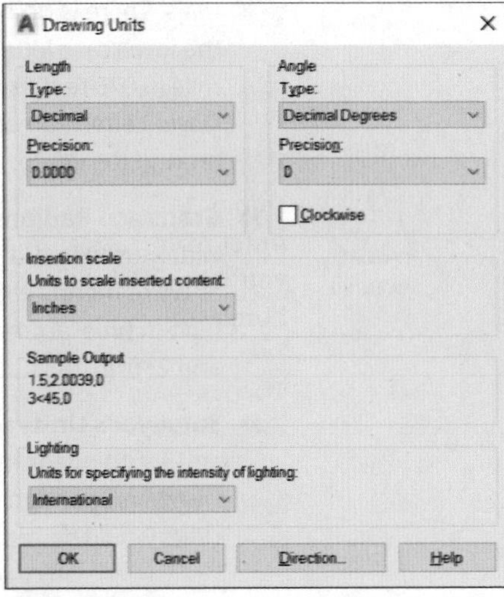

FIGURE 4-1:
The Drawing Units dialog box.

- **Architectural** units are in feet and inches and use fractions to represent partial inches: for example, 12'3 1/2". The base unit is the inch, unless otherwise specified, so when you enter a number like 147.5, AutoCAD will understand it to be 147 1/2" or 12'3 1/2".
- **Decimal** units are *unitless* — that is, they're not based on any particular real-world unit, although this is the type used for metric drawings. With decimal units, each unit in the drawing could represent an inch, a millimeter, a parsec, a furlong, a fathom, a cubit (should you be into building arks in case that super rainy day comes), or any other unit of measure you deem suitable, from Danish alens to the Prussian zoll. An example would be 15.5.
- **Engineering** units are in feet and inches and use decimals to represent partial inches: for example, 12'3.5".
- **Fractional** units, like decimal units, are unitless and show values as fractions rather than decimal numbers: for example, 15 1/2.
- **Scientific** units, which are unitless and display values as exponents, are used for drawing really tiny or really large things. If you design molecules or galaxies, this is the unit type for you. Examples are 15.5E+06 (which is 15,500,000) and 15.5E–06 (which is 0.0000155).

AutoCAD's angle unit types are as follows:

- **Decimal Degrees** show angles as decimal numbers and are by far the easiest to work with, if your type of work allows it.

- **Deg/Min/Sec** is based on the old style of dividing a degree into 60 minutes and minutes into 60 seconds. Seconds aren't fine enough to display AutoCAD's precision capabilities, though, so seconds can be further divided into decimals. There is no degree symbol on a standard keyboard, so AutoCAD uses the lowercase letter *d*. An example would be 45d30'10.7249". One nautical mile (6,076 feet) is approximately 1 minute of arc of longitude on the equator. David Letterman once said that the equator is so long that it would reach once around the world.
- **Grads** and **Radians** are mathematically beautiful (so we're told) but are not widely used in drafting. Apparently, the French artillery uses grads but as long as we're friends with them we shouldn't have to worry. Some of AutoCAD's built-in programming language uses radians for angles. There are 400 grads, and 2*pi (6.2831. . .) radians, in a circle.
- **Surveyor's Units** type is similar to Deg/Min/Sec but uses north, east, west, and south quadrants (quarter circles) rather than an entire circle. An angle in Deg/Min/Sec might measure 300d0'.00", while the same angle in Surveyor's Units would be represented as S 30d0'0.00" E.

REMEMBER

The unit types you'll most likely use are Decimal, Architectural, and Decimal Degrees. You'll know or be told if you need to use one of the other types.

TECHNICAL STUFF

AutoCAD always works internally to something like 16-digit accuracy. Changing the type of units changes only how values are *displayed*, suitably rounded off, but not does change the internal accuracy. For example, a line that's 15.472563 decimal units long would be displayed as 12'3 1/2" long when units are changed to Architectural. The actual length — and any calculations based on it — remains at 15.472563 and does not change to 15.5".

When you use dash-dot linetypes (Chapter 11) and hatching (Chapter 15) in a drawing, it matters to AutoCAD whether the drawing uses an imperial (inches, feet, miles, and so on) or metric (millimeters, meters, kilometers, and so on) system of measure. The MEASUREINIT and MEASUREMENT system variables control whether the linetype and hatch patterns that AutoCAD lists for you to choose from are scaled with inches or millimeters in mind as the plotting units. For both variables, a value of 0 (zero) means inches (that is, an imperial-units drawing), whereas a value of 1 means millimeters (that is, a metric-units drawing). When you start from an appropriate template drawing (as described in the section "A Template for Success," later in this chapter), the system variable values will be set correctly, and you won't ever have to think about it. For an explanation of system variables and how to set them, see Chapter 26.

So why are there two variables? Simple. MEASUREINIT (short for *measurement initial*) sets the default value for new drawings, while MEASUREMENT sets the value for the current drawing.

AutoCAD automatically sets the measurement system according to your country when it's being installed, but you can change it.

Weighing up your scales

Somewhat surprisingly, you don't need to consider scale when setting up a new drawing. In fact, rule number 1 in AutoCAD is to always draw everything full size. Hmmm, that makes quite a few rule number 1s, doesn't it?

"Wait a minute! I want to draw a map of the known universe! If I draw full size, where am I going to find a sheet of paper big enough to print it — and who will help me fold it?"

Trust me. All will become clear shortly.

DRAFTING ON PAPER VERSUS ELECTRONICALLY

If you've ever done paper-and-pencil drafting (and there are fewer and fewer of us left who have), you'll find that AutoCAD's electronic paper works backward from dead-tree paper.

- **Dead-tree paper:** In the (distant) past, we considered the approximate size of the object that we were drawing by hand and the views we wanted, selected a suitable sheet of paper from a set of standard sizes, and then scaled the drawing of the object to suit the sheet of paper. We were constantly translating sizes between the real-world object and our drawing of it. The height of text and the size of the components in dimensions were fixed. Again, scales were selected from a list of preferred values, such as 1:2. You rarely saw a paper drawing at a scale of, say, 1:2.732486921.
- **Electronic paper:** Now we draw everything full size and then tell the PLOT command to grow or shrink things accordingly. This approach is much easier because we never have to translate sizes. Chapters 13–16 show you how AutoCAD now makes text and dimension sizing extremely simple.

"Okay," you're saying, "I understand that I need to print my drawings at a *scale* acceptable to the printer and discipline I work in. But when I'm drawing stuff full size, when do I need to worry about the *scale factor?*" Grab yourself a nice mug of cocoa and settle down 'round the fire because I'm going to tell you. By now you know (because I've told you so) that you draw real things full size, but drawings contain other things that are *not* real, such as text, dimensions, hatch patterns, title blocks, dash-dot linetypes, and so forth. And those nonreal things need to be legible on your plotted drawing.

Say, for example, you draw a plan of your big garage, and now you want to plot it on an 11-x-17-inch sheet of paper. No problem; just tell the PLOT command to scale everything down by a scale factor of 1:24, which architects would commonly represent as 1/2"=1'0".

Oops, problem. Text annotations are typically about 3/32" or 1/8" high. Now, if you draw your 6-inch–wide wall full size, put a 1/8"-high title beside it, and then print the drawing at a scale of 1:24, the wall itself will measure 1/4" on the sheet, and the note will be an illegible little speck beside it. You fix it by making the text 24 times larger, or 3 inches tall, so that it scales down to the correct size when plotted. See, sometimes two wrongs do make a right. Unusually, it takes three, although three rights make a left and two Wrights made an airplane.

DRAWING SCALE VERSUS DRAWING SCALE FACTOR

CAD users employ two different ways of talking about a drawing's intended plot scale: drawing scale and drawing scale factor.

Drawing scale is the traditional way of describing a scale: "traditional" because it existed long before CAD came to be. Drawing scales are expressed with an equal sign or colon: for example, 1/8" = 1'0", 1:20, or 2:1. You can translate the equal sign or colon as "corresponds to." In all cases, the measurement to the left of the equal sign or colon indicates a paper measurement, and the number to the right indicates a real-world measurement. A metric drawing scale is usually expressed without units, as a simple ratio. Thus, a scale of 1:20 means 1 unit on the plotted drawing corresponds to 20 units in the real world.

Drawing scale factor is a single number that represents a multiplier, such as 96, 20, or 0.5. The drawing scale factor for a drawing is the conversion factor between a measurement on the plot and a measurement in the real world.

The "Drawing scale versus drawing scale factor" sidebar explains how you arrive at the scale factor, but this isn't an issue until it comes time to annotate your drawing. I cover noncontinuous linetype scaling in Chapter 9, text sizing in Chapter 13, dimensions in Chapter 14, and crosshatching in Chapter 15. The good news is that AutoCAD can now take care of all this automatically, but a well-placed rumor has it that there may still be a few drawings out there that did it the old way, so I cover both methods in the appropriate chapters.

WARNING

You shouldn't just invent some arbitrary scale based on what looks okay on whatever size paper you happen to have handy. Most industries work with a small set of approved drawing scales that are related to one another by factors of 2 or 5 or 10. If you use other scales, you'll be branded a clueless newbie, at best. At worst, you'll have to redo all your drawings at an accepted scale.

WARNING

The SCale command covered in Chapter 11 has nothing to do with ladders, fish, or setting drawing scales or scale factors!

Table 4-1 lists some common architectural drawing scales, using both imperial and metric systems of measure. The table also lists the drawing scale factor corresponding to each drawing scale and the common uses for each scale. If you work in industries other than those listed here, ask drafters or coworkers what the common drawing scales are and for what kinds of drawings they're used.

TABLE 4-1 Common Architectural Drawing Scales

Drawing Scale	Drawing Scale Factor	Common Uses
1/16" = 1'-0"	192	Large-building plans
1/8" = 1'-0"	96	Medium-sized building plans
1/4" = 1'-0"	48	House plans
1/2" = 1'-0"	24	Small-building plans
1" = 1'-0"	12	Details
1:200	200	Large-building plans
1:100	100	Medium-sized building plans
1:50	50	House plans
1:20	20	Small building plans
1:10	10	Details

TIP

After you choose a drawing scale, engrave the corresponding drawing scale factor on your desk, write it on your hand (don't reverse those two, okay?), and put it on a sticky note on your monitor. You eventually need to know the drawing scale factor for many drawing tasks, as well as for some plotting. You should be able to recite the drawing scale factor of any drawing you're working on in AutoCAD without even thinking about it.

WARNING

Even if you're going to use the Plot dialog box's Fit to Paper option (rather than a specific scale factor) to plot the drawing, you still need to choose a scale to make the nonreal things (such as text, dash-dot linetypes, hatch patterns, and so on) appear at a useful size. I cover plotting in Chapter 16.

Thinking about paper

You don't normally need to worry about the size of the paper that you want to use for plotting your drawing until much later in the drawing process. And that's the beauty of CAD: You can easily move views around to suit after you get the basic object drawn, and you don't need to worry about scale factors until you're ready to add annotations. I cover this in excruciatingly more (just kidding, it's actually quite simple) detail in Chapters 12–16. Here again, most industries use a small range of standard sheet sizes.

Here are the two ways of laying out a drawing so it's ready to be plotted:

- **In model space:** In this process, everything is drawn in model space, which at one time was the only known universe in AutoCAD. The drawing is created full size, while text and dimension component sizes, hatch pattern scaling, title blocks, and borders are all created at the inverse of the final plotting scale. This was the only way for many years, so you'll encounter many drawings that were made this way.
- **In paper space:** Finally, Autodesk programmers figured out how to tunnel through into a parallel universe called paper space, which revolutionized drawing production. Current preferred practice is to draw the object full size in model space, cut a viewport in paper space so you can look through to the model space, and then apply documentation such as dimensions and text in paper space.

I cover model space versus paper space in Chapter 16.

Defending your border

The next decision to make is what kind of border your drawing needs. The options include a full-blown title block, a simple rectangle, or nothing at all around your drawing. If you need a title block, do you have one, can you borrow an existing one, or will you need to draw one from scratch? Although you can draw title block geometry in an individual drawing, you'll save time by reusing the same title block for multiple drawings. Your company or client should already have a standard title-block drawing ready to use, or someone else who's working on your project may have created one for the project.

The most efficient way of creating a title block is as a separate DWG file, drawn at its normal plotted size (for example, 36" x 24" for an architectural D-size title block, or 841mm x 594mm for an ISO A1-size version). You then insert (see Chapter 17) or xref (see Chapter 18) the title block drawing into each drawing sheet.

A Template for Success

When you start in either the Drafting & Annotation workspace (as I do throughout this book) or the old AutoCAD Classic workspace, AutoCAD creates a new, blank drawing configured for 2D drafting. Depending on where you live (your country, not your street address!) and the dominant system of measure used there, AutoCAD will base this new drawing on one of two default drawing templates:

- `acad.dwt` for the imperial system of measure, as used in the United States
- `acadiso.dwt` for the metric system, used throughout most of the rest of the galaxy

In AutoCAD LT, the two default templates are `acadlt.dwt` and `acadltiso.dwt`. When you create a new drawing by using the big New icon, the new drawing is based on the default template. On the other hand, if you use the NEW command in the Application (red A) menu or from the NEW icon in the Quick Access menu, the Select Template dialog box appears, as shown in Figure 4-2, so you can choose a template on which to base your new drawing.

You may be familiar with Microsoft Word or Excel template files, and AutoCAD drawing templates work pretty much the same way because Autodesk stole the idea from them, encouraged, of course, by Microsoft, although one could argue that it was the other way around.

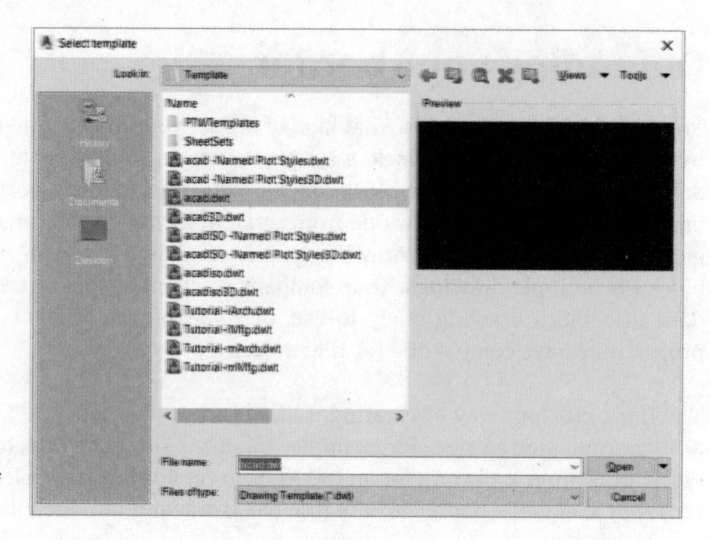

FIGURE 4-2: A toolbox of templates.

A *template* is simply a drawing whose name ends in the letters *DWT*, which you use as the starting point for another drawing. When you create a new drawing from a template, AutoCAD makes a copy of the template file and opens the copy in a new drawing editor window. The first time you save the file, you're prompted for a new filename to save to; the original template file stays unchanged.

Using a suitable template can save you time and worry because many of the setup options are already set correctly for you. You know the drawing will plot correctly; you just have to worry about getting the geometry and text right. Of course, all this optimism assumes that the persons who set up the template knew what they were doing.

The stock templates that come with AutoCAD are okay as a starting point, but you'll want to modify them to suit your purposes or create your own from scratch. In particular, the stock AutoCAD templates are probably not set up for the scales or layers you'll want to use. The instructions in the rest of this chapter tell you how to specify scale-dependent setup information.

So, the only problems with templates are creating good ones and then later finding the right one to use when you need it. Later in this chapter, in the "Making Templates Your Own" section, I show you how to create templates from your own setup drawings. Here I show you how to use an already-created template — say, one of the templates that comes with AutoCAD or one you get from a CAD-savvy colleague. If you're lucky, someone in your office has created suitable templates that you can use to get going quickly.

Follow these steps to create a new drawing from a template drawing:

1. **Run the NEW command by pressing Ctrl+N or clicking the Application button and choosing New.**

   The Select Template dialog box appears.

2. **Click the name of the template you want to use as the starting point for your new drawing and then click the Open button.**

   A new drawing window with a temporary name, such as `Drawing2.dwg`, appears. The template you opened remains unchanged on your computer's hard drive.

   Depending on which template you choose, your new drawing may open in a paper space layout, not in model space. If that's the case, click the Model button on the status bar before changing the settings described in the next section. I describe how to set up and take advantage of paper space layouts in Chapter 16.

3. **Press Ctrl+S or click the Application button and choose Save to save the file under a new name.**

   Take the time to save the drawing to the appropriate name and location now.

4. **Make needed changes.**

   With most of the templates that come with AutoCAD, consider changing the units, limits, grid and snap settings, linetype scale, and dimension scale. See the next section for instructions.

5. **Save the drawing again.**

   If you'll need other drawings in the future similar to the current one, consider saving your modified template as a template in its own right. See the section "Making Templates Your Own," later in this chapter, for the lowdown on saving templates.

TIP

A few of the remaining templates that come with AutoCAD include title blocks for various sizes of sheets. In addition, most templates come in two versions: one for people who use color-dependent plot styles and one for people who use named plot styles. You probably want the color-dependent versions. Chapter 16 describes the two kinds of plot styles and why you probably want the color-dependent variety.

Making the Most of Model Space

After you decide on drawing scale and sheet size, you're ready to set up your drawing. Most drawings require a two-part setup:

1. Set up model space, where you'll create most of your drawing.
2. Create one or more paper space layouts for plotting.

As I explain in Chapter 2, model space is the infinitely large (for all practical purposes), three-dimensional environment in which you create the real objects you're drawing. You can set up your model space as described in this section; Chapter 16 introduces you to setting up your paper space layouts.

Setting your units

Follow these steps to set the linear and angular units that you want to use in your new drawing:

1. **Click the Application button and then choose Units from the Drawing Utilities group.**

   The Drawing Units dialog box appears, as shown in Figure 4-3.

2. **Choose a linear unit type from the Length Type drop-down list.**

   Choose the type of unit representation that's appropriate for your work. Engineering and Architectural units are displayed in feet and inches; the other types of units aren't tied to any particular unit of measurement. You decide whether each unit represents a millimeter, centimeter, meter, inch, foot, or something else. Your choice is much simpler if you're working in metric: Choose Decimal units.

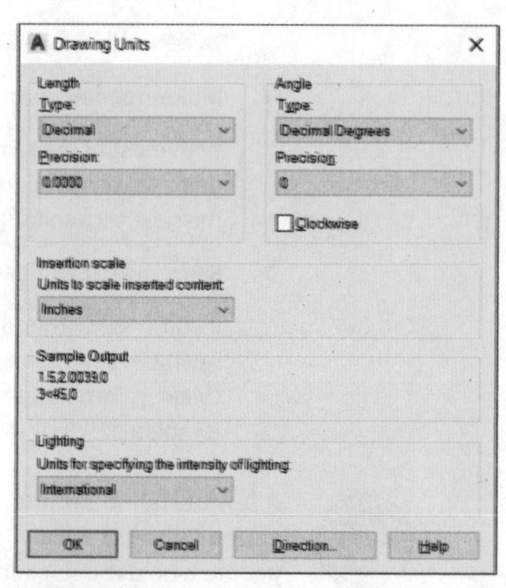

FIGURE 4-3:
Set your units here.

TIP

AutoCAD *can* think in inches! If you're using Engineering or Architectural units (feet and inches), AutoCAD interprets any distance or coordinate you enter as that many *inches*. You must use the ' (apostrophe) character on your keyboard to indicate a number in feet instead of inches. Using the " symbol to indicate inches isn't required but is acceptable.

3. **From the Length Precision drop-down list, choose the level of precision you want when AutoCAD displays coordinates and linear measurements.**

   The Length Precision setting controls how precisely AutoCAD displays coordinates, distances, and prompts in some dialog boxes. For example, the Coordinates section of the status bar displays the current coordinates of the cursor, using the current precision.

REMEMBER

   The linear and angular precision settings affect only AutoCAD's display of coordinates, distances, and angles on the status bar, in dialog boxes, and in the command line and Dynamic Input tooltip areas. For drawings stored as DWG files, AutoCAD *always* uses maximum precision to store the locations and sizes of all objects that you draw, regardless of how many decimal places you choose to display in the Drawing Units dialog box. In addition, AutoCAD provides separate settings for controlling the precision of dimension text. (See Chapter 14 for details.)

4. **Choose an angular unit type from the Angle Type drop-down list.**

   Decimal Degrees and Deg/Min/Sec are the most common choices.

   The Clockwise check box and the Direction button provide additional angle measurement options, but you'll rarely need to change the default settings: Unless you're a land surveyor, measure angles counterclockwise and use east as the 0-degree direction.

5. **From the Angle Precision drop-down list, choose the degree of precision you want when AutoCAD displays angular measurements.**

6. **In the Insertion Scale area, choose the units of measurement for this drawing.**

   Choose your base unit for this drawing — that is, the real-world distance represented by one AutoCAD unit. I discuss the significance of this in Chapter 17.

TECHNICAL STUFF

   The AutoCAD (but not the AutoCAD LT) Drawing Units dialog box includes a Lighting area where you specify the unit type to be used to measure the intensity of photometric lights. I introduce lighting as part of rendering 3D models in Chapter 23.

7. **Click OK to exit the dialog box and save your settings.**

Making the drawing area snap-py (and grid-dy)

For the first three decades, AutoCAD's grid consisted of a set of evenly spaced dots that served as a visual distance reference. You can still configure a dot grid, but starting with AutoCAD 2011, the default is a snazzy graph-paper–like grid made up of a network of lines.

AutoCAD's snap feature creates a set of evenly spaced, invisible hot spots, which make the cursor move in nice, even increments as you specify points in the drawing. Both Grid mode and Snap mode are like the intersection points of the lines on a piece of grid paper, but the grid is simply a visual reference — it never prints — whereas Snap mode constrains the points that you can pick with the mouse. You can set grid and snap spacing to different values.

REMEMBER

You don't have to toggle Snap mode on and off. It's on only when you're using a command that asks you to specify a point. For example, when Snap is on, you can move your crosshairs freely around the screen, but when you start the Line command, Snap mode kicks in, and your cursor jumps to the closest snap point.

Set the grid and the snap intervals in the Drafting Settings dialog box by following these steps:

1. **Right-click the Snap Mode or Grid Display button on the status bar and choose Settings from the menu that appears.**

   The Drafting Settings dialog box appears with the Snap and Grid tab selected, as shown in Figure 4-4.

   The Snap and Grid tab has six sections, but the Snap Spacing and Grid Spacing areas on that tab are all you need to worry about for most 2D drafting work.

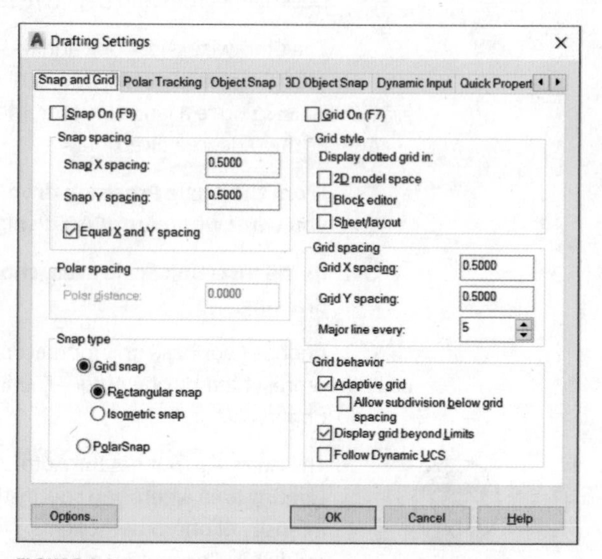

FIGURE 4-4:
Get your drafting settings here!

2. **Select the Snap On check box to turn on Snap mode.**

   This action enables default snaps half a unit apart (ten units apart if you're working with the default metric template).

REMEMBER

AutoCAD usually has several ways of doing things. You can also click the Snap Mode button on the status bar to toggle snap on and off; the same goes for the Grid Display button and the grid setting. Or you could press the function keys: F7 toggles the grid off and on, and F9 does the same for Snap mode.

3. **Enter the snap interval you want in the Snap X Spacing text box.**

   Use the information in the sections preceding these steps to decide on a reasonable snap spacing.

TIP

   If the Equal X and Y Spacing check box is selected, the Y spacing changes automatically to equal the X spacing, which is almost always what you want. Deselect the check box if you want to specify snap spacings that are different vertically and horizontally.

4. **Select the Grid On check box to turn on the grid.**
5. **Enter the desired grid spacing in the Grid X Spacing text box.**

   Use the information in the sections preceding these steps to decide on a reasonable grid spacing.

   As with snap spacing, if the Equal X and Y Spacing check box is selected, the Y spacing automatically changes to equal the X spacing. Again, you usually want to leave it that way.

TECHNICAL STUFF

   X measures horizontal distance; Y measures vertical distance. The AutoCAD drawing area normally displays an X and a Y icon in the lower-left corner in case you forget.

TIP

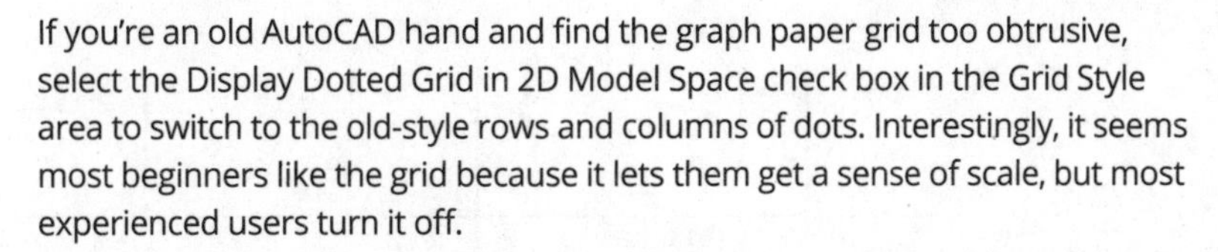

   If you're an old AutoCAD hand and find the graph paper grid too obtrusive, select the Display Dotted Grid in 2D Model Space check box in the Grid Style area to switch to the old-style rows and columns of dots. Interestingly, it seems most beginners like the grid because it lets them get a sense of scale, but most experienced users turn it off.

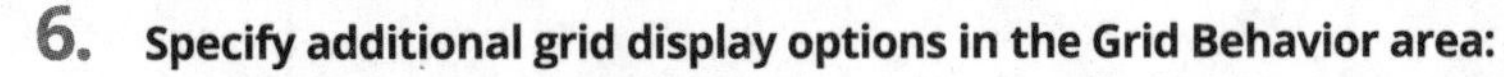

6. **Specify additional grid display options in the Grid Behavior area:**
   - *Adaptive Grid:* AutoCAD changes the density or spacing of the grid lines or dots as you zoom in and out.
   - *Allow Subdivision Below Grid Spacing*: Select this check box in conjunction with the Adaptive Grid check box. The spacing can go lower than what you've set, and it may go higher if you're zoomed a long way out of your drawing. (If it didn't, you wouldn't be able to see your drawing for the grid!)

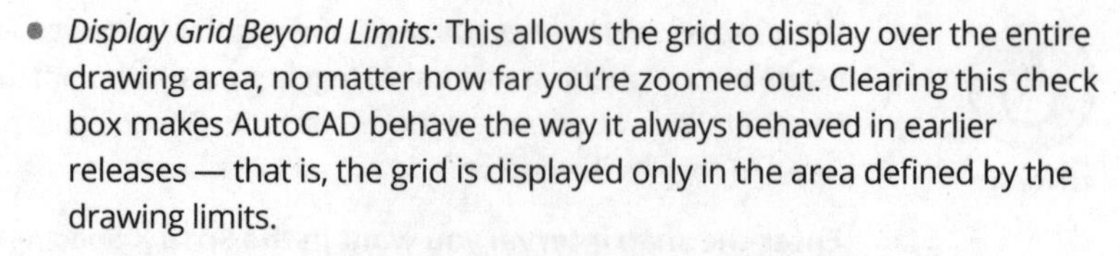

- *Display Grid Beyond Limits:* This allows the grid to display over the entire drawing area, no matter how far you're zoomed out. Clearing this check box makes AutoCAD behave the way it always behaved in earlier releases — that is, the grid is displayed only in the area defined by the drawing limits.

TECHNICAL STUFF

- *Follow Dynamic UCS:* This option (not available in AutoCAD LT) is a 3D-specific feature that changes your drawing plane as you mouse over 3D objects. I cover this feature in Chapter 21.

7. **Click OK to close the Drafting Settings dialog box.**

Setting linetype, text, and dimension scales

Even if you've engraved the drawing scale factor on your desk and written it on your hand — not vice versa — AutoCAD doesn't know the drawing scale until you enter it. Keeping AutoCAD in the dark is fine as long as you're just drawing continuous lines and curves representing real-world geometry because you draw these objects at their real-world size, without worrying about plot scale.

However, as soon as you start using text, dimensions, and noncontinuous dash-dot *linetypes* (line patterns that contain gaps in them), you need to tell AutoCAD how to scale the gaps in the linetypes based on the plot scale. If you forget this, the dash-dot linetype patterns can look waaaay too big or too small. Figure 4-5 shows what I mean.

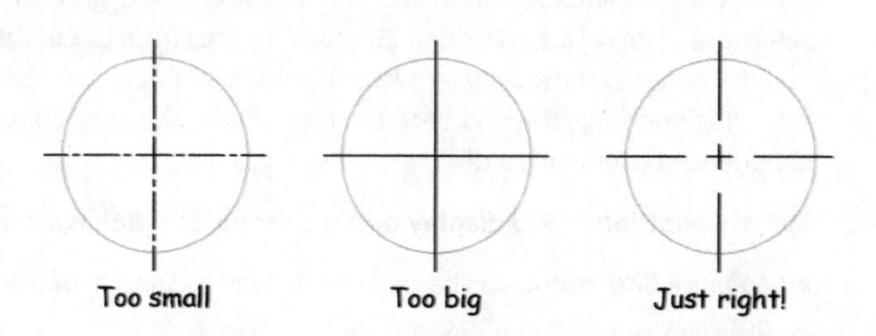

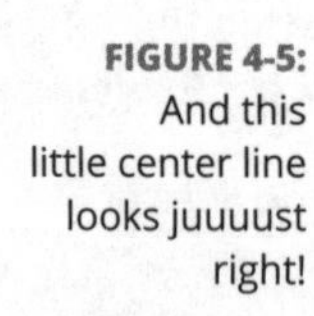

FIGURE 4-5: And this little center line looks juuuust right!

The process of setting the three relevant scale factors was messy in earlier releases but is now trivial. The Annotation Scale button near the right-hand end of the status bar displays the current drawing scale, which by default is 1:1/100%. Click it to display a list of 25 standard drawing scales.

TIP

You can always change the current drawing scale later if the need arises.

At some point in your career, you may encounter a drawing that was created in an earlier release. In the appropriate chapters, I explain how to deal with scale factors in these older drawings.

Entering drawing properties

I recommend one last bit of housekeeping before you're finished with model space drawing setup: Enter summary information in the Drawing Properties dialog box, as shown in Figure 4-6. Click the Application button; in the Drawing Utilities section, choose Drawing Properties to open the Drawing Properties dialog box; then click the Summary tab. Enter the drawing scale and the drawing scale factor you're using in the Comments area, plus any other information you think useful.

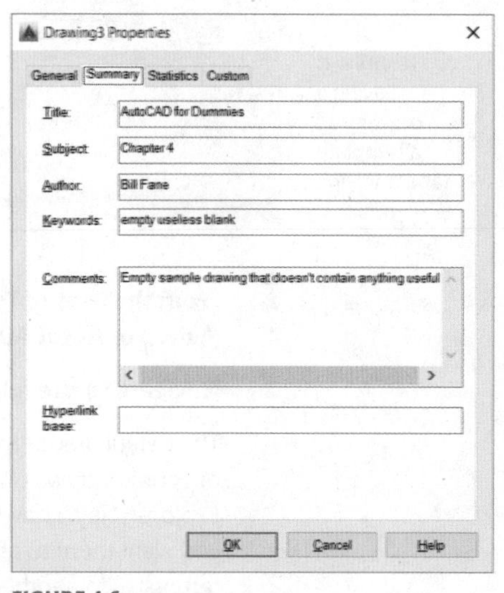

FIGURE 4-6: Surveying your drawing's properties.

REMEMBER

Don't confuse *drawing* properties (which are really file properties) with your drawing's *object* properties — they're different things. The properties you enter here can help you or others you love when they open your drawing and wonder how you set it up. *Object* properties are a big enough topic to merit their own chapter. See Chapter 9.

TIP

Want to find a specific file hidden on your hard drive? You can search for values you've entered in the Drawing Properties dialog box by using Windows Explorer's search functions.

Making Templates Your Own

You can create a template from any DWG file by using the Save Drawing As dialog box. Follow these steps to save your drawing as a template:

1. **Click Save As on the Quick Access toolbar.**

   The Save Drawing As dialog box appears, as shown in Figure 4-7.

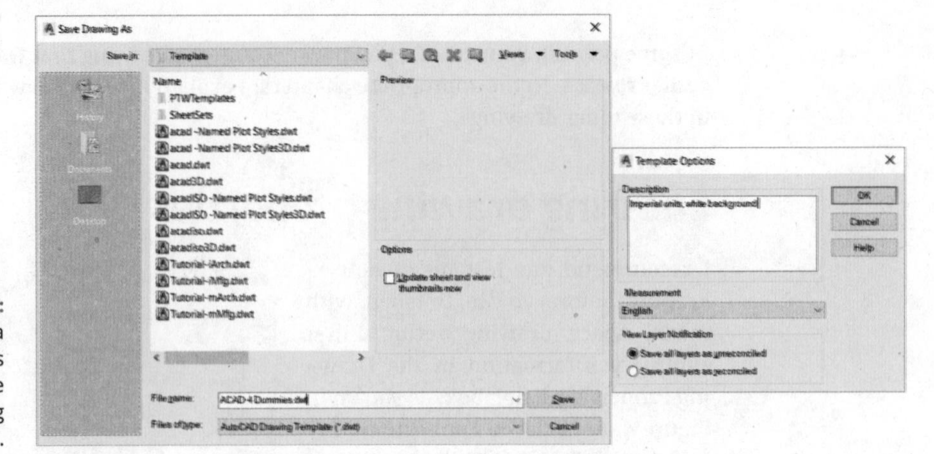

FIGURE 4-7: Saving a drawing as a template and applying options.

2. **From the Files of Type drop-down list, choose AutoCAD Drawing Template (*.dwt) or AutoCAD LT Drawing Template (*.dwt).**

3. **Navigate to the folder where you want to store the drawing template.**

   The default folder for AutoCAD drawing templates is buried deep in the bowels of your Windows user profile, which by default isn't visible in Windows Explorer. Hey, I never said this made any sense! Save your templates there if you want them to appear in AutoCAD's Select Template list. You can save your templates in another folder, but if you want to use them later, you have to navigate to that folder every time you want to use them unless you change your Options settings. I explain how to do this later in the chapter.

4. **Enter a name for the drawing template in the File Name text box and then click Save.**

   A dialog box for the template description and units appears.

5. **Specify the template's measurement units (English or Metric) from the drop-down list.**

   Enter the key info now; you can't do it later unless you save the template to a different name. Don't bother filling in the Description field; AutoCAD doesn't display it in the Select Template dialog box. Don't worry about the New Layer Notification area shown in Figure 4-7 for now; I tell you all about drawing layers in Chapter 9.

6. **Click OK to save the file.**

   The Template Options dialog box closes, and the template is saved to your hard drive.

7. **To save your drawing as a regular drawing, click Save As on the Quick Access toolbar.**

   The Save Drawing As dialog box appears again.

8. **From the Files of Type drop-down list, choose AutoCAD 2018 Drawing (*.dwg).**

   Choose the AutoCAD LT equivalent, if that's your version. Through the years, AutoCAD has changed the file format several times. A drawing produced with an older release can always be opened by a newer release, but the reverse is not always true. Chapter 24 explains which file versions can be opened by which release.

9. **Navigate to the folder where you want to store the drawing.**

   Use a different folder from the one with your template drawings.

10. **Enter the name of the drawing in the File Name text box and click Save.**

   The file is saved. Now, when you save it in the future, the regular file — not the template file — is updated.

REMEMBER

Okay, do you want to know the real secret behind template files? One letter. That's right, the only difference is that a drawing file has the extension DWG, and a template's extension is DWT. You can convert a drawing to a template just by using Windows Explorer to rename the extension. The further secret to this is that a DWT can hold any information that you can put in a DWG. For example, you may often need to produce a drawing of a new machine shaft. All details are the same as previous shafts except for the length. Draw it once and convert it to a template. Now, when you need a new shaft drawing, just start it from the shaft template, change the length, and you're done!

TIP

In Chapter 3, I warn you to not use the New buttons on the Application menu or the Quick Access toolbar. These buttons launch the QNEW (Quick NEW) command, which bypasses the Select Template dialog box and creates a drawing from a default template. The good news is that you can easily put the Quick into QNEW so that it uses your favorite template as the default. Just do this:

1. **Click the Application button and then click the Options button at the lower-right corner of the Application menu.**
2. **On the Files tab, click + next to Template Settings, and then choose Default Template File Name for QNEW.**

   The QNEW default file name setting is None, which causes QNEW to act just like NEW (that is, QNEW opens the Select Template dialog box). Specify the name of your favorite template here, and you get a new drawing file based on it every time you click QNEW.

FINDING YOUR USER FOLDER

Microsoft insists that any program that uses files unique to or modifiable by individual users must be kept in a separate folder system for each user so that different people sharing a computer won't mess up each other's settings. There are just two minor flaws with this edict:

- Hardly anyone shares a computer these days. In fact, my tech editor claims that a typical user today has multiple computers, and brags (or complains) that he has two, not including Android models. I have three, and my wife has one.
- By default, this is a hidden system folder that you can't see in Windows Explorer, making it difficult to copy, rename, or move files in it.

The solution:

Windows 7

1. Start Windows Explorer.
2. Choose Organize ➪ Folder and Search Options.
3. Under the View tab, click the Show Hidden Files and Folders button and then deselect the Hide Extensions for Known File Types check box.
4. Click OK.

Windows 8 and 8.1

1. In File Explorer, choose the View tab.
2. Click the Options button.
3. Follow the remaining steps as for Windows 7 (in the preceding list).

Windows 10 and 11

1. In File Explorer, choose the View tab.
2. In the Show/Hide panel, turn on the Hidden Items option.

Now you can see what's happening in your private folder.

AutoCAD stores drawing templates and many other support files under your Windows user folder. To discover where your template folder is hiding, open the Options dialog box. On the Files tab, choose Template Settings and then Drawing Template File Location, as shown in Figure 4-8.

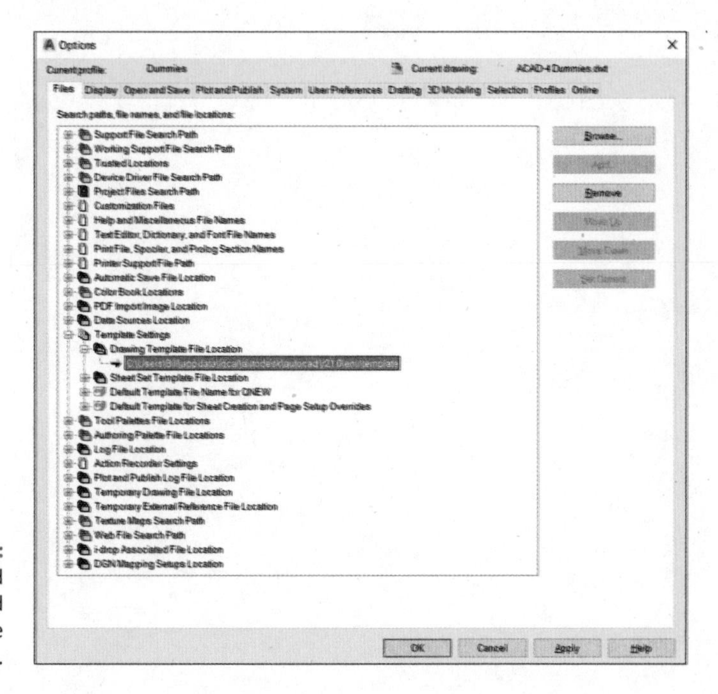

FIGURE 4-8: Seek and you shall find your template folder.

TIP

You don't have to keep your template files where that bossy Mister Gates told you. Create a folder that you can find easily (for example, `C:\Acad-templates` or `F:\Acad-custom\templates` on a network drive), put the templates that you actually use there, and change the Drawing Template File Location setting so that it points to your new template folder. How many Microsoft employees does it take to change a light bulb? None. Dark simply becomes the new standard.

In an office environment in particular, the template folder should live on a network drive so that everyone starts from the same set of templates.

REMEMBER

Once you get everything set up properly, you don't need to do the setup again. Setting up a drawing requires about 20 minutes, on average, but this time drops to almost zero if you configure your templates properly. By using templates, you can save enough time in a year or so to pay for a tropical vacation.

» Naming and restoring views

» Zooming and panning in paper space layouts

» Regenerating the display

Chapter 5

A Zoom with a View

One advantage that AutoCAD has over manual drafting is its capability to show you different views of drawings.

You move the viewpoint in, or *zoom in,* to see a closer view of objects in the drawing; you move the viewpoint out, or *zoom out,* to see a more expansive (not *expensive*) view. If you watch TV or movies or own a camera, you should understand zooming.

Panning refers not to looking for gold but to looking at a different part of a drawing without changing the magnification of the view. If you zoom in so that part of the drawing is no longer shown onscreen, you'll *pan* around in the drawing to see other parts, without zooming in and out. Think of the monitor as a window through which you look at part of the drawing. Now reach through the window and slide the drawing around until you see a different portion of it through the window.

REMEMBER

Panning and zooming do not change the size or position of objects in the drawing. The actions change only how you see them.

In fact, you not only *can* zoom and pan in the drawing but also, in most kinds of drawings, you *must* zoom and pan frequently to be able to draw and edit effectively.

Why do you need to pan and zoom often? For starters, though many architectural drawings plot out at 3 feet by 2 feet, you probably aren't fortunate enough to own a monitor of that size with sufficient resolution to be able to see every little detail.

Early releases of AutoCAD came with a sample drawing of the solar system done to scale. When first opened, it showed circles for each planet's orbit. Zooming in revealed the moon's orbit around Earth, then a crater on the moon, then the lunar landing module, and finally writing on the plaque mounted on a leg of the lunar lander.

In addition, technical drawings are jam-packed with lines, text, and dimensions. As discussed in Chapter 8, drawing with precision is essential to following best practices for AutoCAD drawings. Frequent zooming and panning enables you to better see detail, to draw more precisely because you can see what you're doing, and to edit more quickly, because object selection is easier when the screen isn't cluttered with objects. This chapter describes the most useful display control features in AutoCAD.

Panning and Zooming with Glass and Hand

AutoCAD makes panning easy, by offering scroll bars and real-time panning. In *real-time* panning (as opposed to pretend-time panning?), you can see objects moving on the screen as you drag the mouse upward and downward or back and forth with the middle button held down. Of course, the viewpoint is moving, not the objects.

Both panning and zooming change the *view* — the current location and magnification of the AutoCAD depiction of the drawing. Every time you zoom or pan, you establish a new view. You can give names to specific views so that returning to them is easy, such as a title block or a bill of material, as I demonstrate in the later section "A View by Any Other Name."

You can gain a better sense of panning and zooming in a drawing when you're looking at a drawing. Draw some objects on the screen, or open an existing drawing, or launch a sample drawing in AutoCAD.

If you haven't done so already, you can download sample files from `www.autodesk.com/autocad-samples`. (You can ignore the version numbers.) The AutoCAD LT sample files are also online at `www.autodesk.com/autocadlt-samples`. Note that LT drawings can be opened by standard AutoCAD and vice versa.

The wheel deal

Later in this chapter, I cover in detail various commands and options in AutoCAD for panning and zooming — if you have a wheel mouse, however, you'll rarely need to use the other methods, especially when working in 2D drawings. If you don't have a wheel mouse, run out and buy one now because the small cost will be more than offset by your increased productivity. The following three actions usually suffice for almost all panning and zooming needs:

- **Zoom in, zoom out:** Roll the scroll wheel forward and backward.
- **Pan:** Hold down the scroll wheel (or middle button) as you move the mouse. (The scroll wheel is also considered a button.)
- **Zoom to the extents of the drawing:** Double-click the scroll wheel. This method is particularly useful when you accidentally press Enter at the wrong time during a Move or Copy operation, as described in Chapter 11.

TECHNICAL STUFF

Using the scroll wheel (or middle button on a mouse that has three buttons) for zoom and pan operations depends on the setting of the obscure AutoCAD system variable named MBUTTONPAN. When MBUTTONPAN is set to its default value of 1, you can use the scroll wheel or middle button to pan and zoom. If you change MBUTTONPAN to 0, clicking the scroll wheel or middle mouse button displays the Object Snap menu at the cursor, as it did in earlier AutoCAD releases. If you can't zoom or pan using the scroll wheel or middle mouse button, set MBUTTONPAN to 1. When MBUTTONPAN is set to 1 you can press Shift+right-click to display the Object Snap menu at the cursor. I discuss Object Snaps in Chapter 8.

TIP

If you used the software that came with the mouse to change the function of the middle button, you might not be able to pan with it. To fix this, redefine the function of the middle button to its default.

TIP

Before using the mouse wheel to zoom in, position the cursor over the area you want to zoom into. AutoCAD uses this as the center of the zoom. This way, the area of interest doesn't disappear as you get in close.

Navigating a drawing

You may believe that AutoCAD is all about drawing and, occasionally, even about erasing. If so, you may be surprised to read that two of the most frequently used commands in all of AutoCAD are Pan and Zoom although the hands-down favorite is Undo. You can find these two commands in a couple of convenient places in AutoCAD:

- **On the Navigation bar:** The Navigation bar contains both Zoom and Pan buttons. Figure 5-1 shows the upper-right corner of the AutoCAD window with the Navigation bar in its default location, linked to the ViewCube. Because the ViewCube is more useful in 3D drawing, I tell you about it in Chapter 21.

» **In the Navigate and Navigate 2D panels on the View tab on the Ribbon:** These two panels contain a Pan button and a drop-down set of Zoom tool buttons. This location is not the most convenient for frequently used commands — and to make matters worse, this panel may not even be visible. To display it you must right-click anywhere in the View tab, click Show Panels, and then click Navigate.

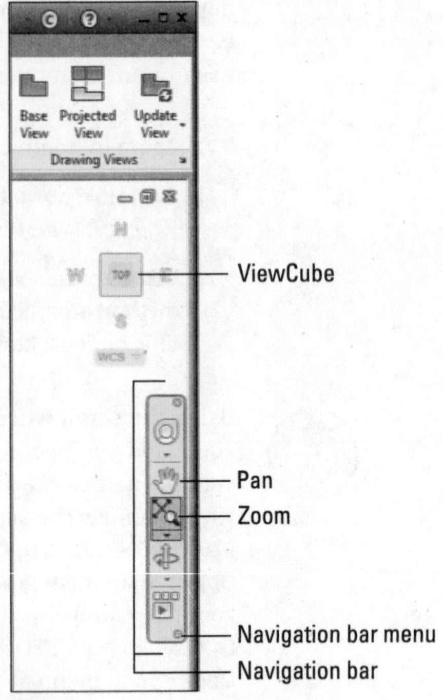

FIGURE 5-1:
Belly up to the Navigation bar.

If you're primarily creating 2D drawings, you can remove some of the 3D related viewing tools from the Navigation bar or turn off the ViewCube itself, either for the drawing session or permanently. Choose a method:

» **Turn off navigation buttons.** Open the Navigation bar menu by clicking the down arrow in its lower-right corner (refer to Figure 5-1) and deselect SteeringWheels, Orbit, and ShowMotion.

» **Turn off the ViewCube or the Navigation bar (or both) in the current viewport.** From the User Interface panel on the View tab, click User Interface and deselect ViewCube or Navigation bar or both.

» **Turn off the ViewCube permanently.** Open the Options dialog box (choose Options from the Application menu or type **OPtions**) and select the 3D Modeling tab. In the Display Tools in Viewport section, under the Display the ViewCube line, deselect the two options 2D Wireframe Visual Style and All Other Visual Styles.

In AutoCAD, unlike in most other Windows programs, panning and zooming are usually more convenient and faster than scrolling. If you want the traditional Windows scroll bars on the right and bottom edges of the drawing window, you can turn them on. Choose Options from the Application menu (or type **OPtions**) to display the Options dialog box. On the Display tab, select or deselect the Display Scroll Bars in Drawing Window check box. The default, and most users' preference, is not to use scroll bars, thereby regaining the screen space they occupy. That's why you don't see them in most figures in this book.

Zoom, Zoom, Zoom

Because zooming is a frequent activity in AutoCAD, you should know some alternative ways to zoom.

In addition to using the mouse's roller wheel and the Zoom button on the Navigation bar (as I describe earlier in this chapter), you'll find tool buttons for all the Zoom options in the Navigate panel on the View tab.

REMEMBER

As mentioned, the Navigate panel on the View tab is turned off by default.

Click the small down arrow in the lower-right corner of the panel and a menu with the other options opens, as shown in Figure 5-2.

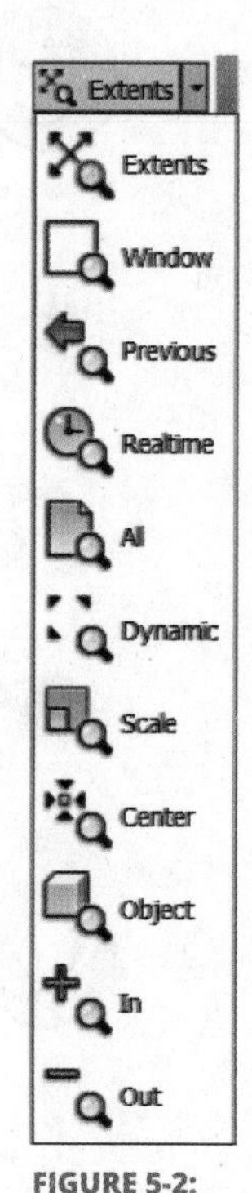

FIGURE 5-2: A menu for magnifying.

The Zoom command has 11 options. The most important ones are described in this list:

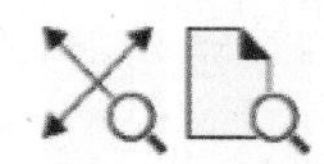

- **Extents and All:** The Zoom Extents button (with the four-headed arrow and the magnifying glass) zooms out just far enough to display all objects in the current drawing. The Zoom All button (the sheet with the folded-over corner) does almost the same thing: It zooms to display the rectangular area defined by the drawing limits set with the LIMITS command, or it zooms to show the extents — whichever is larger. These two options are especially useful when you zoom in too close or pan off into empty space and want to see the entire drawing again.

TECHNICAL STUFF

Limits and extents are slightly different. Limits are set by the Limits command. You can configure AutoCAD to not allow you to draw anything outside the limits, but this usually causes more problems than it solves. Extents are defined by the smallest imaginary rectangle that will just surround every object in your drawing. A good April Fool's trick is to sneak up to another person's computer and draw a tiny circle by typing the location coordinates as something like 10000,10000. The next time the person does a Zoom Extents, the drawing will seem to disappear.

TIP

Use Zoom All or Zoom Extents and then save the drawing before you close it to ensure that

- The next person who opens the drawing can see the full drawing as soon as the person starts working.

- Any objects that you accidentally copied beyond the drawing limits show up so that you can delete or move them.
- The drawing preview that appears in the Select File dialog box and on the Start screen displays the full drawing rather than a tiny, unidentifiable corner of it.

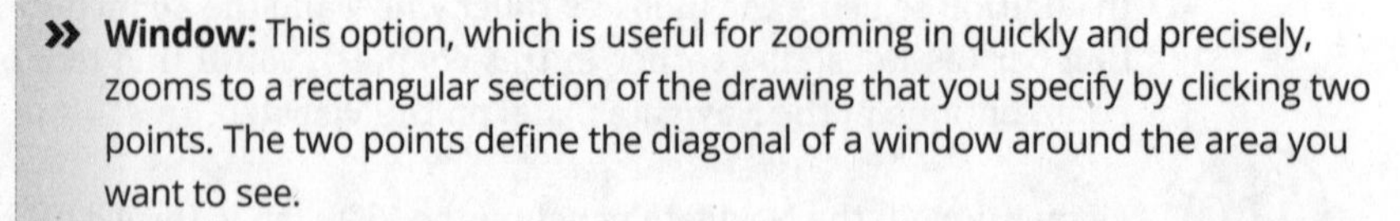

» **Window:** This option, which is useful for zooming in quickly and precisely, zooms to a rectangular section of the drawing that you specify by clicking two points. The two points define the diagonal of a window around the area you want to see.

REMEMBER

The Zoom command's Window option is not a click-and-drag operation, unlike in most other Windows programs and — confusingly — unlike in the Zoom/Pan Realtime Zoom Window option. To use the Zoom command's Window option, you click to specify one corner, release the mouse button, and then click another corner.

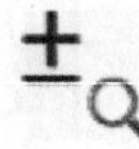
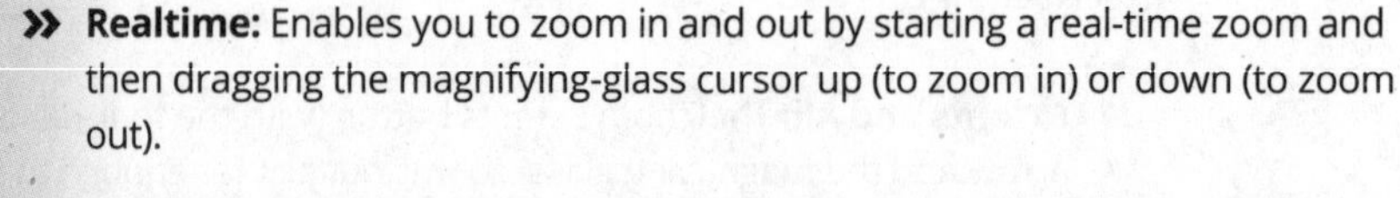

» **Realtime:** Enables you to zoom in and out by starting a real-time zoom and then dragging the magnifying-glass cursor up (to zoom in) or down (to zoom out).

» **Previous:** Use this option to undo the last zoom or pan sequence (or both), returning to where you started. You can repeat this option to step through previous views, even if you create or edit objects after changing the view.

» **Object:** This option zooms in close enough to show selected objects as large as they can be displayed onscreen. Using Zoom Object is similar to examining selected objects under the AutoCAD microscope.

TIP

Entering the keyboard command sequences **Z A**, **Z P**, or **Z W** (note the mandatory space between the letters) is usually faster than trying to find the corresponding zoom tools on the Ribbon or various toolbars. **Z** is the command alias for the Zoom command, while **A**, **P**, and **W** start the **A**ll, **P**revious, and **W**indow options, respectively. These three command sequences, in combination with the wheel mouse, typically account for more than 99 percent of your panning and zooming needs. (I almost never use any Ribbon or navigation bar options.)

I cover another important Zoom option when discussing paper space viewports in Chapter 12.

A View by Any Other Name

If you repeatedly have to zoom or pan to the same area, you can quickly see what you want to see by using a named view. After you name and save a view of a particular area of a drawing, you can return to that area quickly by restoring the view with its name. You use the View command, which displays the View Manager dialog box, to create and restore named views.

Follow these steps to create a named view:

1. **Zoom and Pan until you find the area of the drawing to which you want to assign a name.**
2. **Start the View command from the keyboard. Or on the View tab on the Ribbon, click View Manager in the Named Views panel.**

   If the Named Views panel is not visible, right-click anywhere in the View tab, click Show Panels, and then click Views.

   The View Manager dialog box opens.
3. **Click New.**

   The New View/Shot Properties (New View in AutoCAD LT) dialog box appears, as shown in Figure 5-3.

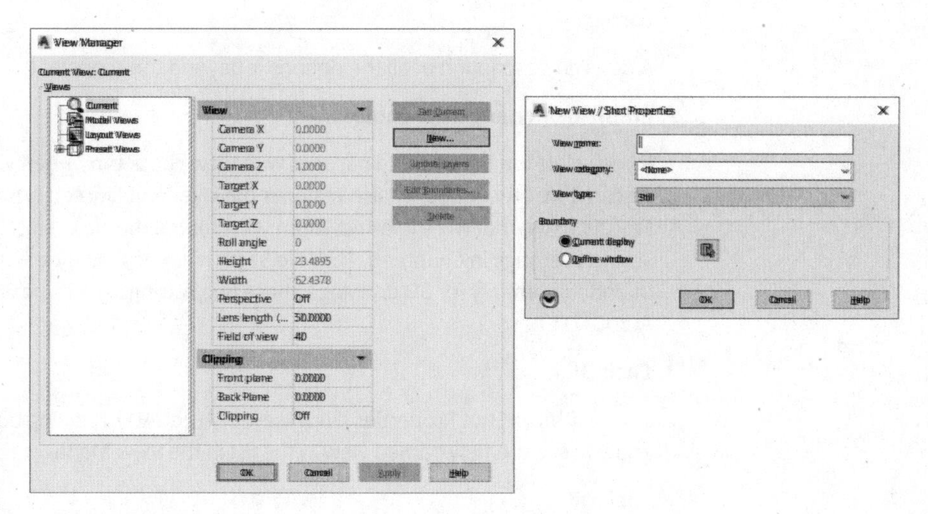

FIGURE 5-3: Save a view in the drawing.

4. **Type a name in the View Name text box.**
5. **(Optional) Type a new category in the View Category text box, or select an existing one from the drop-down list.**

   You create your own view categories to organize views and certain display characteristics of views. This feature is used mainly in sheet sets. If you aren't using them, you can leave the default value, <None>, in the View Category box or clear the current value. (A discussion of sheet sets is beyond the scope of this book.)

6. **Make sure that the View Type drop-down list is set to Still.**

   The Still option creates a static thumbnail image of the objects in the area that is defined by the named view for use in ShowMotion. Visit the online Help system if you want to know more about the Cinematic and Recorded Walk options that you can choose from this drop-down menu, which are used mainly in 3D work.

   AutoCAD LT doesn't support view types, so if you're working with AutoCAD LT, skip Step 6. The New View dialog box in AutoCAD LT omits the View Type list box, most of the Settings drop-down options, and the Background area.

7. **In the Boundary area, select the Current Display radio button, if it isn't selected already.**

   TIP

   If instead you want to save a different view boundary, select the Define Window radio button, click the Define View Window button that appears to the right of it, and pick two corners of the region's boundary (as though you're zooming windows).

8. **If you really want to get fancy, click the down arrow in the lower-left corner.**

   A detailed description of all the options is beyond the scope of this book.

9. **Confirm or change the choices.**

   If you select the Save Layer Snapshot with View check box, when you later restore the view, AutoCAD also restores the layer visibility settings (On/Off and Freeze/Thaw) that were in effect when you created the view. I discuss layer visibility settings in Chapter 9. The Live Section and Visual Style settings are intended primarily for 3D drawings; these two settings aren't included in AutoCAD LT.

10. **Click OK.**

    The New View/Shot Properties (New View in AutoCAD LT) dialog box closes, and you see the new named view in the list in the View Manager dialog box.

11. **Click OK.**

    The View Manager dialog box closes.

To restore a named view, follow these steps:

1. **From the View tab on the Ribbon, click View Manager from the View panel.**

   Alternatively, type View and press Enter. The View Manager dialog box appears.

2. **In the Views list, expand either Model Views or Layout Views (depending on where you saved the view).**

3. **Double-click the name of the view you want to restore, and then click OK to close the dialog box.**

In both AutoCAD and AutoCAD LT, you can set named views to be current without having to open the View Manager dialog box. From the Views panel on the View tab, simply choose a named view to restore from the drop-down list.

You also can plot the area defined by a named view. See Chapter 16 for instructions on plotting views.

Degenerating and Regenerating

As you zoom and pan around the drawing, you may wonder how the image you see onscreen is related to the DWG file that AutoCAD saves on the hard drive. Well, maybe you don't wonder about it, but I tell you in this section anyway.

When you draw and edit objects, AutoCAD stores all their *geometrical properties* (that is, location and size) in a highly precise form — technically, *double floating-point precision*. The program always maintains this precision when you save the DWG file. For reasons of computer performance, however, AutoCAD does *not* use that high-precision form of the data to display the drawing onscreen. Instead, AutoCAD converts the highly precise numbers in the DWG file into slightly less precise integers in order to create the view you see onscreen.

The happy consequence of this conversion is that zooming, panning, and other display changes happen a lot faster than they would otherwise. The unhappy consequence is that the conversion can make circles and arcs look like polygons. AutoCAD solves this problem by regenerating the drawing (also known as performing a **RE**gen operation).

In most cases, AutoCAD regenerates drawings automatically whenever it needs to. You sometimes see command-line messages such as `Regenerating model` or `Regenerating layout`, indicating that AutoCAD is handling the regenerating for you.

If, on the other hand, you see the command-line message `Regen queued`, AutoCAD is warning you that it's *not* performing a regeneration, even though one might be advisable. In addition, you might see the warning message `About to regen -- proceed?`. AutoCAD is saying, "What the drawing looks like onscreen now may not exactly match the real version of the drawing database that's stored when you save the drawing. I'll update the display version at the next regeneration."

You can control whether regenerations happen automatically by using the REGENMODE system variable. See the online Help system for more information on this variable, and see Chapter 23 for general information on system variables.

The REgenAll command, available only from the command line, regenerates all viewports in a paper space layout. If you run the REgenAll command in model space, it has the same effect as the ordinary REgen command.

If you zoom in a long way in a large drawing that has small circles and arcs and you find them converted into hexagons, simply run the REgen command. You can also minimize this effect in the future by running the VIEWRES command and increasing the value from the default of 1,000 to as high as 20,000 although doing so can also slow down a complex drawing. This value is stored in the current drawing, so you can set up the template file (covered in Chapter 4) to cover most of your work.

Almost every new release of AutoCAD has improved graphics performance. The program has come a long way from the so-called high-resolution 640 x 480 16-color Tecmar Graphics Master adapter, which I started with in the 1980s.

2

Let There Be Lines

IN THIS PART . . .

Draw lines and other mostly linear objects, as well as the curvy bits.

Find out how to draw objects with precision.

Manage object properties.

Select and edit objects.

Set up a paper space layout in anticipation of printing.

» Lining up for lines and polylines

» Closing up with rectangles and polygons

Chapter 6
Along the Straight and Narrow

As you may recall from your crayon-and-coloring-book days, drawing is *fun*. Computer-aided drafting (CAD) imposes a little more discipline, but drawing with AutoCAD is still fun. Trust me on this one. In CAD, you usually start by drawing geometry from basic shapes — lines, circles, and rectangles — to construct the real-world object that you're documenting.

For descriptive purposes, I divide the drawing commands into three groups:

- **Straight lines** and objects composed of straight lines, covered in this chapter
- **Curves,** which I explain in Chapter 7
- **Points,** explained in Chapter 7 as well if you're wondering what the point of all this is

After you've created some straight or curvy geometry, you'll probably need to add descriptive elements, like dimensions, text, and hatching, but those come later (in Part 3). Or you may want to use straight and curvy geometry as the basis for some cool 3D modeling. I introduce you to that topic in Part 5. Your first task is to get the geometry right; then you can worry about labeling it.

Drawing geometry properly in AutoCAD depends on the precision of the points that you specify to create the objects. I cover this topic in Chapter 8, so don't start any production CAD drawings until you review that chapter.

Proper geometry creation also depends on creating objects that have the correct appearance. In Chapter 9, I show you how to create hidden lines, center lines, section lines, and other elements.

Drawing for Success

AutoCAD offers a wide range of tools that allow you to create designs in a virtual world that will be accurately manufactured or built in the real world. The 2D and 3D tools that you use require some upfront preparation to ensure that the designs you create are drawn precisely. Nothing is worse than spending time on a great design, only to find out that the objects you drew weren't drawn at the correct size and that lines don't intersect cleanly. Okay, I lied. Worse yet is that your imprecise drawing was used to program a CNC machining center or build a house and now the expensive parts don't fit.

Here are a few important techniques to use when you draw objects:

- **Use precision tools to make sure objects intersect precisely and are drawn at the correct angle.** Precision tools allow you to reference points on existing objects, constrain the cursor to a specific angle or snap distance, and quickly locate a point based on a distance and direction in which the cursor is moved. I cover precision techniques and tools in Chapter 8.
- **Understand the difference between command line and Dynamic Input coordinate entry.** Second and next points entered at the command line are interpreted differently from those entered at the Dynamic Input tooltip. The second and next points entered at the Dynamic Input tooltip are formatted as polar and are relative to the previously entered coordinate value automatically, unlike at the command line where you must first type @ before the coordinate value for relative coordinates.

  AutoCAD's Dynamic Input system displays a lot of the information at the cursor that you used to have to look down to the command line to see. To use Dynamic Input, press F12 to toggle it on and off.
- **Watch the command line.** The command line at times displays additional information about the current command that's not displayed in the Dynamic Input tooltip. This information might be the current text style or justification, or the active extend mode. I cover using the command line in Chapter 2.

» **Organize objects on layers.** All the objects you draw should be placed on specific layers. For example, annotation objects that communicate design information might be placed on a Notes layer, and the lines that are used to represent a wall might be placed on a Walls layer. I cover layers in Chapter 9.

Introducing the Straight-Line Drawing Commands

As I harp on elsewhere in this book, CAD programs are designed for precision drawing, and you'll spend a lot of time in AutoCAD drawing objects composed of straight-line segments. The rest of this chapter covers these commands, all of which are found on the Draw panel of the Home tab on the Ribbon. The icons in the left margin match those on the Ribbon:

» **Line:** Draws a series of one or more visually connected straight-line segments; that is, the end of the current segment has the same coordinates as the start of the next segment. Although the lines look like they're physically connected, each *segment* (piece of a line with endpoints) is in fact a separate object with its own start and endpoints.

» **PLine:** Draws a *polyline,* which is a series of straight- or curved-line segments (or both) connected as a single object. I'm cheating slightly here because I cover curvy components in Chapter 7, but I don't want you to have to read about one command in two different places.

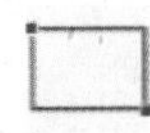

» **RECtang:** Provides a convenient way to draw rectangles in a variety of ways as four right-angle polyline segments. AutoCAD has no actual rectangle object.

» **POLygon:** Gives you a convenient way to draw a polyline of many sides in the shape of a *regular polygon* (a closed shape with all sides and all angles equal).

TIP

The POLygon command may be hidden under the Rectangle button on the Ribbon or vice versa, depending on which button you used last.

The following additional straight-line drawing commands, also available in Auto-CAD LT, are found in the drop-down list below the Draw panel:

» **XLine:** Draws a line (known as an *ex-line* or a *construction* line) that passes through a point at a specified angle and extends to infinity in both directions.

» **RAY:** Draws a line (known as a *ray*) that starts at a point and extends to infinity at a specified angle away from the start point.

TIP

You use the RAY and XLine commands to draw construction lines that guide the construction of other geometry. Using construction lines is less common in AutoCAD than in certain other CAD programs and is far less common than in pencil-and-paper drawings. The many precision techniques in AutoCAD usually provide methods for creating new geometry that are more efficient than adding construction lines to a drawing. In particular, object snap tracking (discussed in Chapter 8) and parametrics (discussed in Chapter 19) usually eliminate the need for construction geometry or even for a mitre line when creating orthographic views.

TECHNICAL STUFF

Although xlines and rays are infinitely long, they don't increase the extents of your drawing to infinity. The question of the day is, "Which is longer: an XLine that extends to infinity in both directions or a RAY that extends to infinity in one direction?"

Drawing Lines and Polylines

The Line command works well for many drawing tasks, but the PLine command works better for others. Experience can help you choose which one works best for your design needs. As the PLine command draws *polylines*, you often hear CAD drafters refer to polylines as a *p-lines* (rhymes with "bee-lines," not to be confused with queues in busy restrooms).

Here are the primary differences between the Line and PLine commands:

- **The Line command draws a series of separate line segments.** Even though they appear to be connected onscreen, each one is a separate object. If you move one line segment, none of the other segments you drew at the same time move with it.
- **The PLine command draws a single, connected, multisegmented object.** A polyline is what a line appears to be; each segment is connected with the others to form a single object. If you select any one segment for editing, the change affects the entire polyline, in most cases. Figure 6-1 shows how the same sketch drawn with the Line and the PLine commands responds when you select one of the objects. A polyline is not a pickup line used by parrots in a bar.

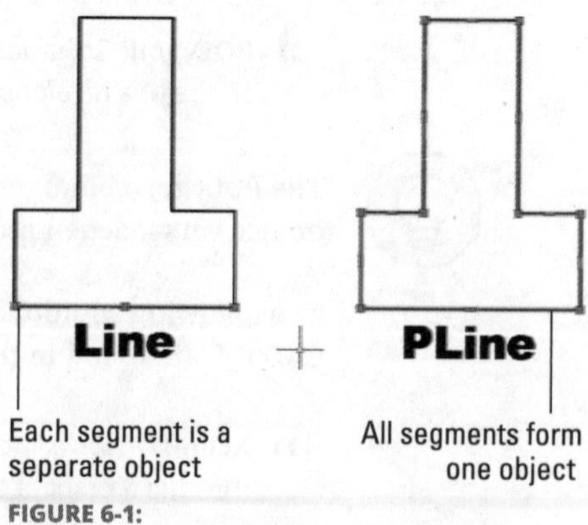

FIGURE 6-1: Results of drawing with the Line and PLine commands.

TIP

Use the PLine command rather than the Line command in most cases where you need to draw a series of connected line segments. If you're drawing a series of end-to-end segments, those segments may well be logically connected. For example, they might represent the outline of a single object or a continuous pathway. If the segments are connected logically, it makes sense to keep them connected in AutoCAD. The most obvious practical benefit of grouping segments into a polyline is that many editing operations are more efficient when you use polylines. When you select any segment in a polyline for editing, the entire polyline is selected.

TIP

As covered in Chapter 19, using parametrics on line segments can often produce the same results as using the PLine command.

- **The PLine command can draw curved segments as well as straight ones.** If you want a combination of separate linear and curved segments, you must switch between the Line and Arc commands. (I cover arcs in Chapter 7.) Using the PLine command, though, you can switch between linear or circular-curve segments in a single polyline.
- **A polyline can have width.** Polyline segment width is visually similar to the lineweight object property in AutoCAD (which I discuss in Chapter 9) except that polyline width can vary from segment to segment, and individual segments can be tapered. Polylines are often used for the copper conductors on electronic printed circuit boards.
- **Polylines can be smooth curves.** After drawing a polyline, you can use the PEdit command to automatically reshape the polyline into a smooth, flowing curve based on the vertex points you selected. I touch on this topic in Chapter 11.

The following sections show you how to create a line and a polyline.

Toeing the line

Unlike many AutoCAD drawing commands, Line offers limited options. It has a Close option only to create one more segment back to the first point you picked in the current run of the command, and it has an Undo option to remove the most recently drawn segment. You can repeat the Undo option back to the start of the current run of the command.

The Line command draws continuous lines. Real-world drawings include several different *looks* of lines, such as hidden, center, and section. Chapter 9 covers how to apply these and several other properties to lines.

Follow these steps to use the Line command:

1. **Start the Line command by clicking the Line button on the Draw panel on the Ribbon, or by entering L and then pressing Enter.**
2. **Draw line segments by picking several random points.**
3. **Terminate the command by pressing Enter, Esc, or the spacebar.**
4. **Press Enter or the spacebar to repeat the Line command.**

TIP

   In AutoCAD, pressing Enter or the spacebar when no command is underway always repeats the last command.

5. **Draw more line segments to complete the figure.**

   If you need to undo some line segments, enter **U** and press Enter.

6. **Enter C and press Enter to close the figure, which draws one more segment back to the first point you picked.**

   The Close option doesn't appear until you've created the first two line segments. You can't close a single line.

TIP

Pressing Enter or the spacebar whenever AutoCAD asks for a first point automatically selects the last point you picked, regardless of the previous or current command.

Connecting the lines with polyline

Drawing polylines composed of straight segments is much like drawing with the Line command, as shown in the steps in this section. The PLine command has a lot more options, however, so *watch the prompts.* If the Dynamic Input feature is on, press the down-arrow key to see the options listed near the cursor, or right-click to display the PLine right-click menu, or simply read the command line.

To draw a polyline composed of straight segments, follow these steps:

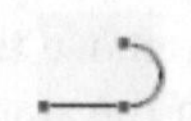

1. **Click the Polyline button on the Draw panel of the Ribbon, or type PL at the command line and press Enter.**

   AutoCAD starts the PLine command and prompts you to specify a start point.

2. **Specify the starting point by clicking a point or typing coordinates.**

   Now you truly do need to read the command line, because the Dynamic Input tooltip at the cursor doesn't display any of the options.

TIP

You can right-click or press the down-arrow key to see a list of the options at the cursor, as shown in Figure 6-2, but it's usually faster to use the command line.

AutoCAD displays the current polyline segment line width at the command line and prompts you to specify the other endpoint of the first polyline segment:

```
Current line-width is 0.0000
Specify next point or [Arc Halfwidth Length Undo Width]:
```

3. **If the current line width isn't zero, you can change it to zero by typing W to select the Width option and then entering 0 as the starting and ending widths, as shown in this command-line sequence:**

```
Specify next point or [Arc Halfwidth Length Undo Width]: W
Specify starting width <0.0000>: 0
Specify ending width <0.0000>: 0
Specify next point or [Arc Halfwidth Length Undo Width]:
```

TECHNICAL STUFF

Despite what you may think, a zero-width polyline segment isn't the AutoCAD equivalent of drawing with invisible ink. *Zero width* means, "Display this segment using the normal, single-pixel width on the screen, and print as thin as possible." How can you tell when your pen has run out of invisible ink? You can see the writing.

4. **Specify additional points by clicking or typing.**

After you specify the second point, AutoCAD adds the Close option to the prompt. The command line shows

```
Specify next point or [Arc Close Halfwidth Length Undo
  Width]:
```

TIP

What you do next can get a little weird. If you invoke the Close option after selecting only the first and second points, PLine doubles back on itself and creates the second segment back over the first, unlike the Line command, which won't. Normal practice would be to not invoke Close until you have created at least two non-collinear segments.

5. **Pick several points to create several line segments, and then enter W to start the Width option. Then set a new width, as shown in this command line sequence:**

```
Starting width <0.0000>: 5
Specify ending width <5.0000>: Press Enter
```

6. **Pick several more points and note the line width of the new segments.**
7. **Set the Width option again:**

```
Starting width <5.0000>: 15
Specify ending width <0.0000>: 0
```

8. **Pick another point.**

   You just created a cool arrowhead!

TIP

Any time AutoCAD prompts for a length or distance, you can either type a value or you can show it what you want by picking two points in the drawing.

9. **After you finish drawing segments, press Enter to leave the figure open, or type C and press Enter to close it back to the start.**

   AutoCAD draws the final segment and miters all the corners perfectly.

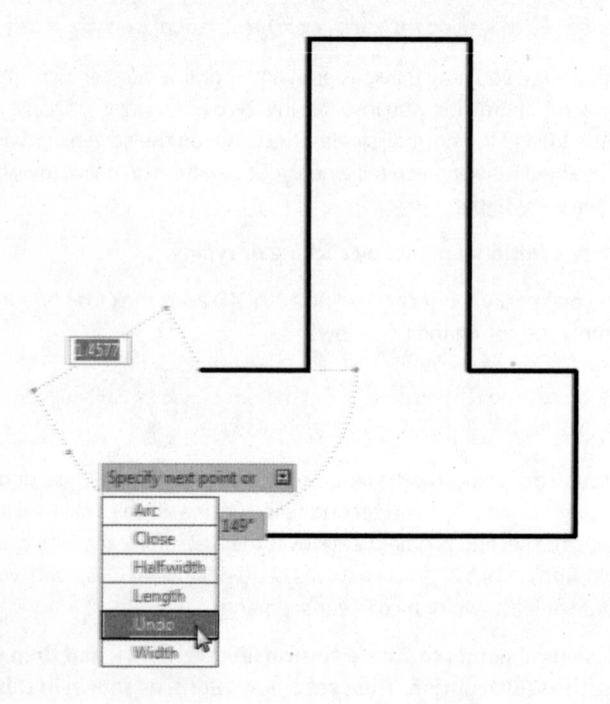

FIGURE 6-2: The optional extras.

REMEMBER

After you create a polyline, you can adjust its segments by grip-editing any of the vertex points. See Chapter 11 for details on grip editing.

In the following steps, I spice things up by adding an arc segment to a polyline.

ALTERNATIVES TO THE LINE AND PLINE COMMANDS

The Line and PLine commands work well for drawing a series of end-to-end single lines, but you may want to draw a series of double lines to represent, for example, the edges of a wall or roadway. Here are some options:

- Use the AutoCAD MLine command to draw *multilines,* which are a series of two or more parallel straight lines. The AutoCAD Multiline feature was full of limitations when it debuted way back in 1994 (in the notorious Release 13), and despite minor tweaks in AutoCAD 2006, it hasn't improved significantly since then. It exists mostly for compatibility with MicroStation files. Look up the MLine and MLStyle commands in the AutoCAD online help system if you want to tangle with this feature, but be prepared to spend time experimenting, struggling, and (possibly) cursing. Don't let small children near you.
- Use the PLine command to draw a single set of connected line or arc segments or both, and then use the OFFset command (described in Chapter 11) to create one or more sets of parallel segments.
- In AutoCAD LT only, use the DLine, or Double Line, command to draw pairs of parallel line or arc segments or both. Unlike MLine, DLine works well. AutoCAD LT doesn't have the MLine command, which is more of a blessing than a limitation. AutoCAD, on the other hand, doesn't have the DLine command. (Score one for the little brother!)

Curved segments in polylines are *circular arcs* that you can create while running the PLine command. AutoCAD can draw other kinds of curves, including ellipses and splines, but not in the PLine command.

To draw a polyline that includes curved segments, follow these steps:

1. **Repeat Steps 1–4 in the preceding step list.**
2. **To add one or more arc segments, type A, and then press Enter to select the Arc option.**

   The prompt changes to show options for drawing arc segments. Most of these options correspond to the many ways of drawing circular arcs in AutoCAD; see the section on arcs in Chapter 7. The command line shows:

   ```
   Specify endpoint of arc or [Angle CEnter/CLose Direction
     Halfwidth Line Radius Second pt Undo Width]:
   ```

3. **Specify the endpoint of the arc by clicking a point or typing coordinates.**

   AutoCAD draws the curved segment of the polyline. The prompts continue to show arc segment options.

   ```
   Specify endpoint of arc or [Angle CEnter/CLose Direction
     Halfwidth Line Radius Second pt Undo Width]:
   ```

   The options at this point are to

   - Specify additional points to draw more arc segments.
   - Choose another arc-drawing method, such as **CE**nter or **S**econd point.
   - Return to drawing straight-line segments with the **L**ine option.

   In this example, you draw straight-line segments.

TIP

   Perhaps the most useful of the alternative arc-drawing methods is `Second pt`. It gives you more control over the direction of the arc but at the cost of losing tangency of adjacent segments. (Sometimes, it's best not to go off on a tangent, anyway.) If you want both ends of the arc segments to be tangent to the adjacent line segments, draw the polyline as straight-line segments and then use the Fillet command (described in Chapter 11) to add the arcs later.

4. **Type L and then press Enter to select the Line option.**

   The prompt reverts to showing straight-line segment options.

   ```
   Specify next point or [Arc Close Halfwidth Length Undo
     Width]:
   ```

5. **Specify additional points by clicking or typing.**
6. **After you're finished drawing segments, either press Enter or type C and press Enter.**

Figure 6-3 shows some elements you can draw with the PLine command by using straight segments, arc segments, varying-width segments, or a combination of all of them.

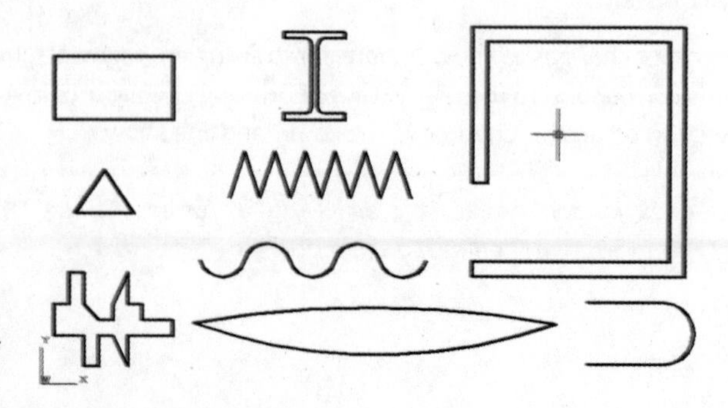

FIGURE 6-3: A plethora of polylines.

Squaring Off with Rectangles

You could use the PLine or Line command to draw a rectangle, segment by segment. In most cases, though, you'll find that using the special-purpose RECtang command is easier. The following steps show you how:

1. **Click the Rectangle button on the Draw panel of the Home tab, or type REC and press Enter.**

REMEMBER

Can't find a command? Some commands may be hidden in slideout panels on the Ribbon interface. Many Ribbon panels display a down-facing arrowhead beside the name of the panel. Click the arrowhead to see a collection of related but less-used commands. Still can't find the command? Some Ribbon command buttons may be hidden under others in the same category. AutoCAD remembers the last one you used. For example, you may find POLygon under RECtangle or vice versa.

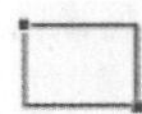

AutoCAD starts the RECtang command and prompts you to specify a point for one corner of the rectangle. The command line displays the prompt:

```
Specify first corner point or [Chamfer Elevation Fillet
  Thickness Width]:
```

TIP

You can add fancy effects by using additional command options. The default options work best for most purposes. Look up RECTANG in the AutoCAD help system if you want to know more about the options.

2. **Specify the first corner by clicking a point or typing coordinates.**

AutoCAD prompts you to specify the corner of the rectangle that's diagonally opposite from the first one.

```
Specify other corner point or [Area Dimensions Rotation]:
```

3. **Specify the other corner by clicking a point or typing coordinates.**

The rectangle is drawn after you specify the second corner point.

TIP

If you know the size of the rectangle that you want to draw (for example, 100 units long x 75 units high), type **D** to enter the Dimensions option, and then type the width and height. Pick a point to indicate which of the four possible alignments you want, and AutoCAD draws the rectangle.

Choosing Sides with POLygon

Rectangles and other closed polylines are types of *polygons*, or closed figures with three or more sides. The AutoCAD POLygon command provides a quick way of drawing *regular* polygons with many sides where all angles are equal. The command has nothing to do with missing parrots.

The following steps show you how to use the POLygon command:

1. **Click Polygon from the Rectangle drop-down list on the Draw panel of the Home tab, or type POL and press Enter.**

   AutoCAD starts the POLygon command and prompts you to enter the number of sides for the polygon.

   ```
   Enter number of sides <4>:
   ```

2. **Type the number of sides for the polygon that you want to draw and then press Enter.**

   Your polygon can have from 3 to 1,024 sides.

   AutoCAD prompts you to specify the center point of the polygon:

   ```
   Specify center of polygon or [Edge]:
   ```

   You can use the Edge option to draw a polygon by specifying the length of one side instead of the center and then the radius of an imaginary inscribed or circumscribed circle. The imaginary circle method is much more common.

3. **Specify the center point by clicking a point or typing coordinates.**

   After you specify the center point, nothing happens in the graphics area but the command line prompts you to specify whether the polygon is inscribed in an imaginary circle whose radius you specify in Step 5 (the corners touch the circumference of the circle) or circumscribed about the circle (the sides are tangent to the circle).

   ```
   Enter an option [Inscribed in circle Circumscribed about
     circle] <I>:
   ```

4. **Type I (for inscribed) or C (for circumscribed), and press Enter.**

   The command line prompts you to specify the radius of an imaginary circle:

   ```
   Specify radius of circle:
   ```

5. **Specify the radius by typing a distance or clicking a point.**

   AutoCAD draws the polygon.

TIP

   If you type a distance or you click a point with Ortho Mode turned on, the polygon will align orthogonally. I cover using Ortho Mode in Chapter 8.

TECHNICAL STUFF

Rectangles and polygons aren't special object types. They're simply regular polylines that have been constructed by special command macros.

Figure 6-4 shows the results of drawing plenty of polygons — a practice known as *polygony*, which, as far as I know, is legal nearly everywhere.

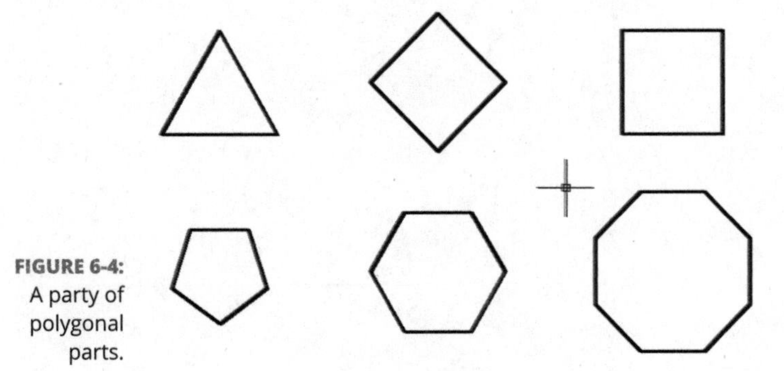

FIGURE 6-4: A party of polygonal parts.

Specify the radius by typing a distance or clicking a point.

» **Employing eccentric ellipses**

» **Dunking for donuts**

» **Making your points**

Chapter 7
Dangerous Curves Ahead

Although straight-line segments predominate in many CAD drawings, even the most humdrum, rectilinear design is likely to have a few curves. And if you're drawing Audi car bodies or Gaudí buildings, those drawings will contain a *lot* of curves! In this chapter, I cover the AutoCAD curve-drawing commands. Your drawings should also have a point; in fact, they may have several points, so at the end of this chapter, I fill you in on creating point objects in AutoCAD.

Throwing Curves

Curves are used to represent many different types of features in a drawing. A circle is used to show a bolt hole in a steel beam, but can also be used for a bicycle tire. After circles, the next most commonly used curve objects are arcs. Arcs can be found in drawings where two pieces of steel meet to show fillets or even in curved lips on countertops. The design of the ark probably included arcs. Other curve types that you can create are filled circles, ellipses, and splines.

AutoCAD supports the following curve commands, all of which can be found on the Draw panel of the Home tab:

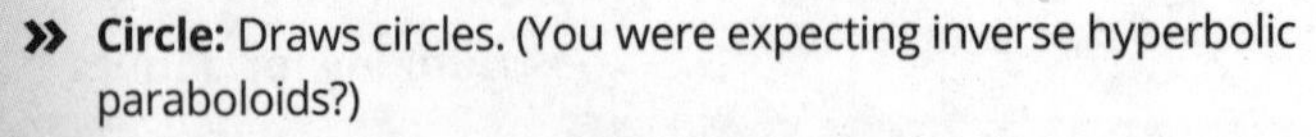

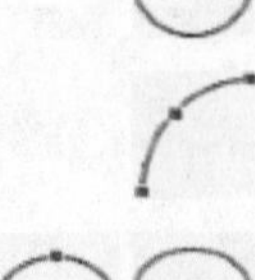

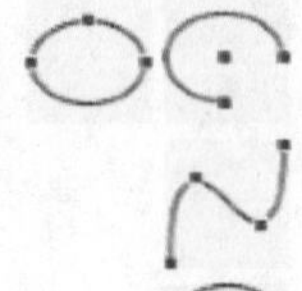

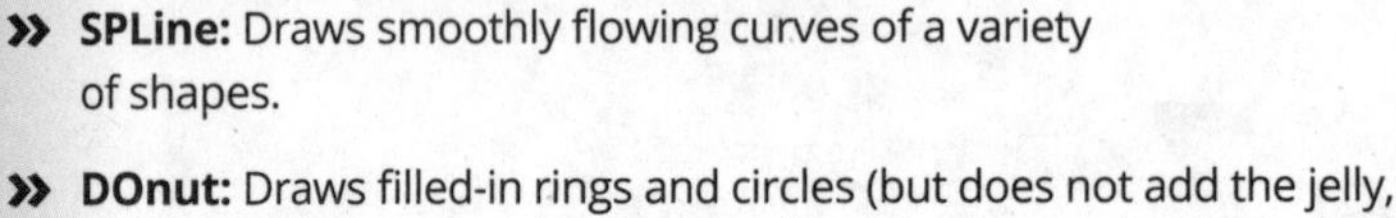

- **Circle:** Draws circles. (You were expecting inverse hyperbolic paraboloids?)
- **Arc:** Draws *circular* arcs with center points and fixed radii, not arcs cut from ellipses, parabolas, or other complex curves.
- **ELipse:** Draws ellipses and elliptical arcs.
- **SPLine:** Draws smoothly flowing curves of a variety of shapes.
- **DOnut:** Draws filled-in rings and circles (but does not add the jelly, unfortunately).
- **REVCLOUD:** Draws freeform clouds, the most common application of which is to indicate revised areas in drawings.

REMEMBER

Can't find a command? Some commands may be hidden in slideout panels on the Ribbon interface. Many Ribbon panels display a down-facing arrowhead beside the name of the panel. Click the arrowhead to see a collection of related but less-used commands. Still can't find the command? Some Ribbon command buttons may be hidden under others in the same category. AutoCAD remembers the last one you used. For example, you may find any one of the 11 Arc options on top of the heap.

Going Full Circle

AutoCAD offers several ways to define circles. Center point/radius and center point/diameter are likely to cover most of your needs. In the following list, the information in parentheses indicates which characters to type at the command line to initiate an option after you start the Circle command. On the other hand, the drop-down list below the Circle button in the Draw panel on the Home tab of the Ribbon starts the Circle command and then feeds it the appropriate option.

You can define circles in AutoCAD using these options:

- **Radius:** This setting is the default. Simply pick the center point, and then specify a radius by either typing a value or dragging the circle to the size you want.
- **Diameter (D):** Pick the center point and then define the diameter. Even though Radius is the default setting, the vast majority of circles you draw are defined by the diameter.

- **2-Point (2P):** Choose this option to draw a circle where the distance between two specified points is equal to the diameter of the circle. If you're using the command line, enter **2P** to choose this option; it's spelled out as *2-Point* on tooltips.
- **3-Point (3P):** Choose this option to draw a circle through any three specified points.
- **Tangent-Tangent-Radius (Ttr):** Pick this option to draw a circle tangent to two existing drawing objects and having a specified radius.
- **Tangent-Tangent-Tangent:** This option lets you draw a circle tangent to three valid existing drawing objects. By *valid,* I mean that it's mathematically possible to construct a circle tangent to the three selected objects.

TIP

Note that you can't create a T-T-T circle by typing a command option at the command line because, weirdly, no such command option exists. This method is performed through a menu macro, so the only way to run it is to choose Tan, Tan, Tan from the Circle button's drop-down list in the Draw panel on the Home tab.

TECHNICAL STUFF

AutoCAD can draw circles tangent to lines, arcs, polylines, or other circles. You can mix and match, so a Ttr circle can be tangent to a line and to an arc.

Figure 7-1 illustrates the six ways to draw circles. Whether these additional methods are useful depends on the kinds of drawings you make and how geometry is defined in your industry. Become familiar with the default center point/radius method, and then experiment with other methods. Or you can always draw circles using the default method and move them into position using object snaps by referencing points on other geometry.

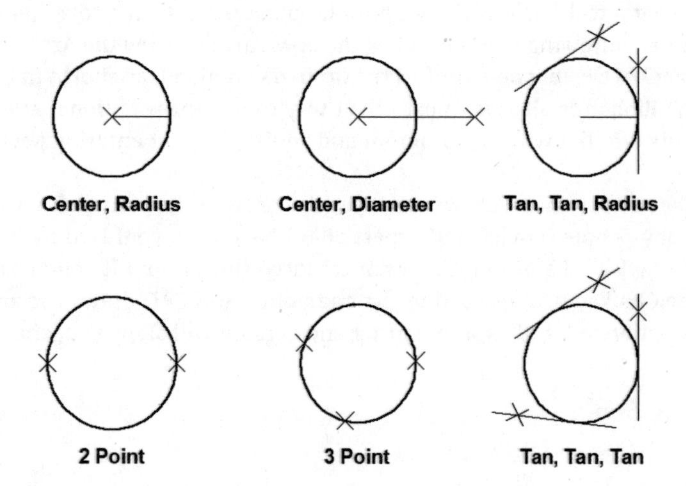

FIGURE 7-1: Pi R squared: Circles R round; pizzas are flat and round.

WARNING

Watch the command line. Pressing Enter repeats the last command; it doesn't repeat the *options* of the last command. If you draw a circle by using the Ttr option, for example, pressing Enter doesn't repeat that method; instead, it repeats the Circle command in its default radius mode.

Arc-y-ology

An *arc* is simply a piece of a circle. On the other hand, if you want a boat full of animals, you would use AutoCAD to design an ark.

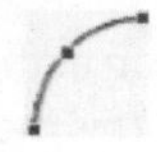

AutoCAD offers you an easy way to define arcs. Just start the Arc command, and then specify three points onscreen to define the arc: where to start the arc, how big a curve it has, and then where to end it. Easy, right?

The trouble is that you nearly always need to specify arcs more precisely than is possible by using this method. AutoCAD helps you specify such arcs by allowing a total of 12 generous option combinations.

To experiment with all the available options, start the Arc command, and specify a start point or choose the Center option. AutoCAD prompts you first for the center point and then for the start point. AutoCAD normally draws arcs in a counterclockwise direction, so pick a start point in a clockwise direction from the endpoint.

On the other hand, after you pick the first point, AutoCAD offers two more options: the included angle and the chord length (the straight-line distance) from the start point to the end point.

TIP

To get a feel for how these permutations can be strung together to create different arc-drawing methods, click the down arrow below the Arc button on the Draw panel of the Ribbon to unfurl the drop-down menu, as shown in Figure 7-2. Using the Ribbon is also the most direct way to use these options, at least until you're truly familiar with the program and you're adept at entering keyboard shortcuts.

TIP

The Arc command draws *counterclockwise* from the start point. Over the years, so many people (novice and expert alike) have been confused by "arcs go counterclockwise" that AutoCAD 2014 introduced this option: If you press and hold down the Ctrl key while selecting the end point, AutoCAD draws the arc in a clockwise direction. A bit of experimenting and cursing will show you what I mean.

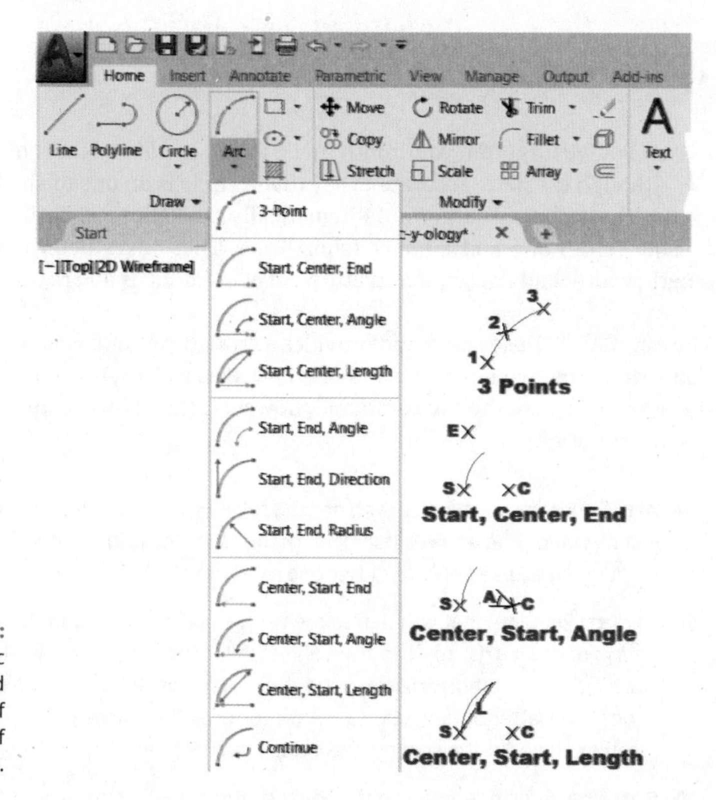

FIGURE 7-2: A deluge of Arc options, and the results of using some of them.

REMEMBER

As usual, whenever AutoCAD asks for the first point, you can simply press Enter to automatically select the last point you specified, even if it's from a different command. Arc has an added twist, however: If the previous point was at the end of a line, a polyline, or another arc, the new arc automatically starts from that point and is tangent to the previous object.

The converse is also true. Draw an Arc and start the Line command. When the command line asks for a starting point, press Enter to make the new line tangent to the end of the arc.

TIP

You probably don't need to use the Arc command often. It might be easier to draw a circle and then remove the part you don't need to arrive at the arc. The Fillet, TRim, and BReak commands (discussed in Chapter 11) cover many alternative arc-creation needs.

Solar Ellipses

In case you've forgotten your ninth grade math, an *ellipse* is sort of a squished circle, although it's more accurate to say that a circle is an unsquished ellipse. Please excuse the technical jargon. Mathematically, an ellipse is defined by two axes: a *major* (long) one and a *minor* (short) one. These axes determine the ellipse's length, width, and degree of curvature. An *elliptical arc* is an arc cut from an ellipse.

The AutoCAD ELlipse command provides a straightforward way to draw an ellipse: You specify the two endpoints of one of its axes and then specify an endpoint on the other axis. Like the Arc command, however, the ELlipse command offers several other options:

- **Arc:** Generates an elliptical arc, not the full ellipse. You define an elliptical arc just as you define a full ellipse. The methods I discuss in this section for creating an ellipse apply to either one.
- **Center:** Requires that you define the center of the ellipse and then the endpoint of an axis. You can then either enter the length of the other axis or specify that a rotation angle around the major axis defines the ellipse. If you choose the latter option, you can enter (or drag the ellipse to) a specific rotation angle for the second axis that, in turn, completely defines the ellipse.
- **Rotation:** Specifies an angle that defines the curvature of the ellipse — big angles create thin ellipses and small angles create rounder ellipses (0 degrees creates a circle, in fact). After dinner, when your plate is empty, hold it vertically to look at it straight on. It looks like a circle. Slowly rotate the plate about its vertical axis until you're looking at it edge-on. Between these two views, you'll see every possible ellipse as defined by the rotation angle — and you'll see your dinner guests beginning to wonder about you.

The following command line example creates an ellipse by using the default endpoints of the axes method. Enter **ELlipse** and press Enter or click the Ellipse button on the Ribbon:

```
Specify axis endpoint of ellipse or [Arc Center]: Pick or type
   the first endpoint of one axis.
Specify other endpoint of axis: Pick or type the other endpoint
   of one axis.
Specify distance to other axis or [Rotation]: Pick or type the
   endpoint of the other axis.
```

Figure 7-3 shows an ellipse and an elliptical arc.

TIP

You create *elliptical* arcs (as opposed to the circular arcs that the Arc command draws) by using the Arc option of the ELlipse command; it can draw a respectable representation of a snowball trajectory, neglecting air resistance. Alternatively, you can draw a full ellipse and then use the TRim and BReak commands, discussed in Chapter 11, to cut out a piece from it.

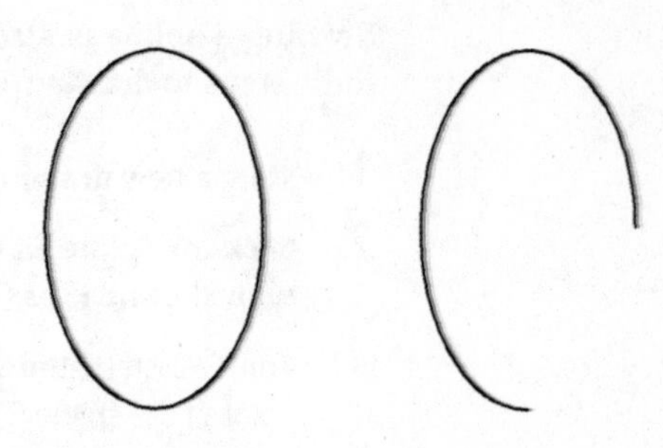

FIGURE 7-3:
To make an omelet, sometimes you have to break a few ellipses.

Splines: Sketchy, Sinuous Curves

Most people use a CAD program because it can perform precision drawing tasks: straight lines, carefully defined curves, and precisely specified points, for example. You may believe that AutoCAD is not the appropriate program to free your inner artist (unless your inner artist is Mondrian) — nonetheless, even meticulously created CAD drawings sometimes need free-form curves. In fact, the advent of processes such as computer numerical control (CNC) machining and 3D printing mean that many consumer goods and even quite a few industrial products are taking on smoother free-form shapes. The AutoCAD *spline object* is just the tool for the job.

You can use AutoCAD splines in one of two ways:

- **Wing it.** Eyeball the location and shape of the curve, and don't worry about making it just so. That's the free-form, sketchy, less-than-precise approach described in this section.
- **Be precise.** Specify the control points and curvature characteristics precisely.

TECHNICAL STUFF

Beneath the easygoing, informal exterior of the AutoCAD spline lies the truly highly precise, mathematically defined entity known as the Non-Uniform Rational B-Spline curve — the *NURBS curve* – wherein a B-Spline is a generalization of the Bézier curve. You can always go to `http://mathworld.wolfram.com/B-Spline.html` to learn far too much about B-splines. Mathematicians and mechanical and industrial designers often care deeply about the precise characteristics of the curves they work with. For these people, the AutoCAD SPLine and SPLINEDIT commands include a number of advanced options. Look up *spline curves* in the AutoCAD online Help system when you need precision splines.

Drawing a spline is straightforward, if you ignore the advanced options. Follow these steps to draw a free-form curve by using the SPLine command:

1. **Start a new drawing.**
2. **Click the Spline Fit button on the Draw panel slideout of the Home tab, or type SPL and press Enter.**

   AutoCAD starts the SPLine command and prompts you to specify the start point of the spline. The command line shows:

   ```
   Current settings: Method=Fit Knots=Chord
   Specify first point or [Method Knots Object]:
   ```

3. **Specify the start point by clicking a point or typing coordinates.**

   AutoCAD prompts you to specify additional points:

   ```
   Enter next point or [start Tangency toLerance]:
   ```

4. **Specify additional points by clicking or typing coordinates.**

   After you pick the second point, press the down-arrow key to display additional options at the Dynamic Input tooltip. You can turn on Dynamic Input by pressing the F12 key. The command line displays:

   ```
   Enter next point or [end Tangency toLerance Undo Close]:
   ```

TIP

   Because you're drawing a free-form curve, you usually don't need to use object snaps or other precision techniques to pick spline points, except perhaps at the start and end of the spline. I discuss precision techniques in Chapter 8.

5. **Press Enter after you choose the endpoint of the spline.**

   AutoCAD draws the spline.

TECHNICAL STUFF

You can specify the start and end tangency of the spline to control the curvature of the start points and endpoints of the spline. If all you want is a swoopy, free-form curve, picking only points works well.

Figure 7-4 shows examples of splines.

TIP

After you draw a spline, you can grip-edit it to adjust its shape. (See Chapter 11 for information about grip-editing.) If you need finer control over spline editing, look up the *SPLine Edit command* in the AutoCAD online help system.

TIP

With its Spline option, the Polyline Edit (PEdit) command can convert a straight-line-segmented polyline into a spline curve form. I cover polylines in Chapter 6, and I touch on PEdit in Chapter 11.

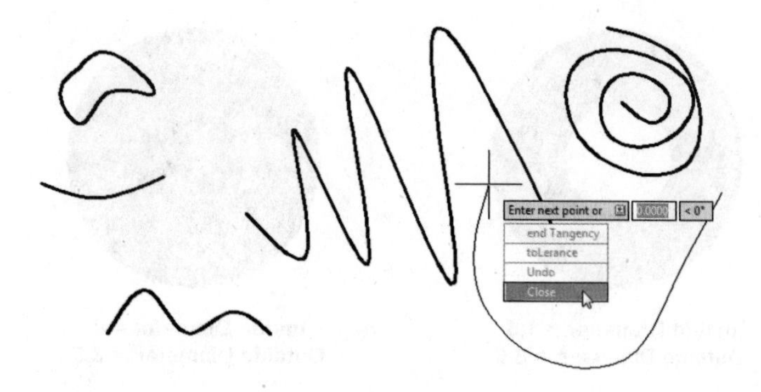

FIGURE 7-4: A slew of splines.

Donuts: Circles with a Difference

Don't tell Homer Simpson, but there's no such thing as a donut in AutoCAD. A donut in AutoCAD is a closed polyline made of two thick 180-degree arc segments that you create with (what else?) the DOnut command. By the way, the rectangles and regular polygons described in Chapter 6 are also polyline objects. A common use for the DOnut command is drawing component-mounting pads on printed-circuit boards; a common use for donuts is as a snack during your coffee break.

When you start the DOnut command, AutoCAD prompts you for the inside diameter and the outside diameter, the size of the hole inside and the size of the donut outside. After you've entered these values, AutoCAD prompts you for the center point of the donut. But one donut is never enough, so AutoCAD continues prompting you for additional center points and drawing donuts until you press Enter. It's the AutoCAD equivalent of saying, "No, really — I'm full now."

The following example draws a regulation-size donut, with a 1.5-inch hole and a 3.5-inch outside diameter. Enter **DOnut** and press Enter, or click the Donut button on the Ribbon. You'll find this button on the slideout panel below the Draw panel of the Home tab of the Ribbon.

```
Specify inside diameter of donut <0.5000>: 1.5
Specify outside diameter of donut <1.0000>: 3.5
Specify center of donut or <exit>: Pick or type the center point
   of one or more donuts.
```

TIP

You can use the DOnut command to create a filled circle, also known as a jelly-filled donut. (Seriously!) Simply specify an inside diameter of 0 (zero). Figure 7-5 shows both kinds of donuts.

FIGURE 7-5: Donuts, plain and jelly-filled.

TIP

If you're Canadian, the DOughnut command also works but *TimHortons* doesn't (unless you use the ALIASEDIT command from Express Tools to create it as a new alias).

Revision Clouds on the Horizon

In many industries, designers and drafters customarily submit a set of drawings at different project milestones or stages of completion, and then submit them again later with revisions, such as corrections, clarifications, and requested changes. Usually, the recipients of these drawings easily locate information that has changed, and a common drafting convention calls attention to revised items by drawing free-form clouds around them. The REVCLOUD command makes quick work of drawing revision clouds.

Drawing revision clouds is easy, after you understand that you click only once in the drawing area. That single click defines the starting point of the cloud's perimeter. Then you move the cursor around — without clicking — and the cloud takes on its curvaceous shape automatically. When you return to a spot near the point you first clicked, AutoCAD closes the cloud automatically.

The following command line example shows you how to draw a revision cloud. Press the REVCLOUD button on the Markup panel of the Annotate tab, or enter **REVCloud** and press Enter:

```
Minimum arc length: 0.5000 Maximum arc length: 0.5000 Style:
  Normal Type: Freehand
Specify first point or [Arclength Object Rectangular
  PolygonalFreehand Style Modify] <Object>:
```

If the Type is indicated as anything other than Freehand, then type **F** and press Enter.

Pick a point along the perimeter of your future cloud:

```
Guide crosshairs along cloud path... Sweep the crosshairs around
   to define the cloud's perimeter.
```

Don't click again. Simply move the crosshairs around without clicking. AutoCAD draws the next arc segment of the cloud when the crosshairs reach the minimum arc length distance from the end of the previous arc segment.

Continue moving the crosshairs around until you return close to the point where you first clicked. AutoCAD will automatically draw one last arc to close the loop.

AutoCAD 2016 added three Type options to revision clouds:

- **O**bject: Pick an existing rectangle, polyline, circle, line, or arc, and AutoCAD turns it into a revision cloud.
- **R**ectangular: Pick two points, and AutoCAD draws a rectangular revision cloud.
- **P**olygonal: Pick a series of points, and AutoCAD connects them with straight-line wavy sections.

TECHNICAL STUFF

If the revision cloud's arcs are too small or too large, erase the cloud, restart the REVCLOUD command, and use the command's Arclength option to change the minimum and maximum arc lengths. The default minimum and maximum lengths are 0.5 (or 15 in metric drawings). If you make the minimum and maximum lengths equal (the default setting), the lobes are approximately equal. If you make them unequal, the lobe size will vary more, and you'll create fluffier clouds. Finally, if you change the Style to Calligraphy, each arc section will taper down its length. Fortunately, these options are more than most non-meteorologists need.

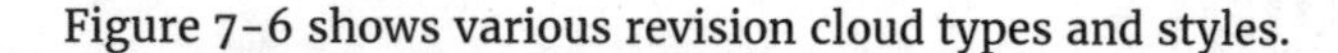

Figure 7-6 shows various revision cloud types and styles.

TIP

Here are a few tips for using revision clouds:

- **Create a separate layer for revision clouds.** Place a revision cloud on its own layer (as described in Chapter 9) so that you can choose to view or plot the drawing with or without the clouds visible. In the days of paper and pencils, revision clouds were sketched on the back of the drafting vellum so that they could easily be erased.

- **Turn off Ortho mode.** To more easily control the shape of a revision cloud, turn off Ortho mode before you start the command.
- **Include the revision number.** Common practice is to delete old revision clouds with each new release of a drawing, but your company's policy may be to leave them on, tagged with a revision number. A good way to handle this requirement is to use a new layer for each new release of revision clouds.

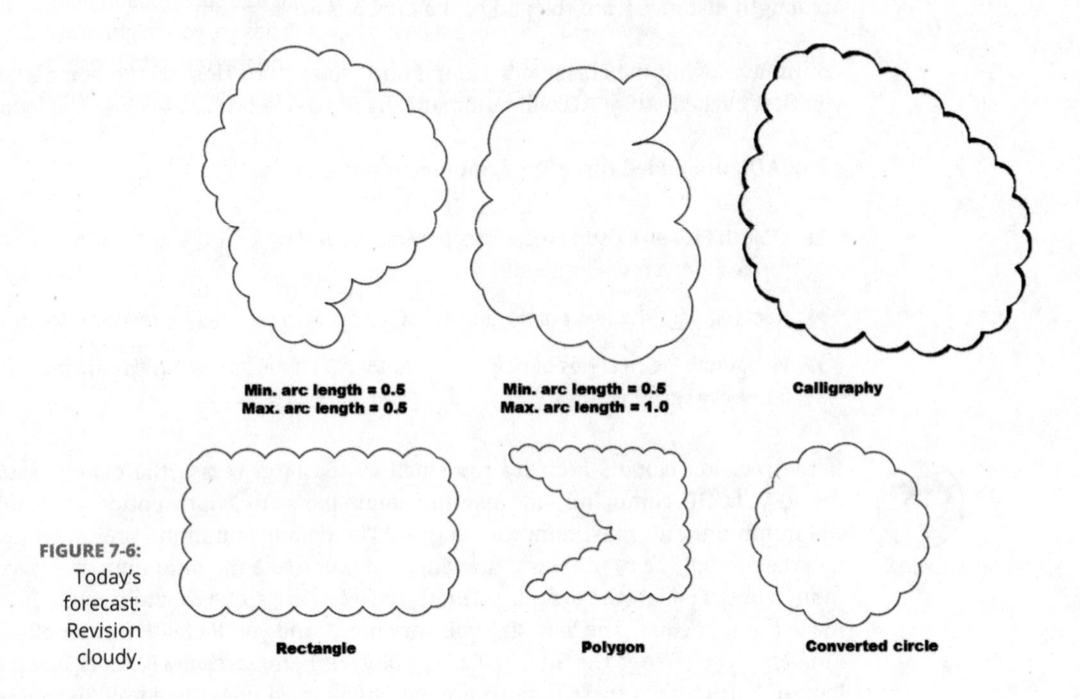

FIGURE 7-6: Today's forecast: Revision cloudy.

Scoring Points

I considered skipping a description of points in this book, but I didn't want you to complain that *AutoCAD For Dummies* was pointless.

The word *point* describes two different elements in AutoCAD:

- A location in the drawing that you specify (by typing coordinates or clicking with the mouse)
- An object that you draw by using the POint command

Throughout this chapter, and most of this book, whenever I tell you to specify points, I'm usually referring to the location. This section tells you how to draw POint objects.

A *POint object* in AutoCAD can serve one of two purposes:

- **Identify a specific location in a drawing to other people viewing the drawing:** A point can be displayed onscreen, as either a tiny dot or another symbol, such as a cross enclosed in a circle.
- **Serve as a precise object snap location:** Think of it as a construction point. For example, when you're laying out a new building design, you might draw point objects at several engineering survey points and then snap to those points as you sketch the building's shape with the PLine command. You use NODe Object Snap mode to snap to AutoCAD point objects. I discuss Snap modes in Chapter 11.

The factor that complicates AutoCAD point objects is their almost limitless range of display options, provided to accommodate these two purposes and possibly some others that I haven't figured out yet. You use the Point Style dialog box, shown in Figure 7-7, to specify how points should look in the current drawing.

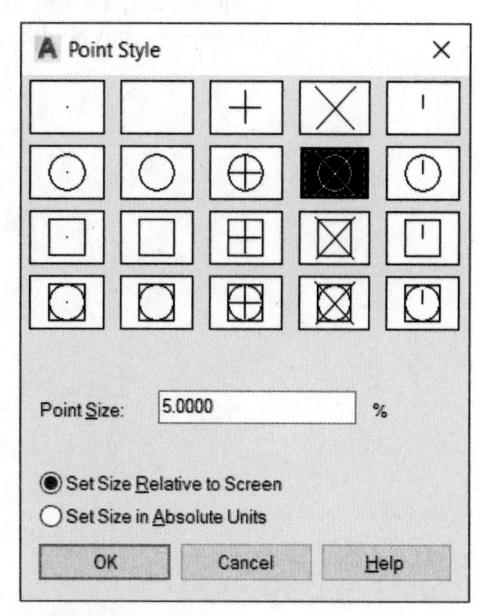

FIGURE 7-7: You use the Point Style dialog box to control how point objects appear onscreen.

To open the Point Style dialog box, open the Utilities panel slideout on the Home tab, or type its command name, DDPTYPE, which has to be one of the least intuitive command names in all of AutoCAD. The upper half of the dialog box shows the available point display styles — just select one.

WARNING

Avoid the first two point style choices if you want point objects to show up onscreen and on plots. The first choice, a single-pixel dot, is hard to see onscreen, and the second choice, which is invisible (a stealth point?), is impossible to see. The single-pixel dot, which is the default display style, works well if you use point objects as object snap locations and want to omit obtrusive points on plots.

The remaining settings in the Point Style dialog box control the size at which points appear onscreen at different zoom magnifications. The default settings often work well, but if you're not satisfied with them, click the Help button in the dialog box to find out how to change them.

After you specify the point style, placing POints onscreen is easy. Simply type **POint** at the command line and press Enter to see this prompt:

```
Current point modes: PDMODE=0 PDSIZE=0.0000
Specify a point: Pick or type the coordinates of a location in
  the drawing.
```

The PDMODE and PDSIZE items listed in the command line are system variables that correspond to the point display mode and display size options in the Point Style dialog box. If you want to know exactly how the system variables correspond to the dialog box choices, you have all the makings of a successful CAD nerd. Click the Help button in the Point Style dialog box to find out more about system variables — not about yourself.

» Getting to know the AutoCAD coordinate systems

» Snapping to object features

» Using other precision drawing and editing techniques

Chapter 8
Preciseliness Is Next to CADliness

Drawing precision is vital to good CAD practice, even more than for manual drafting. Accuracy is, of course, vital to both types of drafting. If you're sketchy on the difference between accuracy and precision, look ahead to the "CAD precision versus accuracy" sidebar in this chapter. If you think CAD managers become tense when you assign properties directly to objects instead of ByLayer (see Chapter 9), wait until someone — I sincerely hope it's not you — fails to use precision techniques when creating drawings in AutoCAD.

REMEMBER

You must draw with precision, at least until you see Chapter 19. Now there's a teaser for you. Typically, the only point you select at random in a drawing is the first one. Then after that one, nearly all line starts and lengths or ends, and all circle and arc centers and radii or diameters, for example, should be specified using exact keyboard input or a suitable object snap.

Controlling Precision

In AutoCAD, a lack of precision makes editing, hatching, and dimensioning tasks much more difficult and time consuming. Keep these facts in mind:

- Small errors in precision in the early stages of creating or editing a drawing often significantly affect productivity and precision later.
- CAD drawings are often used for much more than giving pictures to someone. If the drawings have been properly created, they can be queried for factors such as sizes, areas, and quantities.
- Drawings may end up guiding manufacturing and construction projects; drawing data may drive automatic manufacturing machinery, such as CNC machining and 3D printing. Huge amounts of money, and even lives, can ride on the precision of drawings.

In recognition of these facts, a passion for precision permeates the profession. Precision is one characteristic that separates CAD from ordinary illustration-type drawing work. The sooner you get fussy about precision in AutoCAD, the happier everyone is, including you.

AutoCAD provides a comprehensive package of tools for this task. Table 8-1 lists the more important AutoCAD precision techniques, along with visual cues to the status bar buttons you click to toggle certain features on and off. Note that this list omits status bar buttons that don't directly affect drawing precision directly.

TABLE 8-1 **Precision Tools and Techniques**

Technique	F Key	Status Bar Button	What It Does
Infer constraints	—		Applies geometric constraints at specific pick points (not in AutoCAD LT), as covered in Chapter 19
Snap mode	F9		Forces the cursor to move on an invisible grid of equally spaced hot spots
PolarSnap	—		Forces the cursor to track along specific distances at polar angles
Grid display	F7		Displays a nonprinting reference grid of lines or dots arranged in rows and columns

Technique	F Key	Status Bar Button	What It Does
Ortho mode	F8		Forces the cursor to move horizontally or vertically from the previous point
Polar tracking	F10		Causes the cursor to jump to specified angles
Object snap	F3		Lets you pick specific geometric points on drawing objects without having to initiate a specific snap mode every time
3D Object snap	F4		Lets you pick specific points on existing 3D objects multiple times (not in AutoCAD LT)
Object snap tracking	F11		Causes the cursor to locate new points based on one object snap point (or more)
Object snap overrides	—	—	Lets you pick specific points on existing drawing objects one time only
Coordinate input	—	—	Lets you type exact X,Y or polar coordinates
Direct distance entry	—	—	Lets you locate a point by moving the cursor to show the direction and then typing a precise distance

You can switch many of the buttons — and hence their functionality — on and off by using the F keys. Table 8-1 lists the icons and F keys, where appropriate, for each button.

Precision is especially important when you're drawing or editing *geometry* — the lines, arcs, and other elements that make up whatever you're representing in the CAD drawing. Precision placement usually is less important with notes, leaders, and other annotations that describe geometry.

To use object snap tools, follow these steps:

1. **Make sure all status bar buttons are turned off.**

   They should have gray, not blue, icons.

2. **Start the Line command.**
3. **Pick three points to draw two sides of a triangle.**
4. **Draw the third side of the triangle.**

   Pick a point as close as you can to the start of the first line (that is, the first point you picked). Don't use the Close option.

CAD PRECISION VERSUS ACCURACY

You often hear the words *precision* and *accuracy* used interchangeably, but it's useful to understand the difference. In this book, *precision* refers to controlling the placement of objects so that they lie exactly where you want them to lie in the drawing. For example, lines whose endpoints meet must meet *exactly,* and a circle that's supposed to be centered on the coordinates 0,0 must be drawn with its center *exactly* at 0,0.

Accuracy refers to the degree to which your drawing matches its real-world counterpart. An accurate floor plan is one in which the dimensions of the CAD objects equal the dimensions of the as-built house. In a sense, then, it isn't the drawing that should be accurate — it's the house! Another example is the NASA Mars lander that crashed because of a mix-up in imperial versus metric units. Values may have been very precise, but they weren't accurate.

CAD precision usually, but not always, helps produce accurate drawings. You can produce a precise CAD drawing that's inaccurate because you started with information that was wrong. For example, the contractor gave you incorrect field measurements.

Don't sacrifice accuracy if you want to exaggerate certain distances to convey the relationship between objects more clearly on the plotted drawing. In AutoCAD, you can quite easily show a detail view at a different scale.

5. **Press Enter to complete the command.**

   REMEMBER

   Pressing the spacebar is usually faster than pressing Enter.

6. **Right-click the Object Snap button and choose Settings.**
7. **Deselect all items except for Endpoint and Object Snap On. Click OK.**

   With these two settings, you're turning on object snap functionality and one of the object snap modes (Endpoint).

8. **Start the Line command again.**
9. **Hover the cursor over the first line so that a green square box lights up at the end of the line closest to its starting point. Pick the line at this point.**
10. **Repeat Step 9 for the ending point of the third line and then press Enter.**

    The second green square box may look like it's in the exact location as the first, but it probably isn't.

11. **Enter** Zoom Object Last **and press Enter.**

    AutoCAD zooms in to display a tiny line segment — the last line you drew. The triangle now consists of four lines. The last line may be quite small, but it is a fourth line nonetheless.

12. **Enter** Zoom **P**revious.

13. **Repeat Steps 8–10, but select two lines that form one of the other two vertices of the triangle.**

    Press F2 to expand the command line, where AutoCAD reports that it created a "`zero-length line`". Obviously, the two lines now touch perfectly because the last one snapped like a magnetic attraction to the endpoint of the first line. Okay, it was the start point, but I'm sure that you . . . get the point.

    If you were able to make the first and third lines touch at the limit of AutoCAD's 16-place accuracy, you should throw away this book and buy every lottery ticket you can find.

If you had tried to use the original three lines to control a computer numerically controlled (CNC) milling machine, you would have had problems at the final apex. You need to draw the triangle using the Close option of the Line command or the endpoint object snap.

REMEMBER

Before you draw objects, always check the status bar buttons and set them according to your need for precision.

- **ON:** The button icon is blue.
- **OFF:** The button icon is light gray.

Understanding the AutoCAD Coordinate Systems

Every point in an AutoCAD drawing file can be identified by its X,Y,Z coordinates. In most 2D drawings, the Z-coordinate value is 0. This system of coordinates is referred to in AutoCAD as the *world coordinate system*, or *WCS*.

Keyboard capers: Coordinate input

The most direct way to enter points precisely is to type numbers with the keyboard. AutoCAD uses these keyboard coordinate entry formats:

- **Absolute Cartesian (X,Y) coordinates in the form *X,Y* (for example, 7,4):** This format is the default for input at the command line and shows the distance horizontally (X) and vertically (Y) from the origin at 0,0.
- **Relative X,Y coordinates in the form @*X,Y* (for example, @3,2):** Defines a new point that is X units horizontally (3, in the example) and Y units vertically (2, in the example) away from the previous point.

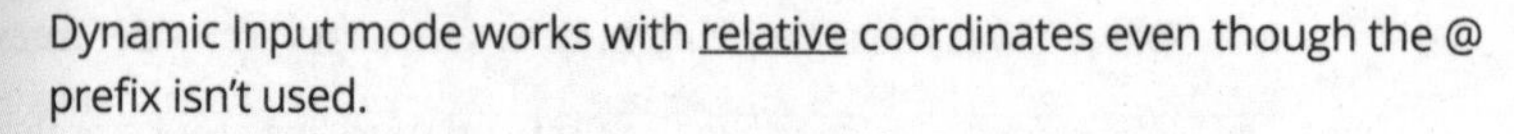

WARNING

Dynamic Input mode works with relative coordinates even though the @ prefix isn't used.

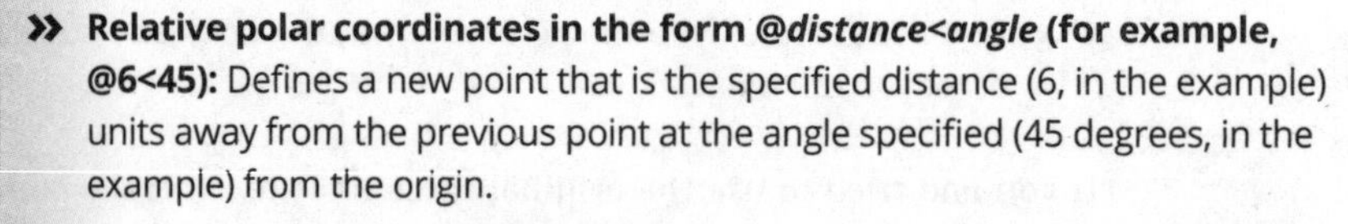

- **Relative polar coordinates in the form @*distance<angle* (for example, @6<45):** Defines a new point that is the specified distance (6, in the example) units away from the previous point at the angle specified (45 degrees, in the example) from the origin.

TIP

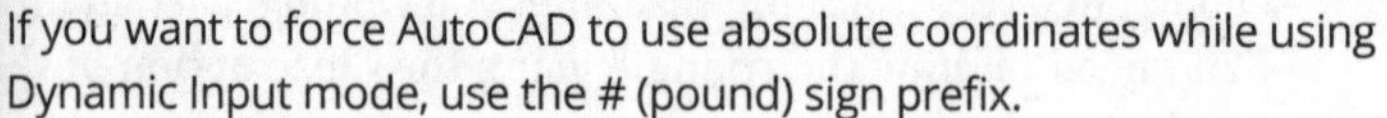

If you want to force AutoCAD to use absolute coordinates while using Dynamic Input mode, use the # (pound) sign prefix.

REMEMBER

AutoCAD nearly always measures angles counterclockwise from the positive X-axis (east), except for surveyors. You know who you are, and you know how your angles work!

Introducing user coordinate systems

Many times, you can conveniently define an additional coordinate system (or "roll your own") to more easily create drawings. This non-world coordinate system is known as a *user-defined coordinate system*.

Why would you want to diverge from the standard world coordinate system (WCS)? Well, the most common reason is that it's much easier to calculate and enter coordinates if they're based on the plane you want to work on in 3D. Suppose that you're modeling an old-fashioned, wedge-shaped rubber doorstop and you want to add the manufacturer's logo to the sloping surface of the wedge. It isn't easy to do if you stay in the WCS, but AutoCAD lets you set a new UCS based on that sloping surface. After the UCS is made current, you draw in it just as you draw in the WCS. I have more to say about UCSs in 3D in Chapter 21.

Think of the real world. A specific latitude and longitude (WCS) always identifies the same point, but the actual location for a description, such as "the corner of 33rd and Main," depends on which city you're in (UCS).

Although originally intended for 3D work, a UCS can be useful in two dimensions as well. The WCS assumes that the north direction is straight up (90 degrees, according to the AutoCAD default settings), but you may be working on a building layout where one wing is at a 37.8 degree angle to the other. No problem: Simply create a UCS that's aligned appropriately. You can look up this process in the online Help system (this book has only so many pages, after all), but here's a quick hint: Click the UCS icon in the lower-left corner of the screen to make grips appear at the origin and the ends of the axis indicators. Then click one of the round grips to turn it red, and drag it to change the angle and thereby set a new UCS. I cover this feature more fully in Chapter 21.

TIP

Creating a UCS by typing options at the command line is generally easier than using the UCS Manager dialog box, but after they're created, they're easier to manage from the UCS dialog box. Type **UCSMAN** and press Enter to display it.

Drawing by numbers

AutoCAD locates *absolute* X,Y coordinates with respect to the 0,0 point (otherwise known as the *origin*) of the drawing — usually, its lower-left corner. AutoCAD locates *relative* X,Y coordinates and *relative* polar coordinates with respect to the previous point you picked or typed. Figure 8-1 demonstrates how to use all three coordinate formats to draw a pair of line segments that start at the absolute coordinates 2,1 and then move to the right 2 units and up 1 unit (@2,1) relative to the first point, and then (relative to that point) move 2 units at an angle of 60 degrees (@2<60). Note in particular how the first two coordinate pairs use the same numbers (2,1) but the second pair defines a different point because of the leading @ symbol.

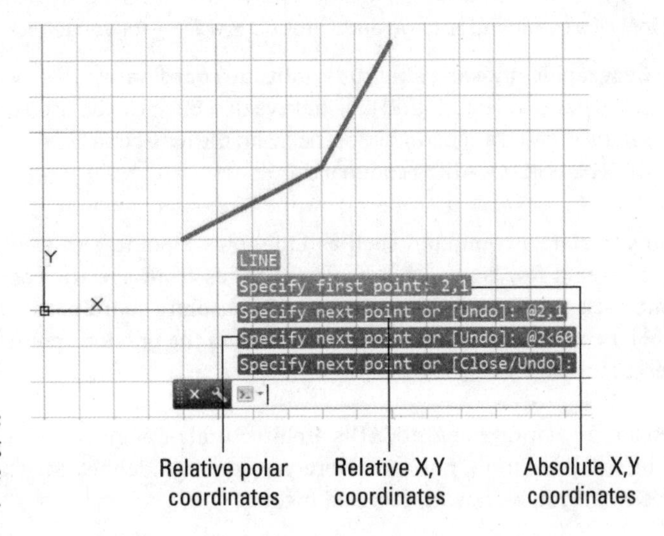

FIGURE 8-1: Coordinating from the keyboard.

TIP

You can find out the X,Y location of the cursor by moving it around in the drawing area and reading the coordinate values in the status bar. The X,Y coordinates should change as you move the cursor. If the coordinates don't change, click the coordinates area until you see <Coords on> in the command line.

REMEMBER

The coordinate display on the status bar is turned off by default. To turn it on, select the three-bar hamburger button at the far right end of the status bar and choose Coordinates.

If you're using the full version of AutoCAD, you may have noticed *three* numbers on the status bar. AutoCAD LT is numerically challenged and only has two. AutoCAD is showing you the X,Y coordinates of the cursor and the current elevation (the *Z value*). However, in 2D drafting, the Z value is or should be 0 (zero), so you can continue calling them X,Y coordinates.

Although it isn't apparent at first, AutoCAD has, in fact, four coordinate display modes. Right-clicking the coordinates reveals the following modes:

- **Specific (<Coords off>):** The status bar coordinate readout is dimmed, and the coordinate values don't update until you pick a point. This behavior is left over from the very old days when computers were too slow to update coordinates in real time.
- **Absolute, showing X,Y coordinates:** The coordinate readout is bright and the absolute X,Y coordinates update continuously as you move the cursor. If no command is active, clicking the coordinates readout alternates between the current mode and off.
- **Relative, showing polar coordinates:** This mode, which displays distance and angle relative to the last point picked rather than absolute X,Y values, appears only if a command is active and AutoCAD is waiting for you to pick a point.
- **Geographic, showing longitude, latitude coordinates:** This mode displays coordinates as latitude and longitude values, but it can be used only after you set the drawing's geographic location with the (what else?) GEOGRAPHICLOCATION command.

When you start a command such as Line, pick a point, and then click the coordinates area a few times, the display changes from specific coordinates (which update each time you pick a point) to live absolute coordinates (X,Y position) to live polar coordinates (distance and angle from the previous point), and then back to specific mode.

REMEMBER

When you're working in AutoCAD's architectural or engineering units, the default unit of entry is *inches*, not feet. Here are some guidelines for entering numeric values when you work with feet and inches:

- **To specify feet,** you must enter the apostrophe (') symbol for feet after the number:

  6' is 6 feet.

- **To separate feet from inches (as architects often do),** enter a hyphen:

  6'-6" is 6 feet, 6 inches.

- **When you enter coordinates and distances,** both the hyphen and the inch mark are optional:

  6'6" and **6'6** are the same as **6'-6"**.

- **To type a coordinate or distance that contains fractional inches,** you *must* enter a hyphen — not a space — between the whole number of inches and the fraction:

  6'6-1/2 (or **6'-6-1/2**) represents 6 feet, 6-1/2 inches.

- **To enter partial inches,** use decimals instead:

  6'6.5 is the same as **6'6-1/2"**, whether you're working in architectural or engineering units.

TIP

To check the accuracy (or precision) of objects that are already drawn, the MEAsuregeom command is a one-stop shop where you can query drawing objects for distances, angles, areas, and other geometric or locational information. You can find the command on the Utilities panel of the Home tab — look for an icon with a yellow ruler. Click the down arrow below the ruler to see a drop list of measuring operations.

The LIst command also tells you all you need to know about selected objects.

Grabbing an Object and Making It Snappy

After you've drawn a few objects precisely in a new drawing, the most efficient way to draw more objects with equal precision is to grab specific, geometrically precise points, such as endpoints, midpoints, or quadrants, on the existing objects. Every object type in AutoCAD has at least one geometric point, and you can snap to them precisely as you draw by using *object snaps* (*osnaps*).

AutoCAD provides two ways to use an object snap:

- **Object snap *override:*** Active for a single pick
- ***Running* object snap:** Remains in effect until you turn it off

AutoCAD (but not AutoCAD LT) has a suite of object snaps specifically for working in 3D. Although I cover them fully in Chapter 21, I point them out here because the 3D Object Snap status bar button sits next to the standard Object Snap button, and it isn't easy to distinguish them. In this section, you use the regular Object Snap button — it shows the square with the little spot in the upper-left corner, not the 3D-looking box with the little green spot on one front corner.

Grabbing points with object snap overrides

Here's how to draw lines precisely by using object snap overrides:

1. **Open a drawing containing some geometry, or start a new drawing and create some.**

2. **Turn *off* running object snap mode by clicking the Object Snap (F3) button on the status bar until the button appears to be dimmed and `<Osnap off>` appears in the command line.**

   TIP

   Although you *can* use object snap overrides while running object snaps are enabled, I recommend that you turn off running object snaps while you're becoming familiar with object snap overrides. After you get the hang of each feature separately, you can use them together.

3. **Start the Line command by clicking the Line button in the Draw panel on the Ribbon or by typing Line (or L) and pressing Enter.**

   AutoCAD prompts you to select the starting point of the line:

   ```
   Specify first point:
   ```

4. **Hold down the Shift key, right-click anywhere in the drawing area, and release the Shift key.**

   The Object Snap menu appears, as shown in Figure 8-2. Object snaps are grouped by type: first ones that apply mostly to lines, then those that apply mostly to curves, and then a group of miscellaneous ones.

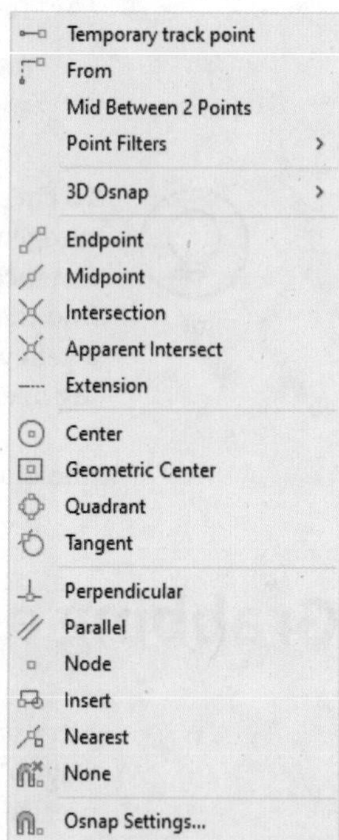

FIGURE 8-2: The Object Snap right-click menu.

5. **Choose an Object Snap mode, such as Endpoint, from the Object Snap menu.**

   The Object Snap menu disappears and the command line displays an additional prompt indicating that you've directed AutoCAD to seek out, for example, the endpoints (endp) of existing objects:

   ```
   _endp of:
   ```

6. **Move the cursor slowly around the drawing, pausing over various lines and other objects. Don't click yet.**

   If you move the cursor near an object with an endpoint, a colored square icon appears at the endpoint, indicating that AutoCAD can snap to that point. When you stop moving the cursor for a moment, a tooltip displaying the Object Snap mode (for example, `Endpoint`) appears, to reinforce the idea.

7. **Click when the ENDpoint object snap square appears on the point you want to snap to.**

   AutoCAD snaps to the endpoint, which becomes the first point of the new line segment that you're about to draw. The command line prompts you to select the other endpoint of the new line segment:

   ```
   Specify next point or [Undo]:
   ```

   When you move the cursor around the drawing, AutoCAD no longer seeks out endpoints, because object snap overrides last for only a single pick. You can use the Object Snap right-click menu again to snap the other end of the new line segment to another point on an existing object.

8. **Display the Object Snap menu again (refer to Step 4). Then choose another Object Snap mode, such as MIDpoint, from the Object Snap menu.**

   The command line displays an additional prompt indicating that you directed AutoCAD to seek, for example, midpoints (mid) of existing objects:

   ```
   _mid of:
   ```

   When you move the cursor near the midpoint of an object, a colored triangle appears at the snap point, as shown in Figure 8-3. Each object snap type (such as ENDPoint, MIDpoint, and INTersection) displays a different symbol. When you stop moving the cursor, the tooltip text reminds you of what the symbol means.

9. **Draw additional line segments by picking additional points. Use the Object Snap right-click menu to specify a single object snap type before you pick each point.**

   Try the INTersection, PERpendicular, and NEArest object snaps. If the drawing contains arcs or circles, try CENter and QUAdrant.

10. **When you finish working with object snap overrides, right-click anywhere in the drawing area and choose Enter from the menu or press the spacebar to end the Line command.**

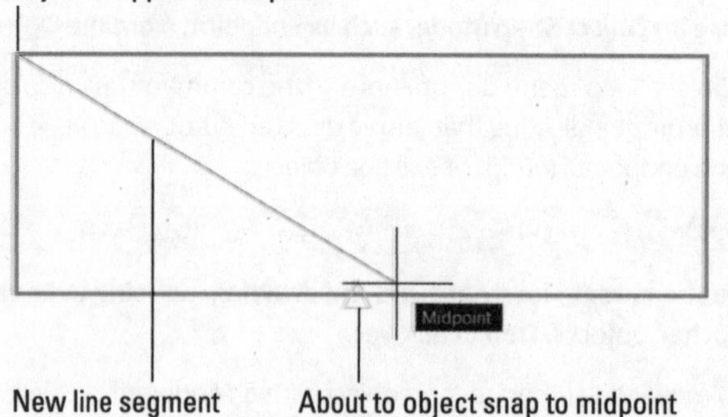

FIGURE 8-3: A snappy line.

TIP

You can activate object snap overrides during a command by entering the first three letters of a mode at the command line whenever AutoCAD is asking you to select a point.

Snap goes the cursor

Table 8-2 explains the subtleties of the different object snap modes.

TIP

Deferred object snaps are turned on automatically when the first point you pick doesn't fully satisfy the current object snap. Following are two examples:

- **Deferred tangent:** You would normally draw a line from a first point to a TANgent point on a circle. On the other hand, if you select the circle first and then select the second point, the line is drawn outward from the circle to the point. This technique can be used to draw a line that is tangent to two non-collinear circles.
- **Deferred perpendicular:** You normally draw a line from a first point to a PERpendicular point on a circle. On the other hand, if you select the circle first and then select the second point, the line is drawn outward from the circle to the point. This technique can be used to draw a line that is perpendicular to two non-collinear circles.

REMEMBER

Right-clicking and Shift+right-clicking do different things in the drawing area:

- **Right-click:** Displays menu options for the current command (or common commands and settings when no command is active)
- **Shift+right-click:** Always displays the same Object Snap menu

TABLE 8-2 **Object Snap Modes**

Mode	What It Does
ENDpoint	Snaps to the nearest end of an open drawing object, such as a line, an arc, a polyline segment, or an elliptical arc.
MIDpoint	Snaps to the nearest midpoint of an open drawing object, such as a line, an arc, a polyline segment, or an elliptical arc.
CENter	Snaps to the center of a circle, an arc, or an elliptical arc.
Geometric CEnter	Snaps to the center of gravity of a closed polyline or a spline.
NODe	Snaps to a point object, or to the definition point of a dimension, or to the insertion point of dimension text.
QUAdrant	Snaps to the quadrant points (0, 90, 180, or 270 degrees in the current UCS) of a circle or an arc, or to the ends of the major or minor axis of an ellipse or an elliptical arc.
INTersection	Snaps to the point where two drawing objects cross or meet.
EXTension	Displays a temporary extension line when you pass the cursor over the end of an object so you can snap to a point along this temporary extension.
INSertion	Snaps to the insertion point of text, mtext, or a block insertion.
PERpendicular	Snaps to the point that forms a right angle between the start point and the object being selected. Perpendicular is not confined to the common image of a horizontal line with a vertical line attached at one end. Lines do not need to be horizontal and vertical nor do they even need to touch to be perpendicular. A line can also be perpendicular to a circle, an ellipse, or even to the sinuous curve of a spline.
TANgent	Snaps to the tangent of an arc, a circle, an ellipse, an elliptical arc, a polyline arc, or a spline.
NEArest	Snaps to the point on the selected object that is the shortest distance from the starting point.
APParent Intersection	Snaps to the point where objects appear to intersect in the current view, even when they don't actually intersect in 3D.
PARallel	Creates a new line segment, polyline segment, ray, or xline that is parallel to an existing linear object. ORTHO mode must be turned off before using the PARallel object snap.
M2P	Snaps to a point midway between two other points.

Running with object snaps

Often, you use an object snap setting (such as ENDpoint) repeatedly. Use running object snaps to address this need.

Follow these steps to set a running object snap:

1. **Right-click the Object Snap button on the status bar.**

   The Object Snap menu appears, as shown in Figure 8-4.

   Many status bar buttons display shortcut menus when you right-click, as shown in Figure 8-4. Object Snap and Object Snap Tracking both display the menu of running object snaps. Many settings, including all running Object Snap types, can be set by clicking a menu item.

2. **Click in the menu to select one or more Object Snap settings.**

   Active running Object Snap settings show a check mark beside the icon (see Figure 8-4).

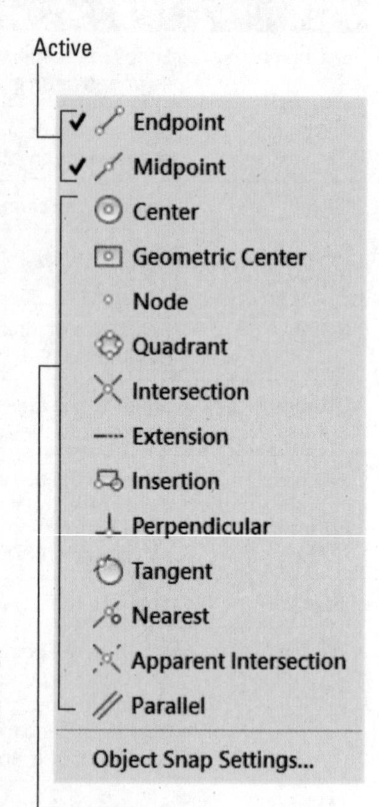

FIGURE 8-4: Grabbing multiple object features is an osnap.

You click the Object Snap button on the status bar or press the F3 key to toggle Running Object Snap mode. After you turn on this mode, AutoCAD hunts for points that correspond to the object snaps you selected on the Object Snap button's right-click menu. As with object snap overrides, AutoCAD displays a special symbol — such as a square for an ENDpoint object snap — to indicate that it has found an object snap point. If you keep the cursor still, AutoCAD also displays a tooltip that lists the object snap point type.

Use object snap overrides or running object snaps to enforce precision by ensuring that new points you pick coincide *exactly* with points on existing objects. In AutoCAD, you can't let points *almost* coincide or look like they coincide. You lose points, both figuratively and literally, if you don't use object snaps or another precision technique covered in this chapter to enforce precision.

When using earlier releases of AutoCAD, applying crosshatching rapidly became impossible if the boundary of the area you were hatching had even microscopic leaks. Current releases have a Hatch Gap option to overcome this problem, but using this option is almost never correct. Although you can crosshatch a leaky area, you will run into far greater problems downstream. For example, leaky geometry will create havoc if you try to use it to program a CNC machining center.

REMEMBER

Most, but not all, object snap overrides have running object snap equivalents. For example, ENDpoint, MIDpoint, and CENter work as either overrides or running object snaps, but M2P (Midway between 2 Points) works only in override mode.

Other Practical Precision Procedures

The following list describes additional AutoCAD precision techniques (refer to Table 8-1):

- **Snap:** If you turn on Snap mode, AutoCAD constrains the cursor to an invisible rectangular grid of points at the spacing that you've specified when AutoCAD prompts you to specify a point.

  When you enable Snap mode, at first it seems to be broken because the cursor doesn't snap to the imaginary grid but travels freely. Snap mode becomes active only when the program asks you to pick a point.

  Follow these steps to turn on Snap mode:

  a. *Right-click the Snap Mode button on the status bar.*

  b. *Choose Snap Settings.*

     The Snap and Grid tab in the Drafting Settings dialog box appears.

  c. *Enter a snap spacing in the Snap X Spacing field and click OK.*

  Click the Snap Mode button on the status bar or press F9 to toggle Snap mode off and on.

TIP

  Usually, you can turn on Snap mode all the time, set at the usual smallest measurement increment. For example, to design doorknobs, I set it to 0.001, whereas an architect might set it to 1/8". Now, I know that even randomly selected points have the degree of accuracy I want.

  You can switch between Grid Snap (snap points in rows and columns) and Polar Snap (snap points based on distances and angles) by using the Snap mode button's shortcut menu. See the Polar Snap bullet for more information.

- **Ortho:** Using Ortho mode forces the cursor to move horizontally or vertically relative to the current coordinate system's X- and Y-axes. To toggle Ortho mode, click the Ortho Mode button on the status bar or press F8. You can also hold down the Shift key to temporarily go in and out of Ortho mode. Because technical drawings often include lots of orthogonal lines, you may use Ortho mode a lot — but take a close look at polar tracking as well.

You know you've been using AutoCAD too long when the police officer asks you to walk a straight line and you reply, "No problem. Can you press F8 please?"

- **Direct distance entry (DDE):** This point-and-type technique is an easy and efficient way to draw with precision. You simply point the cursor in a particular direction, type a distance value at the command line, and press Enter. You can use DDE at any time the cursor is anchored to a point, and the command line or Dynamic Input tooltip prompts you for another point or a distance.

  You usually use DDE with polar tracking turned on to specify distances in particular directions (for example, in angle increments of 45 degrees). You can also combine DDE with Ortho mode to specify a distance in an orthogonal direction (0, 90, 180, or 270 degrees).

- **Object snap tracking:** This feature extends running object snaps so that you can locate points based on more than one object snap point. For example, you can define a point at the center of a square by tracking to the midpoints of two perpendicular sides. Click Object Snap Tracking on the status bar or press F11 to toggle Object Snap Tracking.

  You can locate points based on more than one object snap point. Follow these steps:

  a. *Draw a rectangle with the RECtang command.*

  b. *Right-click the Object Snap button, and turn on MIDpoin.*

  c. *Make sure Object Snap (F3) and Object Tracking (F11) are turned on.*

  d. *Start the Circle command.*

  e. *Move the cursor close to the middle of the bottom line of the rectangle until the green triangle osnap marker appears.* Don't pick this point.

  f. *Move the cursor vertically and note that the green triangle disappears to be replaced by a green + sign, and a vertical dotted green line appears.*

  g. *Move the cursor close to the middle of one of the vertical lines of the rectangle until the green triangle osnap marker appears.* Don't pick this point either.

  h. *Move the cursor horizontally until the two dotted green lines cross and an X appears, along with a midpoint tooltip, as shown in Figure 8-5.*

  i. *Click this point to define the center of a circle that is exactly in the middle of the rectangle.*

  You can use this same technique to transfer points from a front view to a top view or to a side view without having to use construction lines.

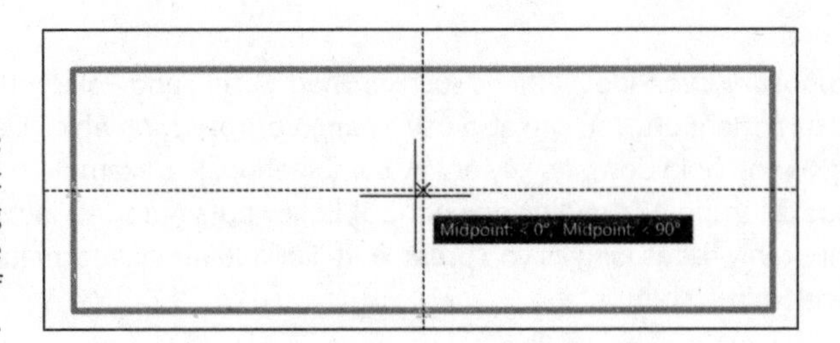

FIGURE 8-5: Using Object Snap Tracking to find the exact middle of a rectangle.

- **Polar tracking:** When you turn on polar tracking, the cursor jumps to increments, such as 15 degrees. When the cursor jumps, a tooltip label starting with *Polar:* appears. Right-click the Polar Tracking button on the status bar and choose an angle. To specify other angles, choose the Tracking Settings option to display the Polar Tracking tab in the Drafting Settings dialog box. Select an angle from the Increment Angle drop-down list and then click OK. Click the Polar Tracking button on the status bar or press F10 to toggle Polar Tracking mode.

  Remember that you can set a predefined polar tracking angle by right-clicking the Polar Tracking button and choosing an angle from the menu. If you want to add an angle that isn't on the list, you have to click Settings to open the Drafting Settings dialog box.

  REMEMBER

  Polar and Ortho modes are mutually exclusive. Turning on one turns off the other.

- **PolarSnap:** You can force polar tracking to jump to specific incremental distances along the tracking angles by changing the snap type from Grid snap to PolarSnap. For example, if you turn on polar tracking and set it to 45 degrees, and then turn on PolarSnap and set it to 2 units, polar tracking jumps to points that are at angle increments of 45 degrees and distance increments of 2 units from the previous point. PolarSnap has a similar effect on Object Snap tracking.

  To switch to PolarSnap, right-click the Snap Mode button and choose PolarSnap from the menu. To specify a PolarSnap distance, follow these steps:

  a. *Right-click the Snap button on the status bar.*

  b. *Choose Snap Settings.*

     The Snap and Grid tab on the Drafting Settings dialog box appears.

  c. *Select the PolarSnap radio button, type a distance in the Polar Distance text box, and then click OK.*

  When you want to return to ordinary rectangular snap, as described at the beginning of this list, right-click the Snap Mode button and choose Grid Snap from the menu.

» **Temporary override:** Settings such as Snap, Ortho, and Polar remain on until you turn them off. You can also use a *temporary override,* which lasts only as long as you hold down its key or key combination. For example, when Ortho mode is turned off, holding down the Shift key puts AutoCAD into Ortho mode temporarily for as long as you press Shift. For additional information, see the online Help system.

TIP

If you're new to AutoCAD, its wide range of precision tools may seem overwhelming. Rest assured that there's more than one way to use the program's precision tools. You can make perfectly precise drawings without using every single implement in the AutoCAD precision toolkit. I recommend these steps:

1. Become comfortable with typing coordinates, Ortho mode, direct distance entry, and Object Snap overrides.
2. Become familiar with running object snaps, and experiment with Snap mode.
3. After you're comfortable using all these precision features, experiment with polar tracking, PolarSnap, and object snap tracking.

REMEMBER

Snap constrains the cursor to locations whose coordinates are multiples of the current snap spacing, and it works whether or not the drawing has objects in it. *Object snap (osnap)* enables you to grab points on existing objects, whether or not those points happen to correspond with the snap spacing.

TIP

What happens if some of your drawing-with-precision choices seem to contradict each other? The answer is simple: AutoCAD uses the one that requires the greater effort to initiate. That is, Snap is overridden by Object Snap, which is overridden by a snap override, which is overridden by the keyboard entry of points or distances.

» Managing layers

» Copying layers and other named objects between drawings with DesignCenter

Chapter 9
Manage Your Properties

Contrary to what you might believe, managing properties has nothing to do with real estate holdings. CAD programs are different from other drawing programs in that you have to pay attention to little details, such as the properties of objects or the precision of points that are specified when you draw and edit objects. If you ignore these details and start drawing, you'll produce a mess of sloppy geometry that's difficult to edit, view, and plot.

This chapter introduces *object properties* — a set of AutoCAD tools and techniques that can help prevent you from making CAD messes. Chapter 8 explains the most important precision drawing techniques that you need to observe to create usable AutoCAD drawings. You must understand the information in Chapters 8 and 9 before you start drawing and editing objects in production drawings (described in Chapters 6, 7, 10, and 11).

When you first start using AutoCAD, the number of property settings and precision controls necessary to draw a simple line can seem intimidating. Unlike in many other programs, you cannot simply draw a line in a more-or-less-adequate location and then slap some color on it. All these settings and controls can inspire the feeling that you have to learn how to drive a Formula 1 car to make a trip down to the corner store. Wait a minute; that could be fun! The benefit is that after you *are* comfortable in the driver's seat, AutoCAD takes you on the long-haul trips and gets you there a lot faster.

Using Properties with Objects

All objects you draw in AutoCAD are like good Monopoly players: They own *properties.* In AutoCAD, these properties aren't physical; instead, they're the characteristics of objects, such as their layer, color, linetype, lineweight, transparency, and plot style. You use properties to communicate information about the characteristics of the objects you draw, such as the kinds of real-world objects they represent, their materials, their relative location in space, or their relative importance. In AutoCAD, you also use properties to organize objects for editing and plotting purposes.

AutoCAD gives you different ways to control object properties:

- **By layer:** Each layer has a default color, visibility, linetype, lineweight, transparency, and plot style property. Unless you tell AutoCAD otherwise, objects always inherit the properties of the layers on which they're created. When objects are selected in a drawing created by using this system, the object properties are shown as ByLayer.
- **By object:** AutoCAD enables you to *override* the property settings that objects inherit from their assigned layers. This gives you independent control over each object's color, linetype, lineweight, transparency, or plot style properties so they differ from the layer's properties. When Autodesk introduced this capability, it gleefully announced that "Now you have total control over an object's properties." But as you will see shortly, the opposite is true; you just lost control.

It is almost never good practice to apply property overrides to individual objects so that, for example, a line on a layer that has the color red as its property is instead displayed as green.

Using the ByLayer approach

In almost all cases, you should create layers, assign properties to each layer, and let the objects on each layer inherit that layer's properties. Here are two benefits of using the ByLayer approach:

- **You can modify properties by layer instead of individually.** You can easily change the property (such as color) of all objects you placed on the layer, not for a bunch of separate objects.

» **You can exchange drawings with other people.** Experienced drafters use the ByLayer approach so that when they work with drawings from other people, the drawing is much more compatible. (You'll also avoid the wrath of irate CAD managers, whose job duties include haranguing any hapless newbie who assigns property overrides to individual objects.)

In a worst-case scenario, you might receive an architectural drawing from someone and everything looks good — even the object colors are correct. You want to plot the drawing with no dimensions showing, so you freeze the Dimension layer. Oh, poo! (Or words to that effect.) Only half the dimensions disappear, but so do some center lines, a couple of walls, the toilets, and other items. This is what I meant when I said that you have lost control of object properties.

If you assign properties with the ByLayer approach, all you have to do is set layer properties in the Layer Properties Manager palette, as shown in Figure 9-1. (I tell you how in this section.) Before you draw any objects, verify that the Color Control, Linetype Control, and Lineweight Control drop-down lists — *and* the Transparency button in the Properties panel on the Home tab of the Ribbon — are all set to ByLayer, as shown in Figure 9-2. Remember that the configuration of panels and drop-down lists may vary according to the resolution of the display. If the drawing is set to use color-based plot styles rather than named plot styles, as described in Chapter 16, the Plot Style Control drop-down list is inactive and displays ByColor.

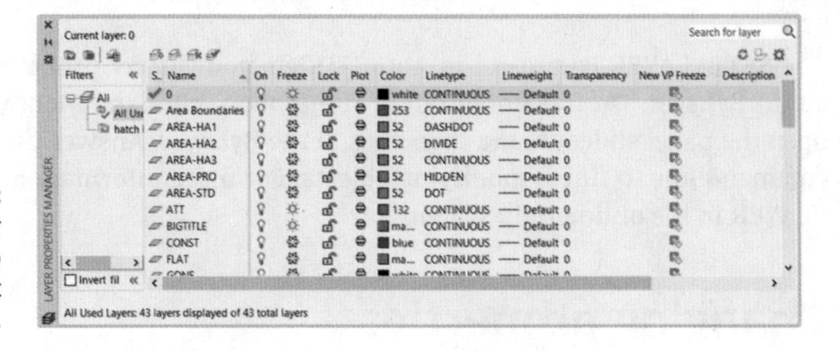

FIGURE 9-1: Use layer properties to control object properties.

If the drawing is set to use named plot styles rather than color-based plot styles (see Chapter 16), the Plot Style control drop-down list should also display ByLayer.

REMEMBER

Don't assign properties to objects by either of these methods:

- Don't choose a specific color, linetype, lineweight, transparency, or plot style from the appropriate drop-down list in the Properties panel on the Home tab of the Ribbon, or from the Properties palette, and then draw the objects. All new objects will be on the current layer but will have property overrides applied.
- Don't draw the objects, select them, and then choose a property from the same drop-down lists.

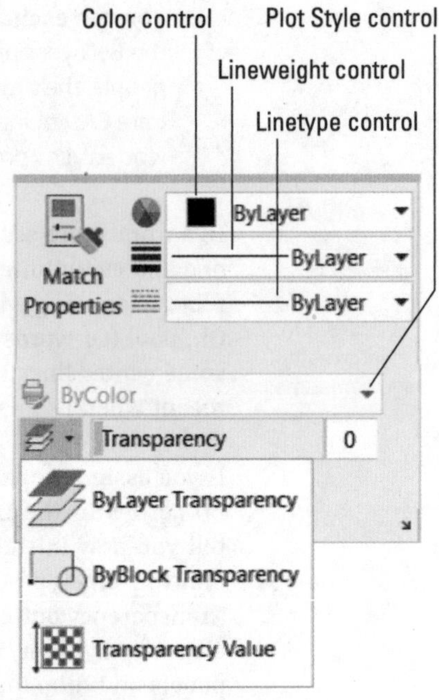

FIGURE 9-2: ByLayer (nearly) all the way.

If you prefer to do things the right way (that is, *my* way), assign these properties ByLayer, as described in the later section "Working with Layers."

TIP

If you want to become a good CAD manager, learn how to edit the Customizable User Interface (CUI) file in AutoCAD, and then delete the property override items on the Ribbon.

TIP

The SETBYLAYER command in AutoCAD lets you correct properties that are not set to ByLayer. On the Home tab of the Ribbon, click the Modify panel label to open the panel slideout, and then click Set to ByLayer. Answer the prompts at the command line to finish modifying objects. For more information, refer to SETBYLAYER in the online Help system.

Changing properties

You can view and change *all* properties of an object in the Properties palette (its icon is shown in the margin), and many of them in the Quick Properties palette (the second icon in the margin). In Figure 9-3, the Properties palette at the left and the Quick Properties palette at the right show properties for the selected line object. Launch the Properties palette by clicking the diagonal down arrow in the lower-right corner of the Properties panel of the Home tab.

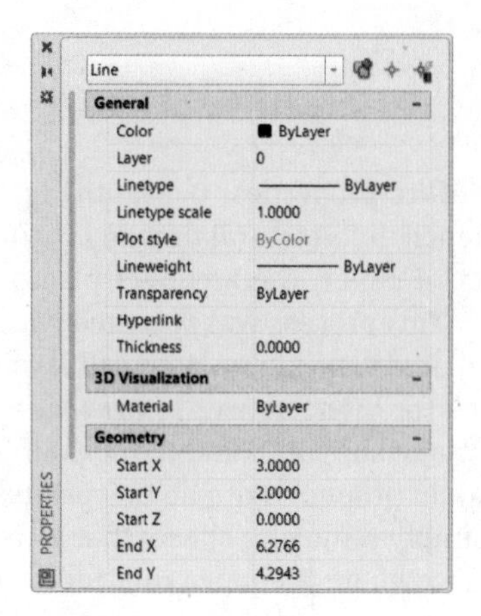

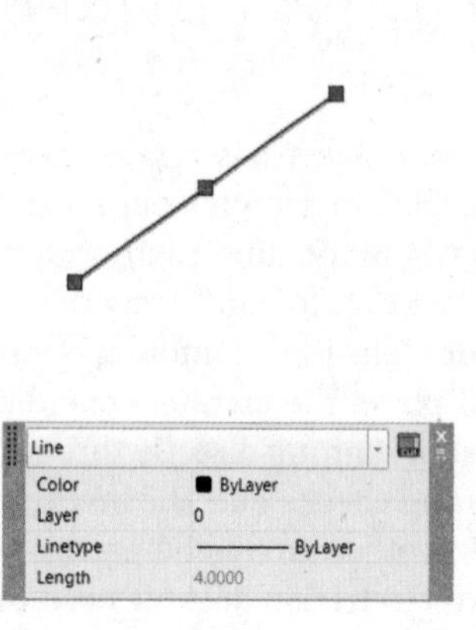

FIGURE 9-3: Comprehensive or quick? Sometimes you need lots of information, and sometimes you don't.

The Properties palette was joined in AutoCAD 2009 by its more streamlined little sibling, Quick Properties (QP). When you turn on the palette by clicking its QP button on the status bar, selecting an object opens a floating palette that displays a small selection of the object's properties, which you can customize.

Handy as it is, the Quick Properties palette has a knack of popping up on top of drawing objects that you need to see. In AutoCAD, you can let Quick Properties mode remain turned off at the status bar and instead use the QuickProperties command. Type its alias (**QP**) and then select an object to display the Quick Properties panel. You can also double-click most objects to display their quick properties.

To toggle the full Properties palette on and off, click the Properties button on the View tab of the Ribbon or press Ctrl+1. Before you select an object, the Properties palette displays the *current properties* — properties that AutoCAD applies to new objects when you draw them. After you select an object, the Properties palette displays the properties for that object. If you select more than one object, the Properties palette displays the properties that they have in common.

WARNING

Properties and Quick Properties let you change nearly any property of one or more selected objects. You should change the main properties (such as color, linetype, and lineweight) only by moving the objects to a different layer.

TIP

If you're a CAD manager, you can easily customize the Quick Properties palettes to remove undesirable properties. For detailed instructions, search for *customize quick palettes* in the Help search window.

Working with Layers

Every object has a layer as one of its properties. Conceptually, layers had their origins in complex paper-and-pencil architectural drawings. After laying out the basic foundation plan on a sheet of paper, the architect placed a sheet of clear plastic on top and drew the walls. This process was repeated with separate sheets for plans like plumbing, electrical, and ventilation. When all layers of plastic were in place, the architect could check for interferences to prevent a ventilation duct from running directly through a major support column, for example. When everything checks out, the architect could produce suitable blueprints for the plumber and other interested parties by simply removing sheets that weren't appropriate. An electrician doesn't need to know where the pipes run, for example. Today, we freeze layers that we don't need to see.

AutoCAD, like most CAD programs, uses the layer as the primary organizing principle for all objects you draw. You use layers to organize objects into logical groups of items. For example, walls, furniture, dimensions, and text notes usually belong on four separate layers, for a couple of reasons:

- Layers give you a way to turn groups of objects on and off, both on the screen and on the plot.
- Layers provide the most efficient way of changing object color, linetype, lineweight, transparency, and plot style.

STACKING UP LAYERS

How do you decide what to name layers and choose which objects to put on them? Some industries have developed layer guidelines, and many offices have created documented layer standards. For example, the American Institute of Architects has a comprehensive set of layer standards. Some projects even impose specific layer requirements. By the way, be careful: If someone says, "You need a brick layer for this project," that statement can have a couple of different meanings. Ask experienced CAD drafters in your office or industry how they use layers in AutoCAD. If you can find no definitive answer, such as through ready-made template files, create a chart of layers for yourself. Each row in the chart should list the layer name, color, linetype, lineweight, transparency, and the kinds of objects that belong on that layer. If you use named plot styles to control the plotted output, add a default plot style to the list. (I cover plot styles in Chapter 16.)

So, to work in AutoCAD efficiently, you create as many layers as the drawing needs and assign them suitable names and properties, such as specific colors and linetypes. Then you draw objects on specific layers. As you draw objects, AutoCAD puts them on the *current layer* automatically — the layer you see in the Layer drop-down list on the Layers panel of the Home tab when no objects are selected. You can also set things up so that dimensions and hatching automatically get put on specific layers. I cover this in Chapters 13 and 14. If a layer already exists in the drawing, you make it the current layer by choosing it from the Layer drop-down list, as shown in Figure 9-4.

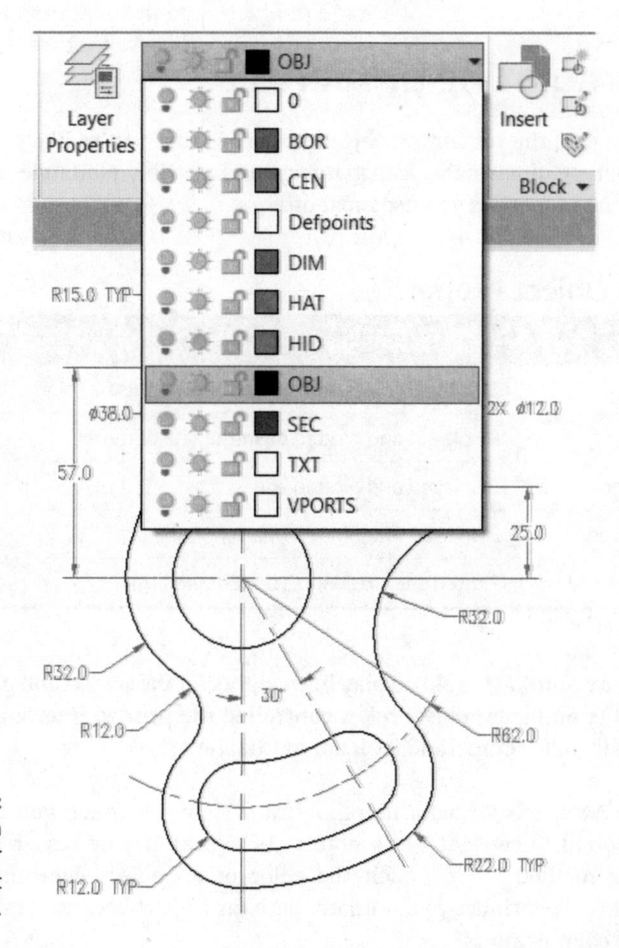

FIGURE 9-4: Setting an existing layer as the current layer.

TIP

You don't have to create *all* layers before drawing an object, but it saves you some time if you start with as many layers as you estimate you need and then add more layers only as needed. You can save even more time by creating all the layers with specific properties that you need in a new drawing, and then saving the drawing as a template file (covered in Chapter 4). Now each drawing you start from that template has all the layers you need. In fact, most companies have (or should have) a standard template file already set up to their layer standards. Similarly, outside clients probably have standard layer sets.

No objects can be selected before you use the Layer drop-down list to change the current layer, so press the Esc key twice to be certain. If objects *are* selected, selecting another layer from the Layer drop-down list changes their layers, which you might not want. When no objects are selected, the Layer drop-down list displays (and lets you change) only the current layer.

Accumulating properties

Besides layers, the remaining object properties that you're likely to use often are color, linetype, lineweight, transparency, and possibly plot style. Table 9-1 summarizes the properties you use most often.

TABLE 9-1 **Useful Object Properties**

Property	What It Controls
Color	Displayed colors and plotted lineweights or colors
Linetype	Displayed and plotted dash-dot line patterns
Lineweight	Displayed and plotted line widths
Transparency	Displayed and plotted opacity of objects
Plot style	Plotted characteristics (see Chapter 16)

Long before AutoCAD could display lineweights on the screen and print those same lineweights on paper, object *colors* controlled the printed lineweight of objects. I explain this indirect method in detail in Chapter 15.

AutoCAD 2000 introduced a more logical system, by which you could assign an actual plotted lineweight to an object. As logical as this newer method seems, the older method — in which the color of an object determines its plotted lineweight — continues to dominate, perhaps in part because of the ready availability of color printers.

You may work this way even in AutoCAD 2023 for compatibility with drawings (and coworkers) that use the old way. Figure 9-5 shows you the idea. The model space view at the left shows objects in different colors but with the same default lineweight. The paper space view at the right (showing what the plotted drawing will look like) shows that although the lines are black, their thicknesses vary, determined by the model space colors. For example, blue is thick, and black is thin.

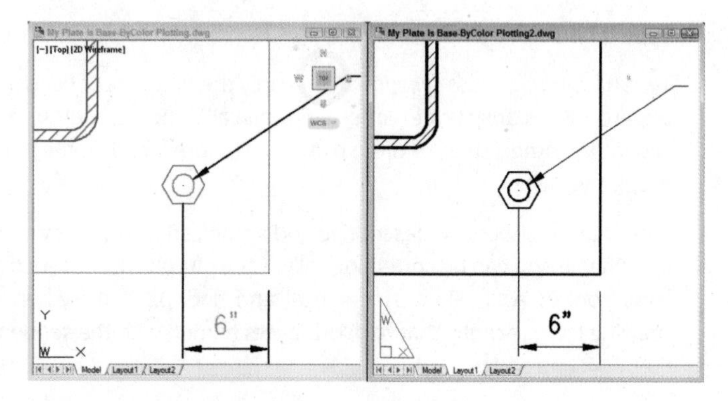

FIGURE 9-5: Change my line thickness, but color me black.

Creating new layers

If no suitable layer exists, you create one in the Layer Properties Manager palette. Follow these steps:

1. **Click the Layer Properties button in the Layers panel on the Home tab of the Ribbon, or type LA at the command line and then press Enter.**

   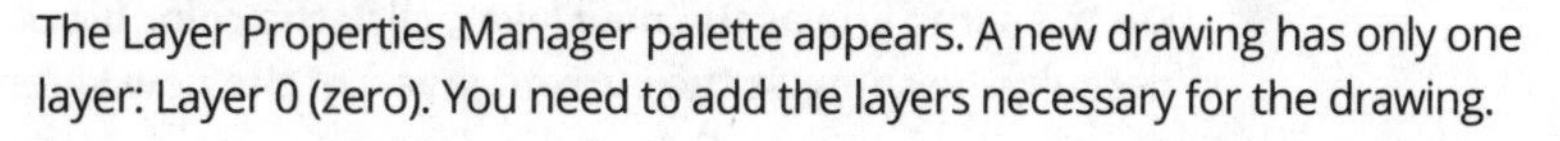

   The Layer Properties Manager palette appears. A new drawing has only one layer: Layer 0 (zero). You need to add the layers necessary for the drawing.

   TIP

   The Layer Properties Manager palette, as well as individual columns within it, can be stretched wider so as to not truncate information. The easy way is to right-click a column header and then choose Maximize All Columns.

2. **Click the New Layer button (it looks like a sheet of paper with a little sunburst on one corner) to create a new layer.**

   A new layer appears. AutoCAD names it Layer1 but highlights the name in an edit box so that you can type a new name to replace it easily, as shown in Figure 9-6.

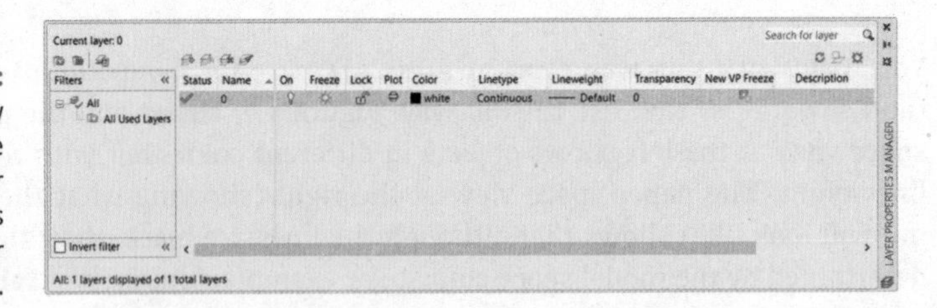

FIGURE 9-6: Adding a new layer in the Layer Properties Manager palette.

3. **Type a name for the new layer.**

TIP

Type a layer name using *initial caps* — only the first letter of each word is in uppercase. Because layer names written in all uppercase letters are much wider than others, they're often truncated (abbreviated) in the narrow Layer drop-down list.

TIP

Layer names should be descriptive and organized so that they're easily identifiable and can be sorted logically. For example, names such as Floor 01 Plan, Floor 01 Walls, Floor 01 Electrical, and Floor 02 Plan are better than a drawing I saw recently that had 132 layers named with the sequence 001, 002, 003, 004, and so on.

REMEMBER

As of 2014, AutoCAD sorts in natural order:

Old Way	New Way
Floor 1 Walls	Floor 1 Walls
Floor 10 Walls	Floor 2 Walls
Floor 2 Walls	Floor 10 Walls

4. **On the same line as the new layer, click the color block or color name (by default, the same as the current layer) of the new layer.**

The Select Color dialog box appears, as shown in Figure 9-7, with Magenta selected from the Color Index list.

The normal AutoCAD color scheme, known as the *AutoCAD Color Index (ACI),* provides 255 colors. That number of choices is overkill for ordinary drafting, but if you truly want to go overboard, the True Color tab provides more than 16 million colors, whereas the Color Books tab gives access to many standard color books used by the design and printing industries, such as Pantone books.

TIP

For now, stick with the first nine colors — the ones that appear in a single, separate row to the left of the ByLayer and ByBlock buttons on the Index Color tab in the Select Color dialog box — for these reasons:

- *These colors are easy to distinguish from one another.*
- *Using a small number of colors makes configuring plot parameters easier.* (I describe plot parameters in Chapter 16.)

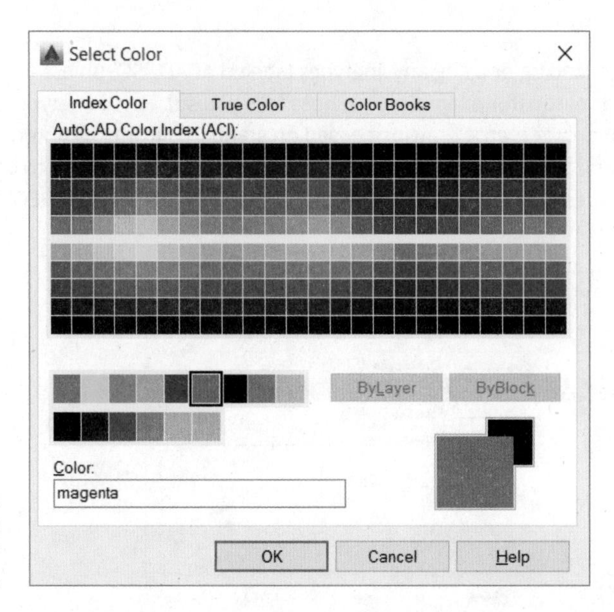

FIGURE 9-7: The Select Color dialog box.

5. **Click a color to select it as the color of this layer and then click OK.**

   The Select Color dialog box closes. In the Layer Properties Manager palette, the Color column now has the new layer color — either the name or the number of the color you selected.

   The first seven AutoCAD colors have numbers, standard names, and initials: 1 = Red, 2 = Yellow, 3 = Green, 4 = Cyan, 5 = Blue, 6 = Magenta, and 7 = White (which appears black when displayed on a white background). You can enter the full name, the initial, or the number. The remaining 248 colors go by numbers only.

6. **On the same line as the new layer, click the Linetype name of the new layer to draw a dashed (hidden) line.**

   The Select Linetype dialog box appears, as shown in Figure 9-8.

   The default AutoCAD linetype is Continuous — lines have no gaps.

   If you already loaded the linetypes you need for the drawing or the initial template file has linetypes loaded, the Select Linetype dialog box displays them in the Loaded Linetypes list. If not, then click the Load button to open the Load or Reload Linetypes dialog box. By default, AutoCAD displays linetypes from the standard AutoCAD or AutoCAD LT linetype definition file — `acad.lin` for imperial-unit drawings or `acadiso.lin` for metric-unit drawings (`acadlt.lin` and `acadltiso.lin` in AutoCAD LT). Load a linetype by selecting its name and clicking OK.

WARNING

Avoid loading or using any linetypes labeled ACAD_ISO, unless you have a truly good reason (for example, your boss tells you so). These linetypes are normally used only in metric drawings based on standards from ISO (International Organization of Standards) — and rarely even then. It's easier to use the linetypes with the more descriptive names: CENTER and DASHED, for example.

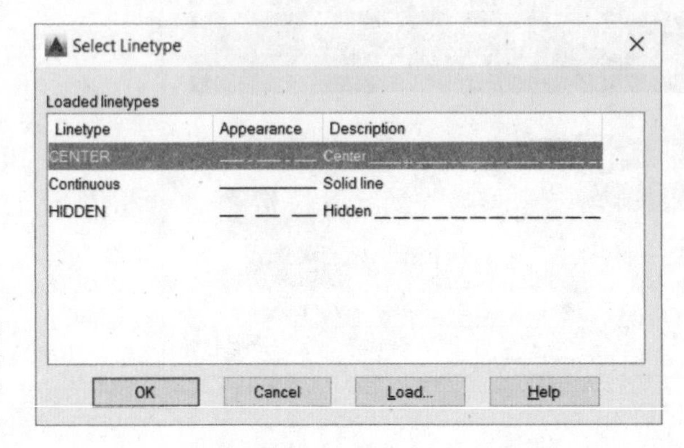

FIGURE 9-8: The Select Linetype dialog box.

7. **Click a linetype in the Loaded Linetypes list to select it as the linetype for the layer; then click OK.**

   The Select Linetype dialog box disappears, returning you to the Layer Properties Manager palette. In the Name list, the linetype for the selected layer changes to the linetype you just chose.

8. **On the same line as the new layer, click the new layer's lineweight.**

   The Lineweight dialog box appears, as shown in Figure 9-9.

9. **Select the lineweight you want from the list and click OK.**

Using the lineweight property is a two-step process. After you assign a lineweight to a layer, you must click the Show/Hide Lineweight button on the status bar to see the effect. By default, AutoCAD doesn't display this button, so you'll need to click the hamburger (Customization) button at the right end of the status bar, and then choose LineWeight.

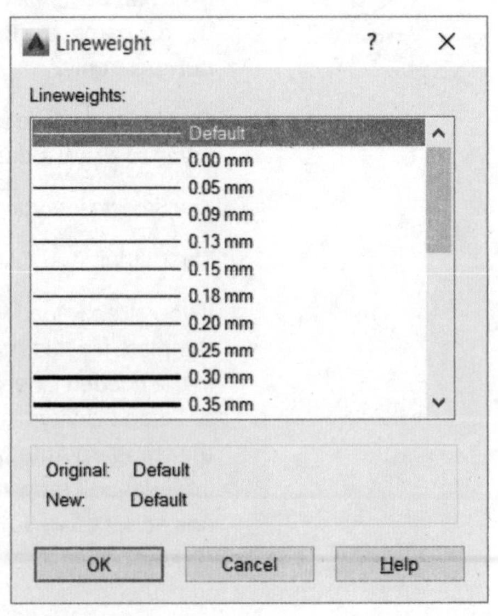

FIGURE 9-9: The Lineweight dialog box.

TIP

A lineweight of 0.00mm tells AutoCAD to use the thinnest possible lineweight on the screen and on the plot. I recommend, for now, leaving the lineweight set to Default, and instead later map screen colors to plotted lineweights, as described in greater detail in Chapter 16.

10. **(Optional) In the same line as the new layer, click the value in the Transparency column.**

    You'll appreciate AutoCAD's transparency property if you're preparing drawings for a presentation. Clicking in the Transparency column opens the Layer Transparency dialog box; type a numeric value between 0 and 90 or use the drop-down list to set a value. Transparency = 0 (no transparency) is the default, which means that objects drawn on a layer set to Transparency = 0 are opaque.

    Set the value to greater than 0, and you start seeing things through the objects you draw. The value of 90 makes objects nearly fully transparent.

11. **(Optional) Set the plot style for the new layer.**

    The Plot Style column's contents depend on whether the drawing uses named plot styles or the traditional color-based plotting. Drawings set to use color-based plotting display an unchangeable plot style name based on the layer's color property. The grayed-out style name changes only when the layer color changes. If, on the other hand, the drawing uses named plot styles, you can assign a named plot style to the layer in this column. (Chapter 16 explains why you might not want to.)

12. **(Optional) Turn plotting on and off.**

    The setting in the Plot column controls whether the layer's objects appear on plots. Click the little Printer icon to turn off this setting (the little printer gets a red bar through it) for any layer whose objects you want to see on the screen but not see on plots. Layout viewports, covered in Chapter 12, are a good use for this feature.

    Another use for this feature is antagonizing coworkers. Go into one of their drawings, start Layer Properties Manager, and turn off plotting on a few random layers. Now go to the top of the Plot column, click the divider between it and the adjacent column, and reduce the Plot column's width to nearly zero. Even a senior experienced AutoCAD user can go nuts trying to figure out why some layers won't plot.

13. **(Optional) To add a description to the layer, scroll the layer list to the right to see the Description column, click in the Description box corresponding to the new layer, and type a description.**

TIP

If you use layer descriptions, you may need to stretch the Layer Properties Manager palette to the right so that you can see the descriptions without having to scroll the layer list.

14. **Repeat Steps 1–13 to create any other layers you want.**

15. **Double-click the name of the layer that you want to make current.**

    A green check mark shows up next to the layer's name.

    Changes you make in the Layer Properties Manager palette are instantaneous. Therefore, if objects in the drawing are already on a particular layer and you change the color for that layer, the existing drawing objects change color immediately and don't wait for you to close the palette.

    The Layer drop-down list now displays the new layer as the current layer — the one on which AutoCAD places new objects you draw.

TIP

As with all palettes in AutoCAD, you can leave Layer Properties Manager open while you do other things in the drawing. Also with other palettes, Layer Properties Manager can be set to automatically minimize itself to its title bar; to be either floating or docked; or to be anchored (I get the sense that some AutoCAD programmers would rather be sailing) to either side of the screen. If your computer has two monitors, you can drag the palettes to the second monitor.

A LOAD OF LINETYPES

When you load a linetype, AutoCAD copies its *linetype definition* — a formula for how to create the dashes, dots, and gaps in that particular linetype — from the `acad.lin` (imperial units) or `acadiso.lin` (metric units) file into the drawing. The files are `acadlt.lin` and `acadltiso.lin`, respectively, in AutoCAD LT. The definitions don't automatically appear in drawings; you have to load into each drawing the linetypes you want to use. If you find that you're loading the same linetypes repeatedly into drawings, add them to template drawing(s) instead. See Chapter 4 for information about templates and how to create them. After you add linetypes to a template drawing, all new drawings you create from that template start with those linetypes loaded automatically.

Don't go overboard on loading linetypes. For example, you don't need to load *all* linetypes in the `acad.lin` file just because you *might* use them all someday. The resulting linetype list would be long and unwieldy. Most drawings require only a few linetypes, and most industries and companies settle on a half dozen or so linetypes for common use. Your industry, office, or project manager may have guidelines about which linetype to use for which purpose.

If you're the techno-dweeb type and don't mind editing a text file that contains linetype definitions, you can define your own linetypes or weed out the ones you'll never use. Press F1 to open the online Help system; search for *custom linetypes*.

Layer 0 (zero) must exist in every drawing. You can assign any of the properties to it, such as color and line type, but you cannot rename or remove it.

TIP

If you set up layers correctly in a template file, you'll almost never need Layer Properties Manager while you're working.

Here are a few things you can do to work effectively with layers:

- **Set a layer to be the current layer.** Make sure that no objects are selected, and then choose the layer name from the Layer drop-down list on the Layers panel on the Ribbon or the Layers toolbar.
- **Toggle properties on and off.** To see through objects, you have to turn on the Transparency button on the status bar. You can also toggle the Lineweight and Transparency buttons on the status bar to see the effect of assigning these properties.
- **Adjust the lineweight properties and scale.** When you finish defining layers and return to the command line, the Lineweight dialog box lets you adjust further lineweight properties, including inches-versus-millimeters and the display scale.

  The display scale affects only how elements *appear onscreen,* not how they're plotted.
- **Let the command line find layers for you.** The command line automatically completes layer names, among other named objects, and it searches for matching partial text in layer names. Follow these steps:
  1. *Open a drawing that contains layers.* If you created the base plate in Chapter 3, open that.
  2. *Make sure that Dynamic Input (F12) is turned off.* Unfortunately, this functionality works only when DYNamic Input is off.
  3. *Start typing letters in the layer name you're looking for, but don't press Enter.* If you're using the base plate, type **BO**. Figure 9-10 shows how the command line displays the three command names that start with *BO* — plus, it shows `Layer: Anchor Bolts` because it found the Bo in Bolts.
  4. *Click the layer name.* It becomes the current layer.

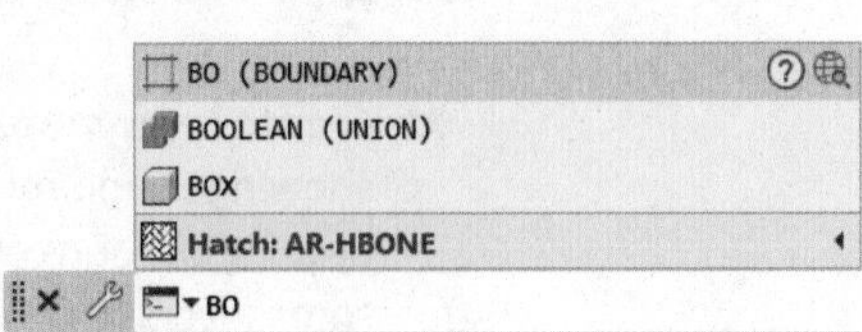

FIGURE 9-10:
Type, and the command line shall find.

Manipulating layers

After you create layers and draw objects on them, you can turn one or more layers off or on to hide or show the objects on layers. In the Layer Properties Manager palette, the first three icons to the right of the layer name control AutoCAD's layer visibility modes:

- **Off/On:** Click the lightbulb icon to toggle visibility of all objects on the selected layer. AutoCAD doesn't regenerate the drawing when you turn layers back on. On the other hand, frozen layers don't regenerate while you're working on the drawing. I give you the lowdown on regeneration in Chapter 5. Current computers are usually fast enough that regens aren't a problem, so Off/On is seldom used anymore.
- **Freeze/Thaw:** Click the sun icon to toggle off visibility of all objects on the selected layer. Click the snowflake icon to toggle visibility on. AutoCAD regenerates the drawing when you thaw layers.
- **Lock/Unlock:** Click the padlock icon to lock and unlock layers. When a layer is locked, you can see (but not edit) objects on that layer.

You can rearrange the column order in the Layer Properties Manager simply by dragging and dropping the column label to a new place. And you can right-click any column label to display a menu from which you can turn columns off and on.

Off/On and Freeze/Thaw do almost the same thing — both settings let you make objects visible or invisible by layer. The difference is that frozen layers effectively don't exist to AutoCAD, so it ignores them when regenerating, printing, and exporting the drawing.

Normally, it's faster and easier to turn layers off and on, freeze and thaw them, and lock and unlock them by clicking the appropriate icons in the Layer drop-down list on the Ribbon instead of using the Layer Properties Manager palette.

Here are a few other useful tricks you can try when working with layers:

- **Create layer states.** Say you have a floor plan of a house that includes a layer showing the plumbing and another layer showing the wiring. You'd probably never show both elements on the same drawing, so you'd need to manage some layers to show the drawing to plumbers or electricians. Rather than turn off a dozen layers and then turn on another dozen layers when you want a different view of the drawing, you can save groups of layers and their settings as a named *layer state.* Click the Layer States Manager button in the Layer Properties Manager to open the Layer States Manager dialog box, or enter **LAYERSTATES** at the command line, or choose Manage Layer States from the Layer State drop-down list in the Layers panel.

- **Fade objects on locked layers.** AutoCAD fades objects on locked layers, giving you a truly effective visual reference without confusing you about which layers might be locked. You can control the amount of fading by setting a nonzero value for the system variable LAYLOCKFADECTL. See Chapter 26 for an explanation of system variables, and check out the online Help system for specific info on this one. You can turn off fading but retain the current setting for future use by adding a minus sign (–) in front of the fade value, such as –50, or you can turn off the fading altogether by setting this value to 0.

  The LAYISO (LAYer ISOlate) command incorporates the same layer-fading feature, plus it locks the layers. Start the command and then set it up the way you want by entering **S** (for Settings) at the command line and pressing Enter; and then type the option letter for the specific settings you want. Look up *LAYISO* in the online Help index for more information.

- **Create layer filters.** If you find that you're using lots of layers, you can create layer filters to make viewing and managing the layer list easier:
  - A *group filter* is a subset of layers you choose (by dragging layer names into the group filter name or by selecting objects in the drawing).
  - A *property filter* is a subset of layers that AutoCAD creates and updates automatically according to layer property criteria that you define (for example, all layers whose names contain *Wall* or whose color is green).

  To find out more, press F1 in the Layers Properties Manager palette, and click the New Property Filter hyperlink.

- **Isolate layers.** On the Layers panel of the Home tab of the Ribbon (shown in Figure 9-11), click the Layer Isolate and Layer Off button to specify the layer to *isolate* (that is, fade all layers except the chosen one) or turn off altogether. You may have to open the panel slideout to see these buttons.

  For more information on layers, open the online Help system and search on *controlling layers manager*.

- **Hide or isolate objects.** Rather than turn off a layer when only a few items are in the way, you can hide or isolate individual objects by using the ISOLATEobjects and HIDEOBJECTS commands while maintaining normal visibility for other objects on the layer. I discuss these commands in Chapter 10.

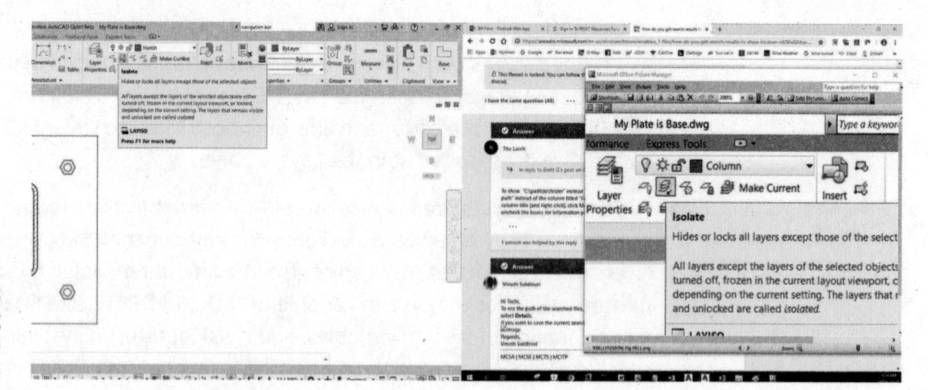

FIGURE 9-11: Tooling through the layer tools.

Scaling an object's linetype

Linetype scale controls the sizing of dash-dot linetypes in a drawing. Some linetypes that you assign to a layer or to an object directly might not look correct on every object based on the drawing's current linetype scale. The problem with the linetypes is that the dashes and gaps are not always noticeable on every object because of the object's size or the way the object was drawn.

To fix this problem, you can change the object's linetype scale, which normally is set to the value of 1. The object's linetype scale is used as a multiplier with the drawing's linetype scale factor. For example, if you change an object's linetype scale factor to 1.5 and the drawing's linetype scale factor is 12, the linetype applied to the object is actually displayed with a linetype scale of 16 (1.5 x 12).

Follow these steps to scale the object's lineweights:

1. **Start a new drawing by using the standard** `acad.dwt` **or** `acadlt.dwt` **template.**
2. **Create a layer that uses the Hidden linetype and set this layer to be current.**

   If you don't feel like creating a new layer, you can always simply set the linetype for Layer 0 (zero) to hidden.
3. **Set the limits to be from 0,0 to 100,100.**
4. **Enter Zoom and press Enter. Then enter A and press Enter to zoom to the limits you set.**

5. **Start the Line command, and type coordinates to draw a line from 10,10 to 70,70.**

   Hey, what happened? That doesn't look like a hidden line! Well, it is, even if it doesn't look like one. The problem is that the line is so long and you've zoomed out so far that the individual spaces and dashes are too small and have all merged into one. The bigger problem is that if you were to plot this drawing to fit an A-size (8½ x 11) sheet of paper, the line would look just as it does onscreen.

1:1 / 100%

6. **Click the Annotation Scale button in the status bar and select 1:20 from the drop list, and then REgenerate the drawing.**

   Ah, that's better! As I mention in Chapter 4, you need to apply a plotting scale of 1:20 to fit the line on the paper, but then everything scales down, including the spaces and dashes in noncontinuous lines. The Annotation Scale setting with a value of 1:20 tells AutoCAD to make the dashes and spaces 20 times longer so that they scale down properly on the screen and when you plot. I cover this topic in more detail in Chapters 13–15.

7. **Start the Line command, and type coordinates to draw a line from 30,10 to 100,80.**

   The new line should look just like the first line you drew.

Using Named Objects

One thing that can make AutoCAD interesting is the somewhat cavalier naming conventions used in the program's documentation. For years, elements such as lines, arcs, and other graphical items were called *entities*, but then they started being called *objects*. Fair enough, but *object* has also long been used to define certain nongraphical components of a drawing — items you'd hardly even consider to be objects — and *those* are the kind of named objects I describe in this section.

TECHNICAL STUFF

Hidden in the innards of every AutoCAD drawing file is a set of *named objects*, which are organized into symbol table. The properties common to all AutoCAD objects are defined in these tables. For example, all line objects in the drawing are stored on one or more layers, so the layer property is common to all lines and is defined in the layer table. But the coordinates that define the start and end points of a given line are unique to that line (or they should be!), so the coordinate properties aren't common to all lines.

A layer is one example of a named object. The layer table in a given drawing contains a list of the layers in the current drawing, along with the settings for each layer including the color, linetype, and on/off setting.

Named objects don't appear as graphical objects in the drawing. They're like the hardworking behind-the-scenes pit crews who keep race cars running smoothly. These named objects are the ones you're likely to use most often (including cross-references):

- **Layer:** They're covered in the section "Working with Layers," earlier in this chapter.
- **Linetype:** They're covered in the section "Working with Layers," earlier in this chapter.
- **Text style:** See Chapter 13.
- **Table style:** See Chapter 13.
- **Multileader style:** See Chapter 13.
- **Multiline style:** This object isn't covered in this book; see the online Help system.
- **Dimension style:** See Chapter 14.
- **Block definition and xref:** See Chapters 17 and 18.
- **Layout:** See Chapter 12.

When you use commands such as LAyer, LineType, and DimSTyle, you're creating and editing named objects. After you create named objects in a drawing, the AutoCAD DesignCenter or Content Explorer gives you the tools to copy them between drawings.

A major real-estate developer might believe otherwise, but you *can* have too many properties (at least in AutoCAD). You may have created layers or loaded linetypes, text, or dimension styles that you don't use. When you suspect that you have some of these superfluously named objects in the drawing, the PUrge command can help you get rid of them. Click the application button to display the Application menu. Choose Drawing Utilities ⇨ Purge to open the Purge dialog box. You can click the plus sign (+) beside a category to purge individual items, or you can click Purge All and get rid of tons of stuff all at one time. Visit the online Help system for more about purging.

Using AutoCAD DesignCenter

DesignCenter is a useful, if somewhat busy, palette. It is launched from the Palettes panel of the View tab. The DesignCenter palette is handy for copying the specifications of named objects: layers, linetypes, block definitions, text styles, and other organizational objects in drawings from one drawing to another.

The DesignCenter palette, shown in Figure 9-12, consists of a toolbar at the top, a set of three tabs below that, a tree view pane on the left, and a content pane on the right. The tree view pane displays a Windows Explorer–like navigation panel, showing drawing files and the symbol tables contained in each drawing. The content pane usually displays the contents of the selected drawing or symbol table.

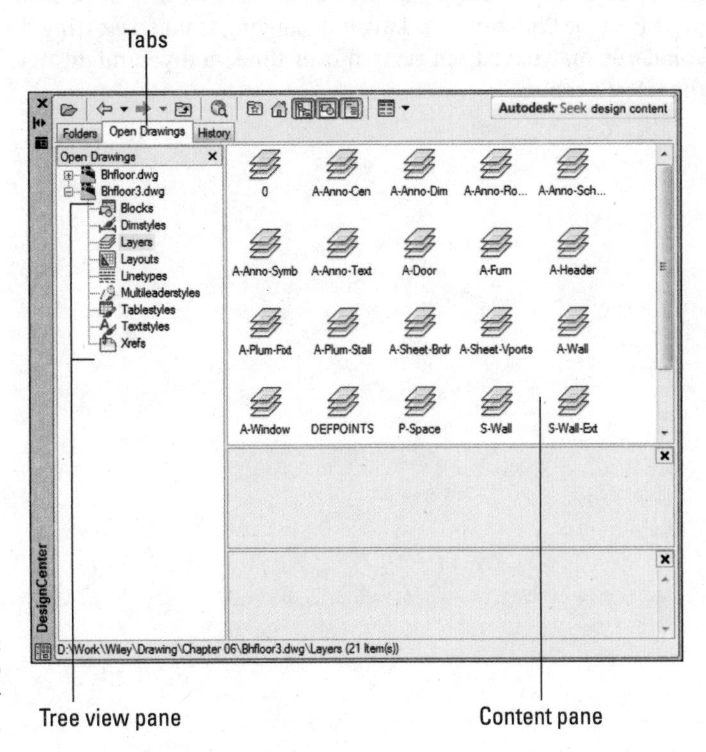

FIGURE 9-12: The AutoCAD DesignCenter palette.

The three tabs just below the DesignCenter toolbar control the elements you see in the tree view and content panes:

- **Folders:** Shows the folders on the local and network drives, just as the Windows Explorer Folders pane does. Use this tab if the drawing you want to copy from isn't open now in AutoCAD.

» **Open Drawings:** Shows the drawings that are open in AutoCAD. Use this tab (it's current in Figure 9-12) to copy named objects between open drawings.

» **History:** Shows drawings that you've recently browsed in DesignCenter. Use this tab to jump quickly to drawings that you've used recently on the Folders tab.

The toolbar buttons further refine what you see in the tree view and content panes. A few of these buttons toggle different parts of the panes.

A detailed description is beyond the scope of this book, but the bottom line is that you can drag and drop the specifications of named objects from one drawing into another. This can be handy at times, but it must be used with care. In particular, if your company or client has defined standards, they should be set up in a template file (see Chapter 4). A danger of copying from an existing drawing is that the standards may have been revised over time, and you might not be copying from the latest version.

» Selecting objects with maximum flexibility

» Editing with grips

Chapter 10
Grabbing Onto Object Selection

Editing objects is the flip side of creating them, and in AutoCAD you usually spend more time editing objects than drawing them from scratch. That's partly because the design and drafting process is, by its nature, subject to changes, and also because AutoCAD lets you edit objects cleanly and easily.

Creating an object and then editing it can sometimes be faster than doing it correctly the first time. For instance, placing a circle and then trimming it is often faster than fighting your way through the Arc command's many options.

Specifying precise locations and distances is as vital to editing objects in AutoCAD as it is when creating them. Become familiar with the precision techniques described in Chapter 8 before you apply the editing techniques described in this chapter to drawings.

Commanding and Selecting

AutoCAD offers three styles of editing, which I list here in order from the most options to the least:

- **Command-first editing:** Enter an editing command, and then select the objects to edit.
- **Selection-first editing:** Select objects, and then enter an editing command.
- **Direct-object editing:** Select an object, and then edit it with grips.

REMEMBER

AutoCAD refers to command-first editing as *verb-noun editing* and to selection-first editing as *noun-verb editing.* When you see this terminology in the Options dialog box or the online Help system, don't worry — you haven't dropped back into fifth-grade English class!

Command-first editing

In *command-first editing,* you start a command and then select the objects with which the command edits. This method is also the *only* way to use certain editing commands (such as Fillet and BReak). It's no surprise that command-first editing is the traditional method in AutoCAD, and the one you should become the most comfortable using.

Selection-first editing

In *selection-first editing,* you perform the steps in reverse order, which is the same order as in most Windows applications: Select an object and then choose a command. Some people claim that selection-first editing tends to be easier to master and makes AutoCAD more approachable when you're a new or an occasional user.

Direct-object editing

In *direct-object editing,* you perform common editing operations by using the mouse to grab an object and then perform an action on it, such as dragging it to a different place in the drawing. No named command is involved; the act of moving the mouse and holding down different mouse buttons performs the edit.

In addition, AutoCAD supports direct-object editing via grip-editing. *Grips* are the small colored squares that appear on objects when you select them. You use grips to edit the objects by stretching, moving, copying, rotating, and scaling them.

Sometimes, grips have other shapes, such as rectangular and triangular, to indicate other kinds of editing. Grips are also known as *handles.* I discuss grip-editing in Chapter 11.

Choosing an editing style

This book emphasizes command-first editing because, in its heart of hearts, AutoCAD is a command-first program. In fact, it started out offering *only* command-first editing and later added the selection-first methods. I stress command-first editing for these reasons:

- **Longevity:** It's the oldest editing style in AutoCAD, and the one with which experienced AutoCAD users are most familiar.
- **Naturalness:** Would you say, "I want to eat my lunch"? or "My lunch I want to eat"? Okay, so Yoda might say, "Eat my lunch I want to!"
- **Consistency:** It works consistently with *all* editing commands — some editing commands are command-first only.
- **Flexibility:** It allows greater flexibility in object selection than the other methods, which is useful when you work on busy complex drawings.

After you know how to do command-first editing, you can simply reverse the order of many editing operations to perform them in selection-first style instead. But if you don't become familiar with command-first editing in the beginning, you'll be bewildered by the useful AutoCAD commands that work only in the command-first style; commands such as these ignore the objects you've already selected and prompt you to select objects before you can continue.

Much of the information in the rest of this book assumes that you're using the AutoCAD's default selection settings. If you find that object selection or grip-editing is working differently from the way it's described in this chapter, click the Application button, choose Options from the bottom of the Application menu to open the Options dialog box, and then check the settings on the Selection tab. The following check box settings should be selected (as shown in Figure 10-1), and all other check box settings should be deselected:

- Noun/Verb Selection
- Object Grouping
- Implied Windowing
- Allow press and drag for Lasso

- Show Grips
- Show Grip Tips
- Show Dynamic Grip Menu
- Allow Ctrl+Cycling Behavior
- Show Single Grip on Groups
- Show Bounding Box on Groups
- Selection Preview when a Command Is Active
- Selection Preview when No Command Is Active
- Command Preview
- Property Preview

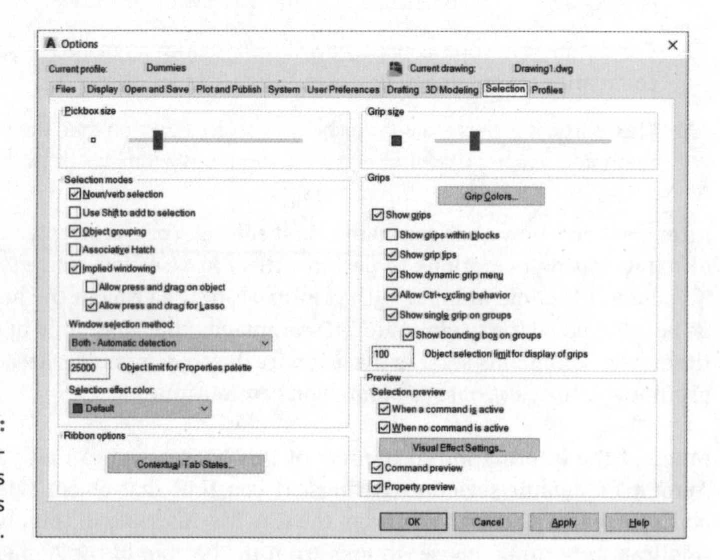

FIGURE 10-1: Setting selection options in the Options dialog box.

For information on what these options do, hover the mouse pointer over an option to display a tooltip with information or else visit the online Help system.

Selecting Objects

Part of AutoCAD's editing flexibility comes from its object selection flexibility. For example, command-first editing offers 16 selection modes, whereas selection-first editing has only six. I describe the most useful ones in this chapter. Don't worry, though: You can squeak by most of the time using only the three that I describe in this section:

- Select a single object by picking it.
- Select multiple objects by enclosing them in a *window* selection box, polygon, or lasso.
- Select multiple objects by enclosing them in a *crossing* selection box, polygon, or lasso.

Most editing commands affect the entire group of selected objects.

One-by-one selection

The most obvious way to select objects is to pick (by clicking) them one at a time. When you select objects (even just one), they are ready for editing and are said to be in a *selection set.* You can build a selection set cumulatively by using this pick-one-object-at-a-time selection mode.

This cumulative convention may be different from the one you're used to. In most Windows programs, if you select one object and then another, the first object is *deselected* and the second one is selected; only the object you selected last remains selected. In AutoCAD, *all* objects you select, one at a time, remain selected and are added to the selection set, no matter how many objects you pick. Hold down the Shift key to remove objects from the selection set.

Selection boxes left and right

Selecting objects one at a time works well when you want to edit a small number of objects, but many CAD editing tasks involve editing lots of objects at the same time. Do you really want to pick 132 lines, arcs, and circles, one at a time? Okay, maybe you do if you're paid by the hour, but after a while your company won't be able to afford you.

Like most Windows graphics programs, AutoCAD provides a *selection window* for grabbing a bunch of objects in a rectangular area. As you may guess, the AutoCAD version of this feature is a bit more powerful than the one in other Windows graphics programs. AutoCAD calls its version *implied windowing*. Here's how you use it:

- **Window object selection:** When you click a blank area of the drawing (not on an object), you're telling AutoCAD that you want to specify a selection by placing a window around the objects. When you move the cursor to the right before picking the next corner of the selection area, you're further implying that you want to select all objects that reside completely within the selection area.

  When you drag the cursor to the right (without lifting the right mouse button), you draw a selection lasso, which also selects all objects completely within the lasso.

- **Crossing object selection:** When you click a blank area of the drawing (not on an object), you're indicating that you want to specify a selection by placing a window around and through the objects. If you move the cursor to the left before picking the next corner of the selection area, you're indicating that you want to select all objects that reside completely within or touch or cross the selection area boundary.

  Dragging the cursor to the left draws a crossing selection lasso.

Fortunately, AutoCAD gives you visual cues about the difference in motion. As you move to the right, the window area appears as an area with a blue fill and a solid border. As you move to the left, the crossing area appears as an area with green fill and a dashed border.

TIP

If your editing method of preference is selection-first, AutoCAD gives you fewer selection options. But here is a selection-first trick most AutoCAD users don't know about: As you begin to drag the mouse, notice that the command line prompts you to press the spacebar. While keeping the right mouse button pressed, tap the spacebar to switch from Crossing Lasso (the default), to Window Lasso, Fence Lasso, and then back to Crossing Lasso selection mode.

Figures 10-2 and 10-3 show a window box and a crossing box, respectively, in action. The fill colors are the same, whether in a selection box, polygon, or lasso.

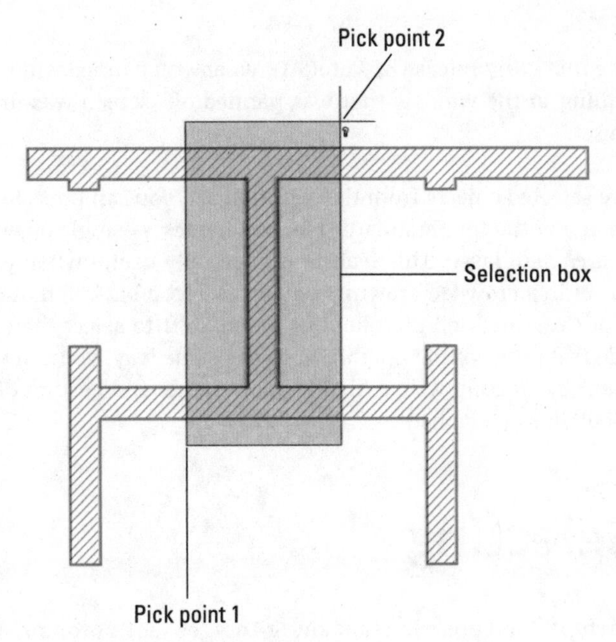

FIGURE 10-2: A window selection box, drawn from left to right, selects the only two objects (the two vertical lines) that are completely within the box.

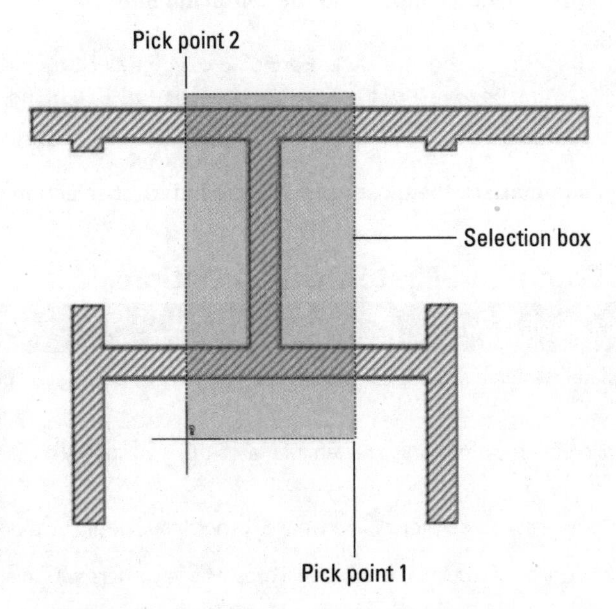

FIGURE 10-3: A crossing selection box, drawn from right to left, selects 11 objects that are completely or partially within the box.

Tying up object selection

You can mix and match selection modes, first selecting individual objects, then specifying a window box, lassoing, and so on. Each selection adds to the current selection set, allowing you to build an enormously complicated selection of objects, and then operate on them with one or more editing commands.

WARNING

Before the mid-2017 release of AutoCAD, when you panned while window selecting, anything in the window that was panned off-screen was dropped from the selection set.

TIP

To *remove* selected objects from the selection set, you can press Shift in combination with any of the four standard selection modes — single object, window area, crossing area, and lasso. This feature is especially useful when you're building a selection set in a crowded drawing; you can select a big batch of objects by using Window or Crossing, and then holding down Shift to select the objects you want to exclude from the editing operation. In the same way, you can carve a statue of an elephant by starting with a big block of marble and cutting off anything that doesn't look like an elephant.

Perfecting Selecting

When you type ? and press Enter at any `Select objects` prompt, AutoCAD lists all command-first selection options on the command line:

```
Window Last Crossing BOX ALL Fence WPolygon CPolygon Group
        Add Remove Multiple Previous Undo AUto SIngle
        SUbobject Object
```

Table 10-1 summarizes the most useful command-first selection options.

TABLE 10-1 Useful Command-First Selection Options

Option	Objects That It Selects
Window	All objects completely within a rectangular area that you specify by picking two points or dragging the cursor
Crossing	All objects within, crossing, or touching a rectangular area that you specify by picking two points or dragging the cursor
WPolygon	All objects completely within a polygonal area whose corners you specify by picking points
CPolygon	All objects within, crossing, or touching a polygonal area whose corners you specify by picking points
Fence	All objects touching a temporary polyline whose vertices you specify by picking points
Last	The last object you drew (whether or not it's visible in the display)
Previous	The selection set that you specified previously
ALL	All objects on layers that aren't frozen or locked and that are in the current space (model space or paper space), including objects that aren't displayed because you've zoomed in

To use any command-first selection options at the `Select objects` prompt, type the uppercase letters (see Table 10-1) that correspond to the option you want, and then press Enter. For example, type **CP** for crossing polygon mode. When you've finished selecting objects, press Enter again to tell AutoCAD that you want to start the editing operation.

REMEMBER

After you finish selecting objects, you must press Enter to tell AutoCAD that you want to start the editing operation. Say . . . is there an echo in here? As a matter of fact, I am repeating myself. Many new AutoCAD users find it difficult to remember the necessity of pressing Enter after they finish selecting objects.

TIP

The *selection preview* features in AutoCAD remove much of the doubt over which objects you're selecting. *Rollover highlighting* displays individual objects with a thick lineweight as you move the cursor over them. *Area selections* display transparent, colored areas when you use any window or crossing option.

The following steps show you how to use the Erase command in command-first mode with several different selection options. The selection techniques used in this example apply to most AutoCAD editing commands:

1. **Open a drawing that contains objects, or start a new drawing and create lines, arcs, or circles.**

   Don't be too particular in drawing them because you blow them away in this step list.

2. **Press Esc to make sure that no command is active and no objects are selected.**

   WARNING

   If any objects are selected when you start an editing command, the command, in most cases, operates on those objects (selection-first editing) instead of prompting you to select objects (command-first editing). For the reasons I describe in the section "Commanding and Selecting" earlier in this chapter, I recommend using the command-first editing style until you're thoroughly familiar with it. Later, you can experiment with selection-first editing, if you like. Just reverse the sequence of commanding and selecting.

3. 

   Click the Erase button in the Modify panel on the Home tab or enter the Erase command.

   AutoCAD displays the `Select objects` prompt at the command line, and if dynamic input is enabled on the status bar, at the Dynamic Input tooltip.

4. **Select two or three individual objects by clicking each one.**

   AutoCAD adds each object to the selection set. The color of the objects you select changes to light gray, and AutoCAD continues to display the `Select objects` prompt.

5. **Specify a window selection box that completely encloses several objects.**

   Move the cursor to a point below and to the left of the objects, click, release the mouse button, move the cursor above and to the right of the objects, and click again.

   All objects that are completely within the box are selected.

6. **Specify a crossing selection lasso that completely encloses a few objects and cuts through several others.**

   Move the cursor to a point below and to the right of some objects, click the mouse button, drag the cursor above and to the left of some of the objects, and release again.

   All objects that are completely within, crossing through, or touching the lasso are selected. AutoCAD continues to display the `Select objects` prompt.

7. **Type WP and press Enter to activate the WPolygon (Window Polygon) selection option.**

   AutoCAD prompts you to pick points that define the selection polygon.

8. **Pick a series of points and press Enter.**

   Figure 10-4 shows an example. After you press Enter, AutoCAD selects all objects that are completely within the polygon. AutoCAD continues to display the `Select objects` prompt until you press Enter.

9. **Press Enter to end object selection.**

   AutoCAD erases all selected objects and returns to an empty command prompt.

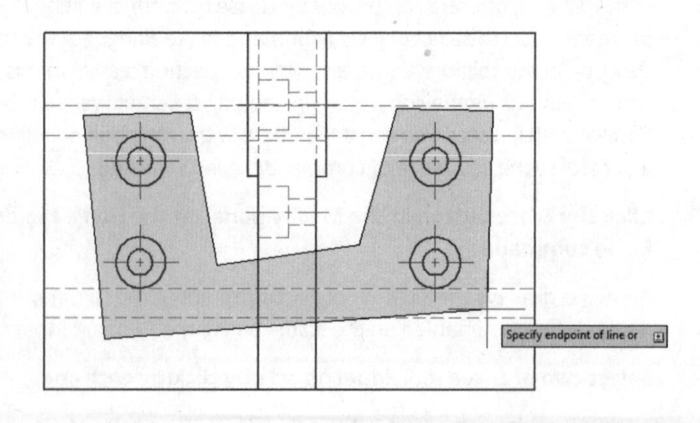

FIGURE 10-4: Lassoing objects by using WPolygon selects the concentric circles and their center lines.

Did you notice how you can use a combination of object selection methods to build a selection set and then press Enter to execute the command on the selected objects? Most AutoCAD editing commands work this way in command-first mode.

TIP

If after erasing a selection set, you immediately realize that you didn't mean to do away with so many objects, enter the **U** command or click the Undo button on the Quick Access toolbar to restore them all. But AutoCAD has one additional unerase trick up its sleeve — the aptly named OOPS command. When you type **OOPS** and press Enter, AutoCAD restores the last selection set that you erased — even when you've run other commands after Erase. This approach works only with objects you erased earlier.

TIP

The Erase command isn't the only way to remove unwanted objects from the drawing. The easiest method in any workspace is to select one or more objects, and then press the Delete key on the keyboard.

Drawing objects on top of other objects is all too easy to do, and after you've done so, AutoCAD recognizes when multiples exist by displaying a tiny icon of overlapping rectangles. If the Selection Cycling button is enabled on the status bar and you pick with overlapping objects, AutoCAD opens a Selection dialog box from which you choose the one object you want.

AutoCAD Groupies

AutoCAD lets you select a bunch of objects and gather them into a group so that when you click one object, everything in the group is selected. You simply select one or more objects and click Group on the Groups panel of the Home tab. If you want, you can name the group as you create it. The buttons on the main Groups panel let you create new groups, toggle group selection off and on, edit groups by adding or removing individual objects, or permanently ungroup a selected group.

The many object-selection modes that I describe in earlier sections — and some that I don't even describe, such as the FIlter command (check out the online Help system for more on that topic) — are useful as far as they go.

Object Selection: Now You See It . . .

AutoCAD lets you control the visibility of individual objects, which is a big deal. Before AutoCAD 2011, the only way to change the display of objects was to turn off or freeze the layer on which they resided. If the layer held other objects that you *did* want to see — too bad. These three commands turn this limitation into ancient history:

- **HIDEOBJECTS:** Prompts you to select the objects that you want to make temporarily disappear from view
- **ISOLATEobjects:** Prompts you to select the objects that you want to see, while temporarily making all others disappear
- **UNISOLATEobjects:** Ends both the hiding and isolating of all objects

The primary method of invoking these commands is via the right-click menu. Using either command-first or selection-first editing as described in the section "Commanding and Selecting," earlier in this chapter, simply right-click and you see the Isolate item, shown in Figure 10-5. A status bar icon — a lightbulb at the lower-right corner of the display — is dimmed when objects are either hidden or isolated. Click this icon to open a menu that lets you turn off the hiding or isolating.

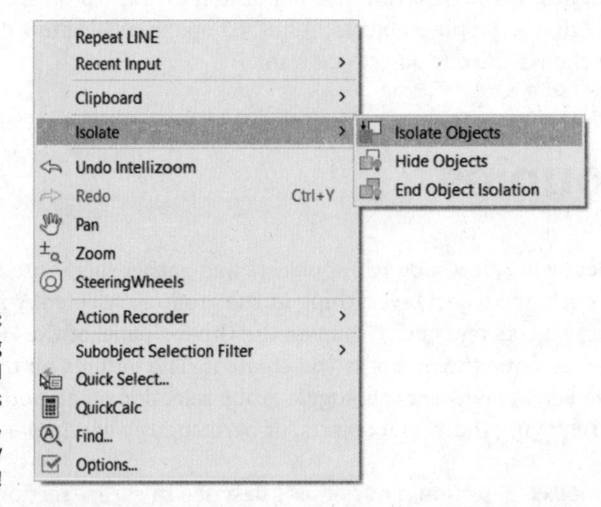

FIGURE 10-5: Making selected objects disappear, but only temporarily!

If you're worried about the possible implications of this concept ("Hmm. I was sure I added those center lines. Do I need to add them again?"), relax. Hiding and isolating objects is temporary — it lasts only as long as the current drawing session. When you close a drawing with objects isolated or hidden, they reappear when you reopen the file.

Rumor has it that the Autodesk programmers stole this cloaking-and-uncloaking concept from the Romulans.

» Manipulating whole objects

» Changing pieces of objects

» Fixing your mistakes — oops, I mean changing your mind

» Grip editing

Chapter 11
Edit for Credit

In Chapter 10, you can see that AutoCAD offers several methods of modifying objects in drawings. You can see how to select objects so that you can edit them. Now it's time to roll up your sleeves and get dirty. In this chapter, I introduce you to the primary editing commands in AutoCAD.

The following sections cover the most important editing commands in AutoCAD, using command-first editing mode. AutoCAD calls commands that edit *modify commands.*

REMEMBER

As I explain in Chapter 10, command-first editing (or verb-noun editing, in AutoCAD-ese) is one of three approaches to modifying objects in AutoCAD. I concentrate on this method, where you start a command and then pick the objects on which the command will act, because it's the only method that works for every editing command in AutoCAD.

Assembling Your AutoCAD Toolkit

Table 11-1 lists AutoCAD's most frequently used editing commands. It shows the tool icons found on the Ribbon's Modify tab and it lists the official command names with their corresponding aliases, where they exist, in case you prefer typing over clicking.

TABLE 11-1 **AutoCAD's Modify (Editing) Commands**

Button	Command
	ARRAYEDIT
	ARRAYPATH
	ARRAYPOLAR
	ARRAYRECT
	BLEND
	BREAK (BR); one point
	BREAK (BR); two points
	CHAMFER (CHA)
	COPY (CO or CP)
	ERASE (E)
	EXPLODE (X)
	EXTEND (EX)
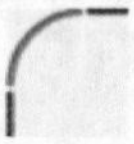	FILLET (F)
	JOIN (J)

Button	Command
	LENGTHEN (LEN)
	MIRROR (MI)
	MOVE (M)
	OFFSET (O)
	OVERKILL
	PEDIT (PE)
	REVERSE
	ROTATE (RO)
	SCALE (SC)
	STRETCH (S)
	TRIM (TR)

The -ARray command underwent a massive revision in AutoCAD 2012; however, I think that there's still a need for simple arrays (that is, copies of objects in regular patterns), so I explain in this chapter how to create those. In addition, you will probably run into a great many existing drawings that were produced using the older process. The four Array commands shown in Table 11-1 run the new associative array functions that I cover in Chapter 18; I include them here because they're grouped with the other Modify commands on the Ribbon.

No matter how you start an editing command, AutoCAD in almost all cases prompts you to select objects, points, distances, and/or options at the command line. Read each prompt during every step of the command, especially when you're figuring out how to use a new one. When all else fails, read the command line!

As I describe in Chapter 8, maintaining precision when you draw and edit is crucial to good CAD work. Nothing ruins a drawing faster than eyeball editing, in which you shove objects around without worrying about precise distances and geometric points.

WARNING

Users of Windows Paint and other bitmap drawing programs may be familiar with the concept of *nudging* — selecting objects and using the arrow keys to move them a certain number of pixels horizontally or vertically. For better or worse, AutoCAD has joined the nudging party. You can move a selected object a pixel's-worth this way or that way by holding down the Ctrl key and pressing an arrow key. When working with AutoCAD, you *always* should use tried-and-true precision drafting techniques because moving objects by the pixel (instead of by real-world units) is the opposite of precise. You can move pieces of text by nudging them, but *never* move actual drawing geometry that way.

The Big Three: Move, COpy, and Stretch

Moving, copying, and stretching are, for many drafters, the three most common editing operations. AutoCAD obliges this need by supplying the Move, COpy, and Stretch commands.

Base points and displacements

The Move, COpy, and Stretch commands each require that you specify how far and in which direction you want objects to be moved, copied, or stretched. After you start the command and have selected objects to be edited, AutoCAD prompts you for these two pieces of information:

```
Specify base point or [Displacement] <Displacement>:
Specify second point or <use first point as displacement>:
```

For the first prompt, you pick a point (AutoCAD calls this the *base point*)_or you enter **D** to specify a displacement, as described next.

In a subtle way, the second prompt says that you have two methods to specify how far and in which direction you want the objects to be copied, moved, or stretched:

- **Pick or type the coordinates of a second point.** AutoCAD calls second coordinate point the, uh, the *second point.* Imagine an arrow pointing from the base point to the second point — the arrow defines how far and in which direction the objects are copied, moved, or stretched.
- **Supply an X,Y pair of numbers that represents a distance rather than a point.** This distance is the absolute *displacement* (distance and direction) that you want to copy, move, or stretch the objects.

How does AutoCAD know whether your response to the first prompt is a base point or a displacement? It depends on how you respond to the second prompt. First, pick a point onscreen or enter coordinates at the `Base` point prompt. Next, there are two possibilities:

- **When you pick or type the coordinates of a point at the second point prompt:** AutoCAD says, "A-ha! — you want a displacement *vector,*" and moves the objects according to the imaginary arrow pointing from the base point to the second point.
- **When you press Enter at the second prompt (without having supplied any information):** AutoCAD says, "A-ha! — you want displacement *distance,*" and uses the X,Y pair of numbers that you supplied at the first prompt as an absolute displacement distance relative to the drawing's origin.

What makes this displacement business interesting is that AutoCAD lets you pick a point at the first prompt *and then* press Enter at the second prompt. AutoCAD still says, "A-ha! — displacement distance," but now it treats the coordinates *of the point you picked* as an absolute distance relative to the drawing's origin at 0,0. If the point you picked has relatively large coordinates, you had better hope that the selected objects are wearing their thermal underwear because they'll end up somewhere in the northeast of Greenland. You probably won't see where they've gone — because you'll typically be zoomed into a normal part of the drawing area, the objects will seem to vanish! Use the **Z A** command to locate them, or press **U** to undo the action.

WARNING

When you start the Move, COpy, or Stretch command, press Enter in response to the second prompt *only* when you want AutoCAD to use your response to the first prompt as an absolute displacement. If you make a mistake, click Undo or press Ctrl+Z to back up and try again. You can use the Zoom All command (described in Chapter 5) to look for objects that have flown off the screen.

Move

The steps in this section demonstrate command-first editing with the Move command, using the base point/second point method of indicating how far and in what direction to move the selected objects. Follow these steps to use precision techniques when you use the Move command:

1. **Press Esc to make sure that no command is active and no objects are selected.**

2. **Click the Move button on the Modify panel of the Home tab.**

   The command line displays the `Select objects` prompt.

3. **Select at least one object.**

   You can use any of the object selection techniques described in Chapter 10.

4. **Press Enter when you finish selecting objects.**

   AutoCAD displays the following prompt:

   ```
   Specify base point or [Displacement] <Displacement>:
   ```

5. **Specify a base point by clicking a point or typing coordinates.**

   This point serves as the tail end of the imaginary arrow that indicates how far and in what direction you want the objects to move. After you pick a base point, it's fairly easy to see what's going on, because AutoCAD displays a temporary image of the object that moves around as you move the cursor. As well, it shows a brown dashed line indicating the move vector. Figure 11-1 shows what the screen looks like.

TIP

   Specify a base point somewhere on or near the object(s) that you're moving. You can (and most of the time should) use an object snap mode to choose a point exactly on one of the objects.

   AutoCAD displays the following prompt:

   ```
   Specify second point or <use first point as displacement>:
   ```

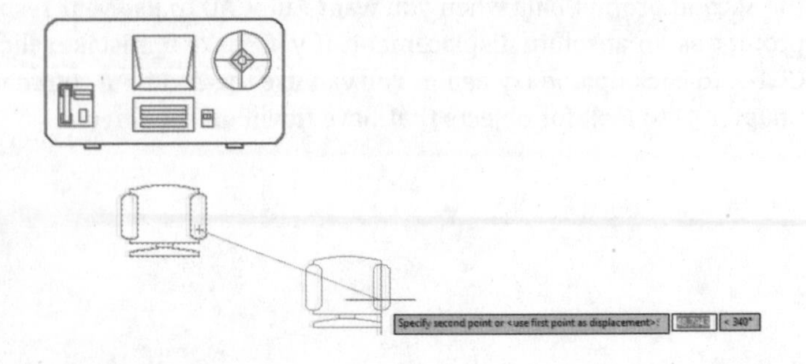

FIGURE 11-1: Dragging objects in the middle of the Move command.

6. **Specify the second point by clicking a point or typing coordinates.**

   The second point serves as the arrow end of the imaginary displacement arrow. After you specify the second point, AutoCAD moves the objects.

   REMEMBER

   Press Enter in response to the second prompt *only* if you want AutoCAD to use your response to the first prompt as an absolute displacement.

TIP

AutoCAD has two common precision techniques for specifying the second point:

» Use an object snap mode to pick a second point exactly on the part of another object in the drawing.

» Use direct distance entry to move objects in an orthogonal or polar-tracking direction. See Chapters 6 and 8 for instructions.

The following steps demonstrate command-first editing with the Move command, using the distance method to indicate how far and in what direction to move selected objects:

1. **Repeat Steps 1–4 in the preceding step list.**
2. **When AutoCAD displays the `Specify base point...` prompt, specify a direction and distance by typing the appropriate values:**
   - Enter **6,2**, for example, to move selected objects 6 units to the right and 2 units up.
   - Enter **3<45**, for example, to move selected objects 3 units at an angle of 45 degrees counterclockwise from the east (3 o'clock).
3. **Press Enter to complete the move.**

COpy

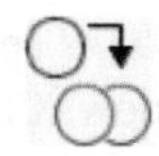

The COpy command works almost identically to the Move command except that AutoCAD leaves the selected objects in place and makes copies of them at the new location.

The COpy command creates multiple copies by default. If you want only one copy, press Enter after placing the first copy in the drawing. Choosing `mOde` at the command prompt or in the Dynamic Input options list lets you switch between making a single copy and multiple copies. Whether you mostly make multiple copies or mostly make single copies, you'll appreciate being able to change the default setting.

The COpy command includes an Array option. In addition to plunking down copied objects just about anywhere, type **A** to choose the Array option and specify spacing for an evenly laid-out linear array, a single row of copies. Using the -ARray command (which I describe later in this chapter and in Chapter 18) is required to create rows and columns or circular patterns of copied objects.

TIP

The COpy command includes the Undo option, with which you can roll back multiple copies while in the COpy command.

Copy between drawings

You can't use AutoCAD's COpy command to copy objects to another drawing. Instead, you can drag objects from one drawing window to another. As well, you can use the most common Clipboard command, COPYCLIP, together with its companion command, PASTECLIP, to copy and paste objects between drawings.

COPYCLIP, CUTCLIP, and PASTECLIP — the three standard Clipboard buttons you find in every Windows program — use the Windows Clipboard to temporarily store drawing objects from one drawing so that they can be pasted into the same drawing or into other documents, such as another AutoCAD drawing or a Word document. The Clipboard panel on the Ribbon's Home tab contains several variation of the Cut, Copy, and Paste tools.

TIP

Windows also offers a Clipboard history (as of the 2018 release of Windows 10). It remembers the last ten pieces of text and images you copied or cut from a document, including from AutoCAD drawings. To access the history, press Windows key+V, and then choose the item to paste into the current document.

You also can paste text into the command line, and if AutoCAD understands it, it will act on it. Examples of valid text include command names, scripts, and pieces of AutoLISP and Diesel code. The last three are examples of programming possible in AutoCAD, a topic I do not cover in this book.

REMEMBER

As you're figuring out where commands lurk on the AutoCAD Ribbon, the standard Windows keyboard shortcuts — Ctrl+X (cut), Ctrl+C (copy), and Ctrl+V (paste) — are available and are often the most efficient ways of using the Windows Clipboard. Better yet, they even work in a single drawing and between drawings. Within certain limitations, you can also Ctrl+X (cut), Ctrl+C (copy), and Ctrl+V (paste) between AutoCAD drawings and many other Windows applications, although this process may not be the best to use in many situations. I discuss this topic a little more in Chapter 24.

AutoCAD has a total of ten Windows Clipboard commands, nine of which are on the Ribbon. Of the remaining ones, PASTEBLOCK (Ctrl+Shift+V) can sometimes be convenient. It pastes a previous Windows Clipboard selection as a block, which I discuss in Chapter 17. A description of the remaining modes, which are rarely used, is beyond the scope of this book.

Stretch

The Stretch command has the same base point and displacement prompts as COpy and Move, and it shifts objects to other locations in the drawing. But it has this important difference: It can move *part* of an object or parts of several objects, leaving the rest unscathed. You can use it to change the length of lines, arcs, and rectangles, for example.

Here are the things you need to know to make Stretch your friend:

- **Select objects to stretch:** To use Stretch to change the length of objects, you must first select objects by using a crossing selection box or crossing polygon, as described in Chapter 10. See Figure 11-2.
- **Define points:** Stretch operates on the defining points of objects (endpoints of lines, vertices of polylines, and the centers of circles, for example) according to this rule: When a defining point is in the crossing selection box that you specify, AutoCAD moves the defining point and updates the object accordingly. For example, if the crossing selection box surrounds one endpoint of a line but not the other, Stretch moves the selected endpoint and redraws the line in the new position dictated by the selected endpoint's new location. It's as though a rubber band is tacked to the wall with two pins and you move one of the pins.
- **Compress and stretch:** Stretch can make lines longer or shorter, depending on the crossing selection box and displacement vector. In other words, the Stretch command combines stretching and compressing.

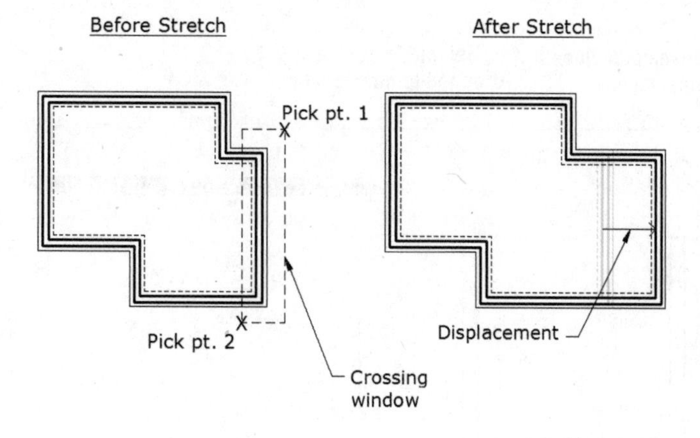

FIGURE 11-2: Use a crossing selection box to select objects for stretching.

» **Get in the mode:** Depending on what you're trying to accomplish, you may want to turn on Ortho mode (F8) or Polar Tracking (F10) mode before stretching. Figure 11-3 shows the results of an Ortho stretch and a non-Ortho stretch.

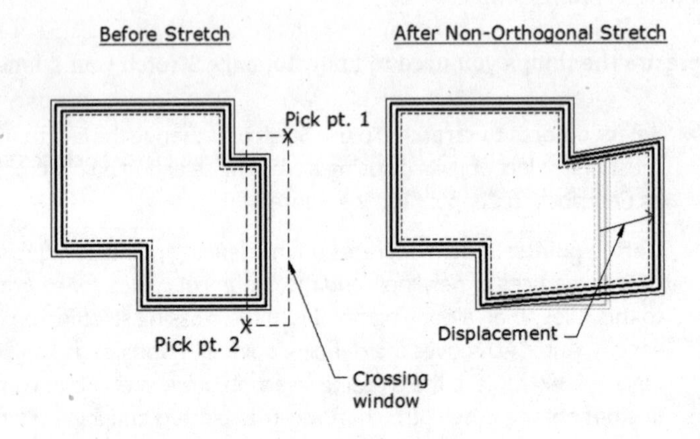

FIGURE 11-3: The hazards of stretching with Ortho mode and Polar Tracking mode turned off.

Follow these steps to stretch some lines:

1. **Draw some lines.**

   Start stretching with simple objects, as shown in Figure 11-4. You can work your way up to more complicated objects, such as polylines and arcs, after you limber up with lines.

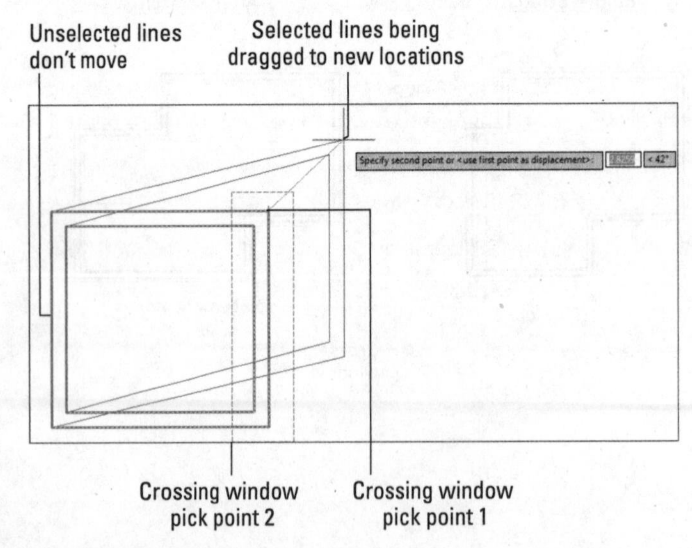

FIGURE 11-4: Dragging objects in the middle of the Stretch command.

2. **Press Esc to make sure that no command is active and no objects are selected.**

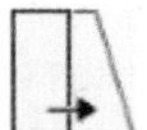

3. **Click the Stretch button in the Modify panel of the Ribbon's Home tab to start the Stretch command.**

   The command line displays the `Select objects` prompt with a reminder to use the Crossing or CPolygon object-selection mode:

   ```
   Select objects to stretch by crossing-window or
     crossing-polygon...
   Select objects:
   ```

4. **Pick points to specify a crossing selection box that encloses some, but not all, endpoints of the lines.**

   Figure 11-4 shows a sample crossing selection box that completely encloses the two vertical lines on the right side of the figure. This crossing selection box cuts through the four horizontal lines, enclosing only one endpoint apiece.

REMEMBER

   You specify a crossing selection box by picking a point, moving the mouse to the *left,* and picking a second point.

5. **Press Enter to end the object selection.**

   AutoCAD displays the following prompt:

   ```
   Specify base point or [Displacement] <Displacement>:
   ```

6. **Specify a base point by object snapping to a point on an existing object or by typing absolute X,Y coordinates.**

   AutoCAD displays the following prompt:

   ```
   Specify second point or <use first point as displacement>:
   ```

TIP

   Toggle Ortho mode on and then off by clicking the Ortho Mode button on the status bar or by pressing F8; move the cursor around first with Ortho mode turned on and then off to see the difference. Refer to Figure 11-3.

   Figure 11-4 shows what the screen looks like as you move the cursor around without the benefit of Ortho mode or Polar Tracking mode.

7. **Toggle Ortho (F3) mode on (if it isn't on already), and then specify the second point — usually, by using direct distance entry.**

   You can also specify the second point by object snapping to a point on an existing object or by typing relative X,Y coordinates.

   After you pick the second point, AutoCAD stretches the objects. Notice that the Stretch command *moved* the two vertical lines because the crossing selection box contained both endpoints of both lines. Stretch lengthened or shortened the four horizontal lines because the crossing selection box enclosed only one endpoint apiece.

Here are some additional tips for using the Stretch command:

- **Practice makes perfect.** Using the Stretch command effectively takes some practice, but it's worth the effort. You can stretch with different crossing selection box locations as well as different base points and second points.
- **Don't try to stretch text.** Create a line of text (I cover text in Chapter 13) that contains the two words *The Truth.* Start the Stretch command, and specify a crossing selection window that starts to the right of the text and ends up passing between the two words. When you finish the command, observe that although you can stretch a great deal in AutoCAD, you can't stretch The Truth. Text doesn't stretch, but it will *move* if its insertion point falls within the crossing window.
- **Use a crossing-polygon selection instead of a crossing-window selection.** The Stretch command prompt says `Select objects to stretch by crossing-window or crossing-polygon`, but picking points doesn't give you the crossing-polygon option. See Chapter 10 for information on crossing-polygon selection. To use a crossing-polygon selection, type **CP** at the `Select objects` prompt and press Enter.
- **Use multiple windows to move objects.** You can use more than one crossing window in a single run of the Stretch command, and you can select objects that are outside a crossing window to have them move if other objects that cross the selection window(s) are stretched. The only objects that change their size are stretchable objects *crossed* by a crossing window or polygon.

More Manipulations

The commands in this section — MIrror, ROtate, SCale, ARray, and Offset — provide ways other than the Move, COpy, and Stretch commands to manipulate objects or create new versions of them. To use these commands, you need to be familiar with the techniques for object selection and precision editing for the Move, COpy, and Stretch commands. See Chapter 8 to find out about object selection and precision editing, and see the discussion earlier in this chapter about Move, COpy, and Stretch.

Mirror, mirror on the monitor

The MIrror command creates a reverse copy of an object. After you select some objects, AutoCAD prompts you to select two points that define a line about which the objects will be mirrored. You can then retain or delete the source objects. Follow these steps to use the MIrror command:

1. **Press Esc to make sure that no command is active and no objects are selected.**

2. **Click the Mirror button on the Home tab's Modify panel, or enter MI and press Enter.**
3. **Select at least one object, and press Enter to end the object selection.**

   AutoCAD prompts you to define the mirror line by picking points:

   ```
   Specify first point of mirror line:
   ```

4. **Specify the start of the mirror line by clicking a point or typing coordinates.**

   AutoCAD prompts you:

   ```
   Specify second point of mirror line:
   ```

5. **Pick a second point.**

   Most of the time, you'll enable Polar Tracking mode or Ortho mode so that you can mirror objects precisely. You can also use object snaps on existing objects, including ones being mirrored, which ensure exact symmetry between the source and the mirrored objects. AutoCAD now prompts you:

   ```
   Erase source objects? [Yes No] <N>:
   ```

6. **Finish the mirror by using one of these options:**
   - *Type* **Y** *at the final prompt.* The source objects disappear, leaving only the new, mirrored copy.
   - *Accept the default No option.* The source objects are retained along with the mirrored copies.

Normally, when you mirror text or dimensions, the words read backward. The system variable MIRRTEXT can handle this little problem. By default, MIRRTEXT is turned off (that is, its value is 0), so the text itself still reads the right way around after it and the other objects are mirrored. When you really want your drawing text to appear in reverse, change the value of MIRRTEXT to 1. I cover system variables in a little more detail in Chapter 26.

ROtate

The ROtate command pivots objects around a point that you specify. Follow these steps to use the ROtate command:

1. **Press Esc to make sure that no command is active and no objects are selected.**
2. **Click the Rotate button on the Home tab's Modify panel, or enter RO and press Enter.**
3. **Select one or more objects and then press Enter to end object selection.**

   AutoCAD prompts you for the base point for rotating the selected objects:

   ```
   Specify base point:
   ```

4. **Specify a base point by clicking a point or typing coordinates.**

   The base point becomes the point around which AutoCAD rotates the objects. You also have to specify a rotation angle:

   ```
   Specify rotation angle or [Copy Reference] </line> <0>:
   ```

5. **Specify a rotation angle by one of these methods:**
   - *Type an angle measurement and press Enter.*
   - *Press Enter to accept the default value shown in angle brackets (which is the last-used value).*
   - *Click a point onscreen.*
   - *Enter **R** to specify a Reference angle.*

   Alternatively, you can indicate an angle on the screen by moving the cursor until the Coordinates section of the status bar indicates the desired angle and then clicking. If you choose this alternative, then use Ortho (F3) mode or Polar Tracking (F10) mode to indicate a precise angle (for example, 90 or 15 degrees), or an object snap to rotate an object so that it aligns precisely with other objects.

Positive angle values rotate objects counterclockwise (counter to what you might expect), and negative values rotate them clockwise. For example, entering 270 degrees or –90 degrees produces the same result, making objects point down.

After you specify the rotation angle by typing or picking, AutoCAD rotates the objects into their new positions. The ROtate command's Copy option makes a rotated copy while leaving the source object in place.

The Reference option can seem a little confusing, but it is actually simple. A typical use for this option is when you want to rotate one or more objects so that they are parallel to an existing object.

The first three steps are the same as what you just did, but enter **R** when you get to Step 4.

```
Specify rotation angle or [Copy/Reference] <0>: r
```

6. **The command line now asks you to specify the reference angle.**

```
Specify the reference angle <0>:
```

Pick two points, usually on the object or objects that you want to rotate. In the angle brackets, AutoCAD suggests an angle that is the last one you used, such as 0.

7. **Now for the slightly tricky part. By default, angles are defined counterclockwise from the X-axis (east). If you want to align the objects with some other angle, then enter P for Points and then define two points.**

```
Specify the new angle or [Points] <0>: p
Specify first point: Pick a point, such as on another
  object
Specify second point: Pick a second point to define the
  angle
```

The objects then rotate so that the angle from the X-axis defined by the first two points matches the angle from the X-axis defined by the second two points.

SCale

If you've read all my harping about drawing scales and scale factors in Chapter 4, then you may think that the SCale command performs magical drawing-scale setup on an entire drawing. No such luck, nor does it have anything to do with ladders or fish. The SCale command uniformly *scales* — enlarges or reduces — objects up or down by a factor that you specify. Here's how it works:

1. **Press Esc to make sure that no command is active and no objects are selected.**
2. **Click the Scale button on the Home tab's Modify panel, or enter SC and press Enter.**

3. **Select at least one object, and press Enter to end the object selection.**

   AutoCAD prompts you for the base point from which it will scale all selected objects:

   ```
   Specify base point:
   ```

4. **Specify a base point by picking a point or typing coordinates.**

   AutoCAD now prompts you for the scale factor:

   ```
   Specify scale factor or [Copy/Reference] <1.0000>:
   ```

WARNING

Don't assume that AutoCAD will scale the objects and just leave them more or less where they are in the drawing. AutoCAD scales the distance *between* objects as well as the objects themselves. For example, when you select a circle to scale, pick a point outside the circle as the base point and then specify a scale factor of 2. AutoCAD not only makes the circle twice as big but also moves the circle twice as far away from the base point that you specified. To make a circle bigger but leave it where it's located, use the CENter object snap to select its center point as the base point of the scale operation.

5. **Type a scale factor and press Enter.**

   AutoCAD scales the objects by the factor that you type, using the base point that you specified. Numbers greater than 1 increase the objects' sizes; numbers smaller than 1 and greater than 0 decrease the objects' sizes. Negative scale factors are invalid.

TIP

Just like the ROtate command, SCale has a Copy option with which you can make enlarged or reduced duplicates of selected objects without altering the source objects. And throughout the drawing session, both SCale and ROtate remember the last scale factor or rotation angle that was entered.

-ARray

The -ARray command is a supercharged COpy command. You use it to create a rectangular grid of objects at regular X and Y spacings or a radial arrangement of objects around a center point at a regular angular spacing. For example, you can use rectangular arrays to populate an auditorium with chairs or use a polar array to populate a bicycle wheel with spokes.

The kind of array I describe in this chapter is the old-style — but still very useful — non-associative array. I cover the new-style array in Chapter 18. Non-associative arrays are simply copies of the source object; they dwell on the same layers as their source, and they can be edited individually, even in older releases of AutoCAD.

The following steps describe how to create an old-style rectangular array, which you'll probably do more often than you create a polar array:

1. **Press Esc to make sure that no command is active and no objects are selected.**
2. **Type** -ARray **(don't forget the leading hyphen) and press Enter.**

   Typing a hyphen in front of an AutoCAD command normally tells the program to run the command line prompts rather than a dialog box. Starting with AutoCAD 2012, however, typing **-ARray** (with the leading hyphen) versus typing **ARray** (without the leading hyphen) runs two different commands. The difference is that ARray creates associative arrays, whereas -ARray creates non-associative ones. I cover associative arrays in Chapter 18.

   AutoCAD prompts you to select the objects you want to array.

```
Select objects:
```

3. **Select one or more objects; then press Enter.**

   AutoCAD prompts you to choose the type of array you want:

```
Enter the type of array [Rectangular/Polar] <R>:
```

4. **Type R and press Enter to create a rectangular array.**

   AutoCAD prompts you to specify the number of rows and columns you want, and the distance between each of them:

```
Enter the number of rows (---) <1>:
```

5. **Enter a value and press Enter.**

   AutoCAD asks for the number of columns, the distance between rows, and (finally) the distance between columns.

   The Rows and Columns numbers include the row and column of the original objects themselves. In other words, entries of 1 don't create any new objects in that direction.

6. **Enter values for the number of columns, and the distances between adjacent rows and columns, and then press Enter.**

   AutoCAD creates regularly spaced copies of the selected objects in neat rows and columns.

To create regularly arrayed objects in rows and columns or arranged around a center point, you must use the command line version of -ARray: Type a hyphen in front of the command name. The hyphen forces AutoCAD to use the command-line version of a command when there is also a dialog box version.

Typing ARray without the hyphen creates new-style associative array objects. Associative arrays, covered in Chapter 18, are more powerful and versatile, but non-associative arrays may still be adequate for many applications and are found all over pre-2012 drawings.

Offset

You use Offset to create parallel or concentric copies of lines, polylines, circles, arcs, or splines. Follow these steps to use Offset:

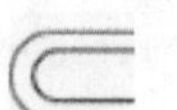

1. **Click the Offset button on the Home tab's Modify panel, or enter Offset and press Enter.**

   AutoCAD displays the current command settings and prompts you for the *offset distance* — the distance from the original object to the copy you're creating:

   ```
   Current settings: Erase source=No Layer=Source
     OFFSETGAPTYPE=0
   Specify offset distance or [Through Erase Layer] <Through>:
   ```

2. **Type an offset distance and press Enter.**

   Alternatively, you can indicate an offset distance by picking two points on the screen. If you choose this method, then you should normally use object snaps to specify a precise distance from one existing object to another.

   AutoCAD prompts you to select the object from which you want to create an offset copy:

   ```
   Select object to offset or [Exit Undo] <Exit>:
   ```

3. **Select a single object, such as a line, a polyline, a circle, or an arc.**

   Note that you can select only one object at a time with the Offset command. AutoCAD asks where you want the offset object:

   ```
   Specify point on side to offset or [Exit Multiple Undo]
     <Exit>:
   ```

4. **Point to one side or the other of the object and then click.**

   It doesn't matter how far away from the object the cursor is when you click. You're simply indicating a direction. As you move the cursor from one side of the source object to the other, AutoCAD previews what the offset object will look like.

AutoCAD repeats the `Select object` prompt, in case you want to offset other objects by the same distance:

```
Select object to offset or [Exit Undo] <Exit>:
```

5. **Repeat Step 3 to offset another object, or press Enter when you're finished offsetting objects.**

Figure 11-5 shows the Offset command in progress. For information on the command options — Multiple, Erase, and Layer are all useful options — look up Offset in the online Help system's Command Reference section.

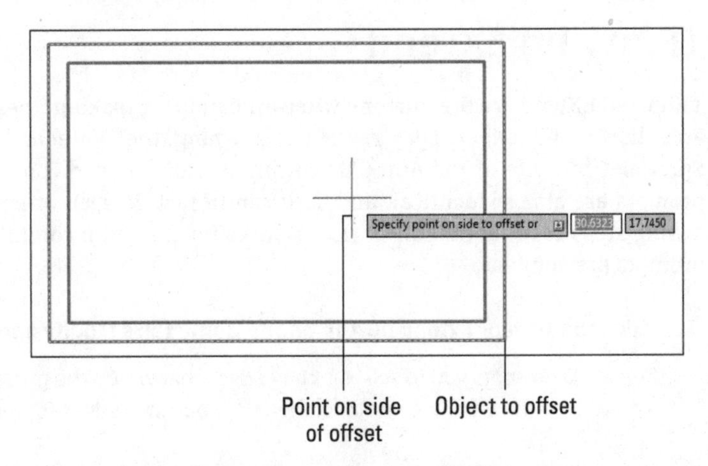

FIGURE 11-5: Offsetting a polyline.

TIP

When you want to offset a series of connected lines, for example a rectangular house plan outline or one side of a pathway on a map, either draw it as a polyline or convert the individual line or arc segments (or both) into a polyline with the Join command. I cover Join later in this chapter. When you draw a series of line segments with the Line command and then try to offset the segments, you have to pick each segment and offset it individually. Even worse, the corners usually aren't finished in the way you expect, because AutoCAD doesn't treat the segments as connected. You avoid all these problems by offsetting a polyline, which AutoCAD treats as a single object. Figure 11-5 shows an offset polyline. See Chapter 6 for more information about the differences between lines and polylines.

TECHNICAL STUFF

Offset is one of several commands during which AutoCAD provides an interactive preview of what the editing action will be. Command Preview gives you real-time feedback for the current editing command so that you can make changes to unintended consequences before finishing the command. The preview function is available during the following commands: BReak, CHAmfer, EXtend, Fillet, LENgthen, MAtchprop, Offset, and Trim. If you are not seeing command previews, then turn them on by setting COMMANDPREVIEW to 1.

AutoCAD also provides previews while placing hatch patterns, new layouts in viewports, properties, and 3D commands that place surfaces and lofts.

Slicing, Dicing, and Splicing

TRim, EXtend, BReak, Fillet, CHAmfer, BLEND, and Join are commands that are useful for shortening and lengthening objects, for breaking them in two, and for putting them back together again.

TRim and EXtend

TRim and EXtend are the mirror-twin commands for making lines, polylines, and arcs shorter and longer. They're the yin and yang, the Laurel and Hardy, the Jack Sprat and his wife of the AutoCAD editing world. The two commands and their prompts are almost identical, and both can in fact do each other's job. The following steps cover both. I show the prompts for the TRim command; the EXtend prompts are similar.

1. **Click the Trim or Extend button on the Home tab's Modify panel.**

   AutoCAD prompts you to select cutting edges that will do the trimming (or, if you choose the EXtend command, then the boundary edges for extending to):

   ```
   Current settings: Projection=UCS, Edge=None, Mode=Quick
   Select object to trim or shift-select to extend or[cuTting
     edges Crossing mOde Project eRase]
   ```

2. **When Quick mode is on, the AutoCAD automatically selects all objects in the drawing as cutting edges for the TRim command (or boundary edges for the EXtend command).**

   When the mOde option is set to Standard, press Enter to accept the default option to select all drawing objects. Alternatively, select individual objects by picking them. Press Enter to end the object selection. The objects you select become the cutting edge of the TRim command or the boundary to which objects will be extended by the EXtend command.

   Figure 11-6 shows a cutting edge (for TRim) and a boundary edge (for EXtend).

   AutoCAD prompts you to select objects that you want to trim or extend (EXtend replaces cutting edges with Boundary edges and doesn't have the eRase option):

   ```
   Select object to trim or shift-select to extend or
     [cuTting edges Crossing mOde Project eRase Undo]:
   ```

Cutting edge (for trim) Boundary edge (for extend)

FIGURE 11-6: Anatomy of the TRim and EXtend operations.

3. **Select a single object to trim or extend. Choose the portion of the object that you want AutoCAD to trim away or the end of the object that's closer to the extend-to boundary.**

   AutoCAD trims or extends the object to one of the objects that you selected in Step 2. If AutoCAD can't trim or extend the object — for example, when the trimming object and the object to be trimmed are parallel —the command line displays an error message, such as `Object does not intersect an edge.`

TIP

   You can select multiple objects to trim and extend by typing **F** and pressing Enter to use the Fence object-selection mode or by entering **C** to use a crossing selection. Even better, you can use implied windowing and drag a selection box to select multiple objects. Refer to Chapter 10 for more on multiple object selection.

   The command line continues to prompt you to select other objects to trim or extend:

```
Select object to trim or shift-select to extend or [Fence
  Crossing Project/Edge eRase Undo]:
```

4. **Choose additional objects or press Enter when you're finished trimming or extending.**

TIP

   Here's a triple-treat tip: You can switch between TRim and EXtend without exiting the command by holding down the Shift key. If you accidentally trim or extend the wrong object and you're still using the TRim or EXtend command, then type **U** and then press Enter to undo the most recent trim or extend operation. And, finally, if a remnant won't trim because it doesn't cross the cutting edge, then type **R** (for eRase) and press Enter to erase the remnant without leaving the TRim command.

The example in Figure 11-6 shows trimming to a single cutting edge, in which the end of each line gets lopped off.

Another common use of the TRim command is to trim out a piece of a line between two cutting edges. In the two-cutting-edges scenario, TRim cuts a piece out of the middle of the trimmed line.

TIP

An object being trimmed or extended can also be a cutting edge or extending boundary, and vice versa, all in the same run of the command.

The LENgthen command provides other useful ways to make lines, arcs, and polylines longer (or shorter). You can specify an absolute distance (or *delta*) to lengthen or shorten by, a percentage to lengthen or shorten by, or a new total length. Look up the LENgthen command in AutoCAD's Help system for more information.

LENgthen is an option on the grip pop-up menus on lines and arcs. I discuss grips later in this chapter. To display the menu, just hover the mouse pointer over an endpoint grip on either of those object types, or over one of the new triangular grips on an elliptical arc, and click Lengthen.

BReak

The BReak command isn't what you use before heading out for coffee. AutoCAD doesn't have a command for that yet, but I keep hoping. The BReak command creates gaps in lines, polylines, circles, arcs, or splines. BReak also comes in handy when you need to split one object in two without actually removing any visible material.

The following steps show you how to break an object (don't worry — in AutoCAD, you won't have to pay for it):

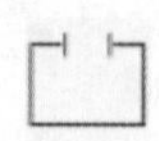

1. **On the Ribbon's Home tab, click the label of the Modify panel to open its slideout, and then click the Break button.**

   AutoCAD prompts you to select a single object that you want to break.

2. **Select a single object, such as a line, a polyline, or an arc.**

TIP

   The point you pick when selecting the object serves double duty: It selects the object, of course, but it also becomes the *default first break point* (that is, it defines one side of the gap that you'll create). Thus, use one of the AutoCAD precision techniques, such as an object snap, to pick the object at a precise point or use the `First point` option (described in Step 3) to repick the first break point.

   AutoCAD prompts you to specify the second break point or to type **F** and press Enter when you want to respecify the first break point:

```
Specify second break point or [First point]:
```

3. **If the point that you picked in Step 2 doesn't also correspond to a break point (see the preceding Tip), then type F and press Enter to re-specify the first break point. Then pick the point with an object snap or another precision technique.**

When you type **F** and press Enter and then re-specify the first break point, AutoCAD prompts you to select the second break point:

```
Specify second break point:
```

4. **Specify the second break point by picking a point or typing coordinates.**

   AutoCAD cuts a section out of the object, using the first and second break points to define the length of the gap.

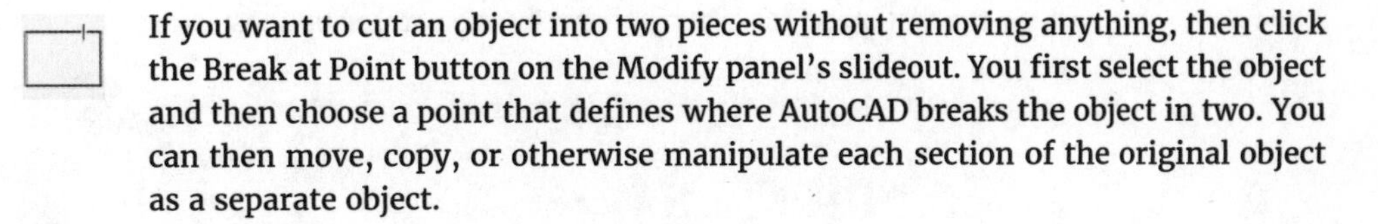

If you want to cut an object into two pieces without removing anything, then click the Break at Point button on the Modify panel's slideout. You first select the object and then choose a point that defines where AutoCAD breaks the object in two. You can then move, copy, or otherwise manipulate each section of the original object as a separate object.

WARNING

Approach the EXPLODE command with caution — it breaks up complex objects into AutoCAD primitive objects. For example, a multisegment polyline explodes into separate line and arc objects. Most of the time, these complex objects are created that way for a reason. The things you can explode but shouldn't unless you have a really, *really* good reason include polylines, blocks, 3D solids, associative arrays, tables, and multiline text. And even though AutoCAD lets you, never, *ever* explode dimensions or leaders. They become extremely difficult to edit.

Fillet, CHAmfer, and BLEND

Whereas TRim, EXtend, and BReak alter one object at a time, the Fillet, CHAmfer, and BLEND commands modify pairs of objects. As Figure 11-7 shows, Fillet creates a curved corner between two lines, whereas CHAmfer creates a beveled corner. In case you wondered, it's pronounced "FILL-et," not "fill-AY." Saying that you know how to "fill-AY" may secure you a job in a butcher shop, but it will only draw strange looks in a design office.

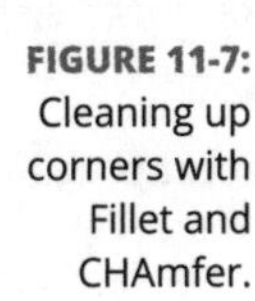

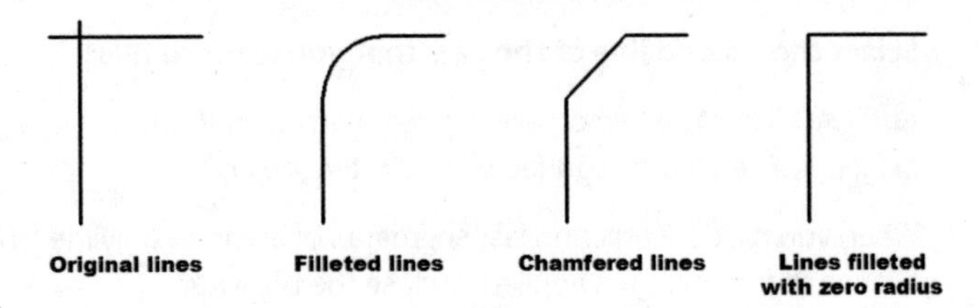

FIGURE 11-7: Cleaning up corners with Fillet and CHAmfer.

Fillet and CHAmfer show a preview of the results of the operation as soon as you select the objects to modify. If the fillet radius or the chamfer distances don't look right in the preview, you can change their values before completing the command.

The following steps describe how to use the Fillet command — the CHAmfer command works similarly except that, rather than specify a fillet radius, you specify either two chamfer distances or a chamfer length and angle:

1. **Click the Fillet button on the Home tab's Modify panel.**

   Fillet and CHAmfer share a flyout button; when you see a straight-line corner instead of a rounded one, click the flyout arrow to select Fillet.

   AutoCAD displays the current fillet settings and prompts you to either select the first object for filleting or specify one of three options:

   ```
   Current settings: Mode = TRIM, Radius = 0.0000
   Select first object or [Undo Polyline Radius Trim
     Multiple]:
   ```

2. **Type R and press Enter to set the fillet radius.**

   AutoCAD prompts you to specify the fillet radius that it uses for the upcoming fillet operation:

   ```
   Specify fillet radius <0.0000>:
   ```

3. **Type a fillet radius and press Enter.**

   The number you type will be the radius of the arc that joins the two lines.

   AutoCAD then asks you to select the first object:

   ```
   Select first object or [Undo Polyline Radius Trim
     Multiple]:
   ```

4. **Select the first line of the pair that you want to fillet.**

   AutoCAD prompts you to select the second object for filleting:

   ```
   Select second object or shift-select to apply corner or
     [Radius]:
   ```

5. **Select the second line of the pair that you want to fillet.**

   AutoCAD fillets the two objects, drawing an arc of the radius that you specified in Step 3. The arc isn't connected to the two objects.

   When you pick the first and last segments of an open polyline, an appropriate fillet or chamfer is then applied to close the polyline.

TIP

Fillet has more tricks up its sleeve:

- You can fillet two lines and specify a radius of 0 (zero) to make them meet at a point. Note that this technique produces *no arc,* which you want, rather than a zero-radius arc.
- If you have lots of lines to fillet, whether with a zero radius or the same nonzero radius, then use the Fillet command's Multiple option to speed the process.
- When you start Fillet and select two parallel lines, you see a nice 180-degree arc joining them.
- You can add fillets between pairs of arcs, splines, and even between an arc and a spline.
- You can fillet between any two combinations of line, arc, circle, polyline, and spline.
- You can fillet all vertices on a polyline at one time by choosing the Polyline option before selecting the object.
- Hold down the Shift key before picking the second line to automatically produce a clean intersection as though you'd set the fillet radius to zero even when the current radius setting isn't zero. The CHAmfer command has the same Shift+select option.

The BLEND command joins Fillet and CHAmfer, offering another method for creating transitions between 2D drawing objects. Whereas CHAmfer creates beveled corners (that is, straight lines), and Fillet creates round corners (circular arcs), the corners BLEND makes are spline objects. Figure 11-8 shows the two types of available blend; the feature is connecting the green lines. Choosing the Tangent option produced the red spline, and choosing the Smooth option produced the blue spline. Unlike Fillet and CHAmfer, both of which modify the source objects to make the radiused or beveled transition, BLEND leaves the source objects intact. Blends appeal to industrial designers and other purveyors of swoopy shapes. If you're a mechanical drafter, then stick with fillets and chamfers.

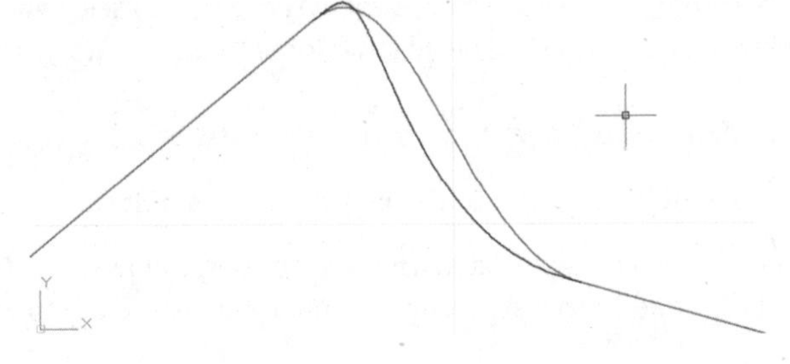

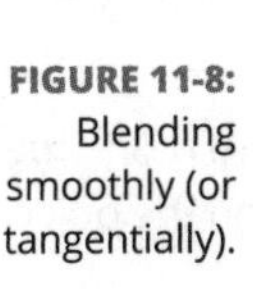

FIGURE 11-8: Blending smoothly (or tangentially).

Join

Use the Join command to fill gaps between lines, arcs, elliptical arcs, open splines, 2D and 3D polylines, and helixes. If the lines are *collinear* (they lie in the same straight line) or the arcs, splines, polylines, or elliptical arcs are on a similarly curved path, then Join creates a single new entity to replace the existing separate pieces, as shown in Figure 11-9.

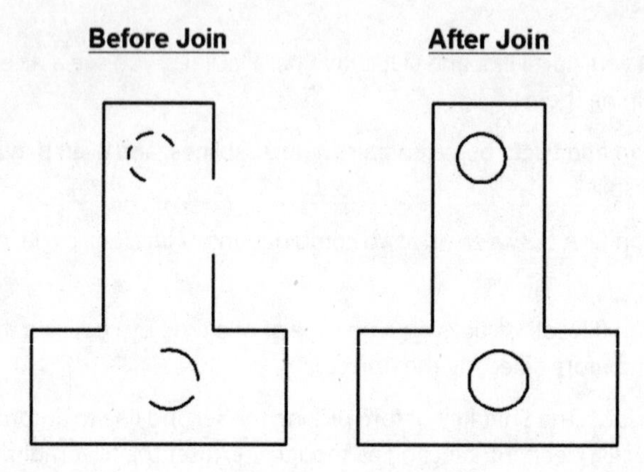

FIGURE 11-9: Joining sundered pieces.

REMEMBER

Objects to be joined must be collinear or else their ends must be coincident. You can't join noncollinear objects that have gaps between them or that cross or overlap.

The following steps describe how to use the Join command:

1. **Click the Modify panel title at the bottom edge of the Ribbon's Home tab to open the Modify panel's slideout, and then click the Join button.**

   When you foresee doing a lot of joinery, you can pin open the slideout. See Figure 11-9. AutoCAD prompts you to select the source object.

2. **Select the *source object* — that is, the object you want to join other objects to.**

   AutoCAD prompts you according to the object type you selected. If you select a line, then the command line or Dynamic Input tooltip shows this message:

   ```
   Select source object or multiple objects to join at once:
   ```

3. **Select valid objects to join to the original source object.**

   For example, when you selected a line as the source object, AutoCAD continues prompting for additional adjoining lines until you press Enter to end the object selection.

4. **Press Enter to end the command.**

   AutoCAD joins the selected objects into a single object. The new object inherits relevant properties, such as the layer or linetype of the first object selected.

POLISHING THOSE PROPERTIES

When you contemplate how to edit objects, you probably think first about editing their *geometry,* such as moving, stretching, and making new copies. That's the kind of editing I cover in this chapter.

Another kind of editing changes the *properties* of objects. As I describe in Chapter 9, every object in an AutoCAD drawing has a set of nongeometrical properties, such as layer, color, linetype, lineweight, transparency, and (perhaps) plot style. Sometimes you need to edit those properties — if you accidentally draw on the wrong layer, for example. Here's a handful of ways to edit an object's properties in AutoCAD:

- **The Properties palette:** It's the most flexible way to edit properties. Select any object (or objects), right-click in the drawing area, and choose Properties from the menu. The Properties palette displays the names and values of all properties relevant to the selected object(s). Click in the appropriate value field to change a particular property.
- **The Quick Properties palette:** When the Quick Properties palette is enabled (click its status bar button), a palette pops up near the cursor when you hover the cursor over an object. When you select the object, a larger Quick Properties palette appears, containing value fields that you can click to change the specific property, just like the Properties palette. You can choose how many lines of information you want Quick Properties to display by right-clicking the Quick Properties status bar button and choosing Quick Properties Settings.
- **Layers and Properties control lists:** Another way to change properties is to select objects, and then choose from the drop-down lists (Layer, Color, and so on) on the Properties palette. It's fine to change layers this way, but don't be slapdash about changing the other properties. See Chapter 9 and the following bullet for more information.
- **Match Properties button:** You can use Match Properties to copy the properties from a source object to other objects. You can find the Match Properties button on the Clipboard panel of the Ribbon's Home tab. Match Properties works similarly to the Format Painter button in Microsoft applications. Match Properties works even when the objects reside in different drawings. Use the command's Settings option to limit which properties are copied.

(continued)

(continued)

- ***Warning:*** As discussed in Chapter 9, it's almost never correct to set the color, linetype, or lineweight to be anything other than ByLayer. To do so invites chaos.
- **Change Space command:** Sometimes, you add some geometry to paper space and then realize that it should have been in model space, or vice versa. I introduce the concepts of model space and paper space in Chapter 12. The CHSPACE command can come to the rescue by moving objects and scaling and aligning them appropriately for the space. You'll find it on the Modify panel slideout on the Ribbon's Home tab. For more information, look up CHSPACE in the Command Reference of the online Help system.

You aren't limited to selecting a single object at the first prompt. If you select multiple objects by using windowing, then Join joins whatever objects it can and rejects those that it can't.

You can turn an arc into a circle or an elliptical arc into a full ellipse with Join's cLose option.

Other editing commands

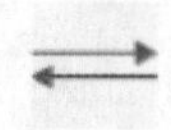

The REVERSE command provides an easy way to reverse the direction of lines, polylines, splines, and helixes. Why does that matter? Well, most of the time it doesn't, but if you happen to be using a complex linetype that uses text or a directional arrow block, then you probably want the text to read right way up and the arrow to point in the correct direction. Lines, polylines, and other entity types have start points and endpoints and, therefore, a direction that runs from the former to the latter. Rather than redraw linework so that the text or symbols appear the right way up, use the REVERSE command to flip the start points and endpoints.

This command also reverses variable-width polylines. The system variable PLINEREVERSEWIDTHS controls how REVERSE handles them. When set to the default of 0 (zero), REVERSE reverses the start and end sequence but doesn't affect the width; when set to 1, the start and end widths of segments are also reversed.

The Delete Duplicate Objects tool (also known as the OVERKILL command) looks for fully or partially overlapping objects, and either combines or deletes them. Overlapping objects, such as one line partially or completely on top of or beneath another line, can mess up editing and other commands.

The PEdit (Polyline Edit) command can perform a great many operations on polylines. The operations fall into two categories:

- They affect the entire polyline, such as its width, how curves are fitted and unfitted, reversing its direction, and defining how noncontinuous line types behave at vertices.
- They affect the polyline at specific vertices, such as setting the start and ending width of individual segments.

PEdit can also be used to turn a regular line into a single-segment polyline, and to join additional line and arc segments to an existing polyline as new segments. Check the Help facility for additional information.

Getting a Grip

At the beginning of this chapter, I say that I cover only verb-noun editing. Grip editing, however, is — by its nature — noun-verb. You grab the selected object and then perform an action on it.

Grips are little square, rectangular, or other shaped handles that appears on objects after you select them. The shape of the grip indicates its function. In its simplest guise, the square, you can simply click on one of the grips displayed by an object, and then drag the grip to a new location. The object tags along, but how it does so depends on which grip you select — sometimes the object moves, sometimes it stretches.

For lines you can

- Drag one end of a line to change the line's length and ending position.
- Drag the midline grip to move a line without changing its length or angle.

For arcs and circles you can

- Drag the center of an arc or circle to a new location.
- Drag the diameter of a circle to a new size.
- Drag either end of an arc to a new location.
- Drag an arc to a new radius.

For more sophisticated operations, you can

» Type in a coordinate pair to specify the new location.
» Use direct distance entry (DDE, discussed in Chapter 8) to specify the distance and direction of the change.
» Use object snaps to existing objects to specify the new location. You can drag one grip until it's close to a grip on another object and snaps to it, even when object snap modes are turned off. Figure 11-10 shows a *hot* (red) endpoint grip of a line being connected to the endpoint grip of another line. The angled line shows the original position of the line being edited, and the thin vertical line shows the new position. Using a grip in this way as a visible object snap offers the same advantage as using Object Snap overrides (as described in Chapter 7): It ensures precision by making objects meet exactly.
» Select multiple objects. Hold down the Shift key while selecting multiple grips. Each one turns red to indicate that it's hot. When you move one grip, all the hot ones move in parallel with the selected one, as shown in Figure 11-11.

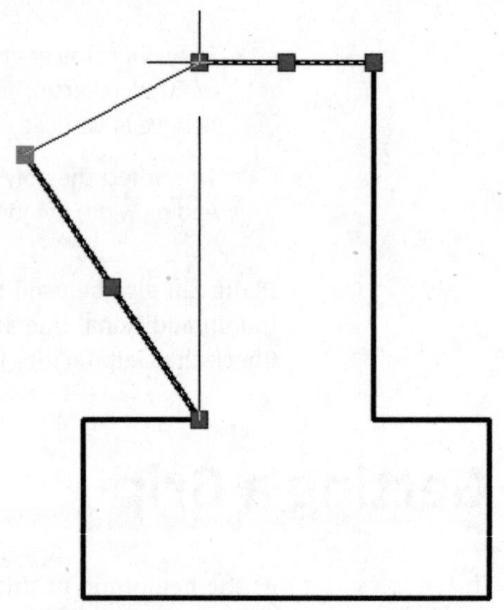

FIGURE 11-10:
Using grips to connect two objects.

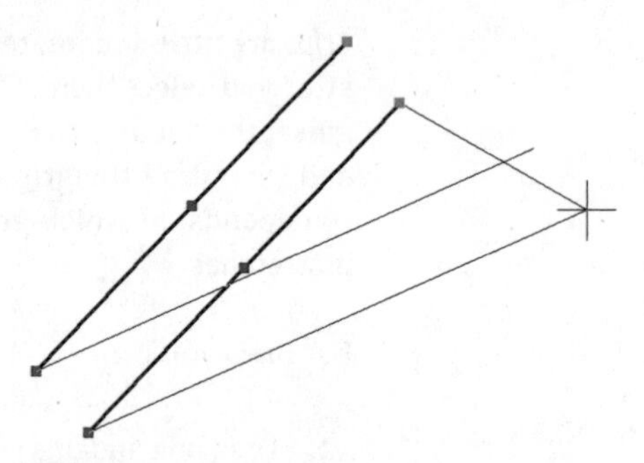

FIGURE 11-11:
Stretching multiple objects with multiple hot grips.

Select one or more objects to activate the grips, and then click a grip. Now take a look at the command line, which displays:

```
** STRETCH **
Specify stretch point or [Base point Copy Undo eXit]:
```

So far, so good. As you move the cursor, AutoCAD displays a ghost image to show you what would happen if you were to click at that moment.

Before you click, however, press the spacebar several times and note the changes to the command line and to the ghost image. The command line scrolls through the full rotation of choices:

```
** STRETCH **
Specify stretch point or [Base point Copy Undo eXit ]:
** MOVE **
Specify move point or [Base point Copy Undo eXit ]:
** ROTATE **
Specify rotation angle or [Base point Copy Undo Reference eXit]:
** SCALE **
Specify scale factor or [Base point Copy Undo Reference eXit]:
** MIRROR **
Specify second point or [Base point Copy Undo eXit]:
```

That's right: The six most-often-used editing options become available in rotation. Wait a minute — only five are in the list! COpy is there, but shown as an option for all of the other five.

Some grips on lines, arcs, and elliptical arcs are multifunctional. Hovering over an endpoint grip displays a pop-up menu that offers you a choice of lengthening or stretching the object.

Hovering over the midpoint grip on an arc offers the choice of stretching the arc by its midpoint (that is, keeping the same endpoints) or changing its radius.

Elliptical arcs display triangular grips at their endpoints that let you increase the length of the arc without changing its other parameters.

Hovering over a rectangular grip on the midpoint of a polyline segment offers the choice of *stretching* (that is, moving the current segment while maintaining its connections to its siblings), adding a new vertex (that is, splitting the current segment in two), or converting a line segment into an arc (or vice versa).

When Editing Goes Bad

Analysis of AutoCAD users' actions suggests that one of the most commonly used functions is undo. It comes in several forms and depths of action. If you're the impatient type, then you can jump ahead to the last two items in this list because they'll probably account for the majority of your undo actions:

- **OOPS:** The OOPS command un-erases the last set of erased objects, even when other editing or drawing actions have taken place in between. This command has no menu option or Ribbon icon, so you have to type it at the command line.
- **UNDO:** The UNDO command is quite versatile and powerful, with several options. Among other things, you can have it undo a specified number of actions, or drop a marker flag and then undo as far back as the marker flag. See the Help facility for more information. You need to type this command on the command line because the only menu option is the typical Windows undo arrow on the Quick Access menu, which undoes just one step and doesn't prompt for the other options.
- **U:** This command is available in two forms:
  - *Type it as a command.* It undoes your last action, and you can use it repeatedly to step back through the current editing session. Just press Enter until you back up to the step you need.
  - *Enter it as a prompt during a number of other commands, such as Line, PLine, and COpy.* In this case, it undoes the last line or polyline segment, or the last objects to be copied — but then the parent command resumes, and you can continue drawing or placing copies.
- **REDO:** This command reverses the last U or UNDO operation. To make REDO work, you must use it immediately after one of these operations. It appears as the typical Windows redo arrow on the Quick Access menu.
- **DOO-DOO:** Or words to that effect. This isn't a command — instead, it's what you say when you undo too many steps and then realize, to your horror, that REDO can only redo the last Undo step.
- **MREDO:** This command reverses multiple Undo actions and pretty much eliminates the need for DOO-DOO. It has several advanced options, though the prompts are self-explanatory.
- **BAK:** This isn't a command — it's a file type. Whenever you save your work in process, AutoCAD creates a file with the extension `.bak` that stores your drawing the way it looked immediately after the previous save operation. You can go back to this version by simply using Windows Explorer to rename the BAK file to one with the `.dwg` extension.

» **SV$:** This one is also a file type. As you work, AutoCAD automatically saves a copy of your drawing every ten minutes and gives it the `.sv$` file extension. You can go back to this version by simply using Windows Explorer to rename the SV$ file to one with the `.dwg` extension. You'll find your SY$ files in the `c:\Users\`*<login name>*`\appdata\local\temp` folder, but you can use the Files tab of the OPtions command to store them in any location you like — as well as change the duration between automatic saves.

» **Undo button:** It invokes the UNDO 1 command and option. The Quick Access menu includes an Undo button, just as many Windows applications do.

» **Redo button:** It invokes the MREDO 1 command and option. The Quick Access menu includes a Redo button, just like many Windows applications do.

Don't you just *wish* that the real world had Undo and Redo buttons?

Dare to Compare

In my engineering classes, I teach a session on document control wherein I emphasize the absolute importance of properly controlling and recording design revisions. Any change, no matter how minor, must be properly recorded. Otherwise, you could have two drawings that say they have the same revision number but are slightly different. You could end up with 52,473 parts that don't quite fit — and you don't want to know how I know that number.

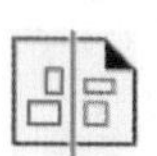

Just in case someone else sends you a suspect drawing file, AutoCAD 2019 introduced the new COMPARE command. It can be started from the big red A (Application) menu's Drawing Utilities, or from the Collaborate tab on the Ribbon menu. It asks you for a drawing to compare with the current one. It then displays something like Figure 11-12.

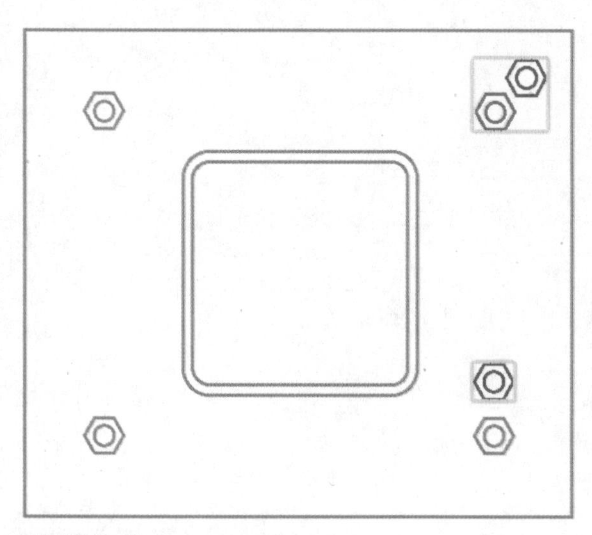

FIGURE 11-12: What's the difference?

In the drawings that I COMPAREd in Figure 11-12, I moved the objects in the upper-right corner and copied the objects in the lower-right corner. The result of a COMPARE operation changes the colors of objects:

- Everything that exists only in the first drawing is red.
- Everything that exists only in the second drawing is green.
- Everything that is exactly identical in both drawings is gray.

Orange revision clouds are placed around the differences. The clouds have a small arc size, so you may have to zoom in to see the arc segments.

AutoCAD 2020 added more functionality. For example, you can import and export changes and differences and you can save a snapshot drawing of the comparison. AutoCAD 2022 added a DWG Compare palette that lets you walk through the differences in the two drawings, one difference at a time, by clicking the forwards and backwards arrows.

The COMPARE functionality could be useful for comparing original design drawings to "as built" revisions returned from the field.

- » **Looking into viewports**
- » **Examining details at other scales**
- » **Working in paper space**

Chapter 12

Planning for Paper

Most of this book revolves around setting up and using the *model space environment* that is infinitely large (or at least 10^{99} units across), a three-dimensional realm in which you create your gleaming towers, tomorrow's wondrous electronic gadgetry — or everyday garden sheds, angle brackets, and doorknobs. You may, however, have picked up a hint here and there that AutoCAD has a whole different parallel universe available to you, known as *paper space.*

The final product of all this setup, remember, is a printed drawing on a piece of paper or in a digital format, such as PDF. In most industries, paper drawings are legal contract documents, so they need to be easy to read and understand. In the first part of the documentation process, you configure the layout of the sheet in paper space, which I explain in this chapter. For the actual process of outputting either model space or paper space layouts to a printer or a file, see Chapter 16.

Chapter 2 introduces you to the two parallel universes of model space and paper space, and Chapter 4 explains how to configure model space for efficient drawing. This chapter explains how to set up layouts in paper space for efficient plotting.

Model space is the drawing environment that's current when the Model tab (not the Model button on the status bar) is active. In model space, you create the "real" objects you're drawing, so these objects are referred to as *model geometry*, whether they're 2D objects or 3D. When the Model tab is active, you see only objects in model space. Anything in paper space is invisible.

Imagine that your screen is displaying your drawing in model space. Now place a sheet of paper in front of your monitor and cut a square hole in it so you can look through it to see all or part of the object's model space. This is like a one-way mirror; you can look through this paper space viewport to see model space, but model space can't look back at you through the hole. You must have at least one paper space viewport in each layout to display anything from model space.

To see what I mean, try this:

1. **Start a new, blank drawing.**

   By default, it starts in model space.

2. **Draw several random lines.**
3. **Click the Layout1 tab in the lower-left corner of the screen.**

   AutoCAD switches to the default layout, which has one viewport into model space. All the lines you just drew fit into the central, rectangular viewport.

4. **Draw a circle in the central, rectangular viewport.**
5. **Click the Model tab in the lower-left corner of the screen.**

   Hey, where did the circle go?

6. **While you're here, draw an ellipse in the general area of the lines.**
7. **Click the Layout1 tab.**

   Like magic, the ellipse appears and the circle reappears.

That's the basic principle of paper space layouts. Everything you draw in model space can be seen from paper space layouts, but nothing you draw in paper space layouts can be seen when you are in model space. This feature is particularly useful for drawing the borders and title blocks in paper space that you need when you plot a drawing, but would only get in the way in model space.

Figure 12-1 shows the arrangement of a default layout tab. It contains the following items:

- The large white rectangular area represents the sheet of paper on which the drawing will be printed.
- The dashed rectangle shows the margin (or boundary) of the actual area which the current printer can print. Anything drawn outside this margin won't print; the margin varies with every printer model.

» The inner rectangle shows one viewport through to model space, which is displaying the grid. There can be more than one viewport, they can be any shape, and they can be located anywhere in paper space.

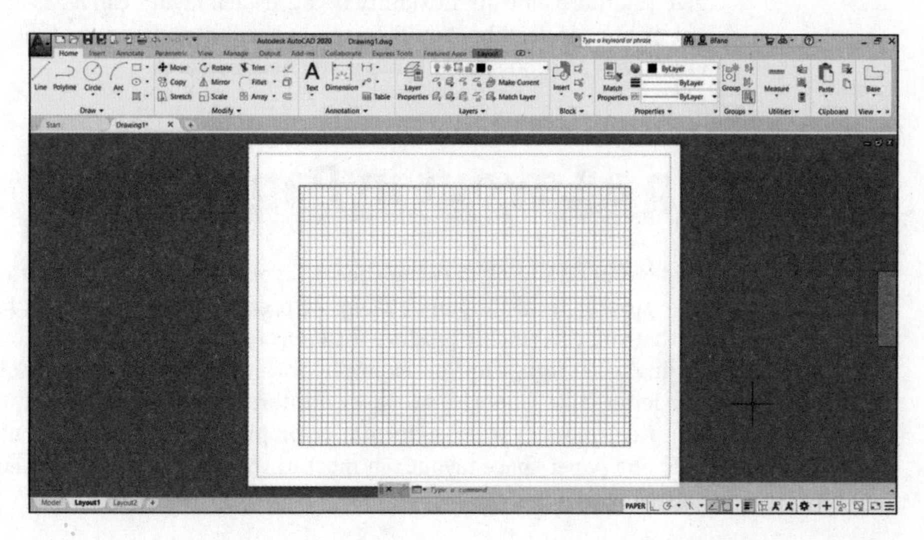

FIGURE 12-1: All spaced out.

I turned on the grid when I did the screen capture to emphasize the fact that it's in model space. As explained in Chapter 3, the grid can be turned on or off, and is never printed.

REMEMBER

An AutoCAD drawing file always contains one and only one model space, but paper space can include more than one layout. For example, a complex drawing might have one layout for a general overview and several other layouts for larger-scale detail views. Each layout can then be plotted on its own sheet of paper. When the model space drawing is edited, all layouts immediately show the changes automatically.

Best practice these days is to draw the real stuff, such as the building, the machine, or the doorknob, in model space and paper-oriented stuff, such as the border, the title block, the view labels, and general notes, in a paper space layout.

As I point out in a number of places in this book, the sacred mantra is "always draw everything full size." The use of paper space layouts greatly simplifies this process. As discussed a little later in this chapter, you can define the scale factor for each viewport, as well as the size of the physical sheet of paper that your printer or plotter produces. Now you can simply plot every paper-space layout as full size and everything will fit.

In AutoCAD, you *can* ignore paper space layouts entirely and do all your drawing *and* plotting in model space. Originally, AutoCAD only had model space, with paper space added to AutoCAD in 1990. But you owe it to yourself to give layouts a try. You'll find that they make plotting more consistent and predictable and they give you more plotting flexibility because each layout can be assigned to a different printer, if need be. You'll certainly encounter drawings that use paper space exclusively, so you should understand both systems.

Setting Up a Layout in Paper Space

Aside from arranging drawing sheets, layouts store plot information. AutoCAD can save several separate plot settings with each layout, as well as for model space, so that you can quickly produce different types of plots for each one, without first having to set things up. For example, with one layout you can do quick check plots to a letter-size monochrome laser printer; with a second layout, you can do full-size final plots to a large-format color plotter. In practice, you'll probably use only one paper space layout tab most of the time, especially when you're getting started with AutoCAD.

The layout two-step

Setting up a layout is a two-step process. The first step is to define the paper, and the second step is to define the viewport or viewports.

Pick a paper, any paper

Defining the paper is a simple process. Follow these steps:

1. **Click the desired layout tab.**
2. **Click the Layout tab of the Ribbon menu.**

   The blue Layout tab is available on the Ribbon only when AutoCAD is in layout mode.

3. **In the Layout panel, click Page Setup.**

   The PAGESETUP command starts, and displays the Page Setup Manager dialog box.
4. **Click New.**

   The New Page Setup dialog box appears.

5. **Type a suitable name and click OK.**

   Next up is the Page Setup dialog box. This dialog box looks remarkably like the Plot dialog box that I discuss in Chapter 16.

6. **Specify the printer you want to use, its paper size, and so on (as per Chapter 16), but leave the Plot Area drop-down list set at Layout.**

   Many of the names in the Printer/Plotter Name list should look familiar because they're the Windows printers (*system printers,* in AutoCAD lingo). Names with the .pc3 extension represent nonsystem printer drivers. See Chapter 16 for details.

7. **Click OK.**

   The Page Setup dialog box closes, and the New Page Setup dialog box returns.

8. **Click Close to apply your new setup to the current layout.**

You can repeat the page setup as often as you like in one drawing. You thus can have multiple layouts or multiple page setups, or both in one drawing. Each layout can have a different page setup, and you can use the PAGESETUP command at any time to switch a given layout to a different setup.

View that port

Now that you've set up the paper properly, it's time to move on to the viewport setup. By default, AutoCAD starts a new drawing with a single viewport each of Layout1 and Layout2, and each viewport has a border that's on layer 0 (zero). A single viewport in a single layout is often appropriate for most drawings. You can add more layouts to the drawing, and each layout can have many viewports.

A viewport border is an actual drawing object. As such, you can move, copy, grip-edit, print, hide, and delete it, and change its properties. You can even use it as the boundary for cross-hatching, but I'm not sure why you would want to.

The default settings are probably not optimal for your needs, so you should follow these steps:

1. **Set the layer.**

   A viewport boundary is a drawing object and so it will print, which you usually don't want. Using Layer Properties Manager, create a new layer, perhaps called VPORTS, and turn off plotting for this new layer. You may also want to change the layer color or line type or both so you can easily identify viewports in drawings. I cover layers in Chapter 9.

2. **Click the viewport object, and then click your VPORTS layer name in the drop-down list below Layers in the Layers panel of the Home tab.**

   The viewport object remains visible, but now it won't plot.

3. **Set the viewport scale.**

   When you did the simple exercise near the beginning of this chapter, did you notice that the first time you clicked the Layout 1 tab, everything you drew in model space just happened to exactly fit within the viewport? To make this happen, AutoCAD automatically adjusted the viewport scale factor. The problem is that it almost never turns out to be a standard scale that anyone would use. For example, when I did the exercise, the scale factor turned out to be 0.11036728. A more logical scale would be 0.1, or 1:10. No problem:

   a. *Click the viewport boundary.*

   1:1 / 100%

   b. *Click the Viewport Scale button, which appears near the right end of the status bar.*

   c. *Select a suitable scale from the list of standard scales that appears.* The list shows every drawing scale registered in the scales list, including metric scales, even if you're working in a drawing using English units and vice versa.

TIP

Most of the time, way too many scales are shown in the list you see in the Viewport Scale button and in the Plot dialog box. AutoCAD has a handy-dandy Edit Drawing Scales dialog box that lets you remove imperial scales if you never work with feet and inches, and vice versa if you never work in metric. To run through the scales, choose Scale List from the Annotation Scaling panel on the Annotate tab, or type **SCALELISTEDIT** and press Enter to open the Edit Drawing Scales dialog box. If (okay — *when*) you make a mistake, click the Add button in the Edit Drawing Scales dialog box to add a lost scale factor, or click the Reset button to restore all default scales.

4. **Lock the viewport scale.**

   When a viewport is selected, a padlock icon appears to the left of the Viewport Scale button. The icon is a toggle that turns viewport scale locking on or off.

WARNING

Always lock the viewport scale immediately after setting it. You'll see why later in this chapter when I discuss details at different scales.

TIP

Practice playing with the paper space and model space layouts in AutoCAD's sample drawings. I don't cover sheet sets in this book, but the individual drawings in the sheet sets all use paper space layouts along with model space. For example, from AutoCAD's Start screen, click the arrow next to Open, and then select Explore Sample Drawings. When the file dialog box opens, click Sheet Sets, and then click Civil, and then click Site Grading Plan. Click Yes in the alert box that appears to see a drawing that has four paper space layouts to print four views of the model space drawing.

Put it on my tabs

The Autodesk documentation sometimes refers to the *Model tab* or to *layout tabs*, as I sometimes do. Out of the box, AutoCAD displays selectable tabs at the lower-left edge of the drawing window, clearly labeled Model, Layout1, and Layout2. Hovering your cursor over a tab pops up a preview of its contents. You can jump to a desired view just by clicking its preview image.

TIP

AutoCAD 2014 added file tabs at the top left of the graphic screen for each open drawing. Hovering the cursor over a file tab also produces quick views of model space and all layouts in the drawing. Figure 12-2 shows a typical arrangement in which AutoCAD has several drawings open. When you hover the cursor over the current file tab, a blue border indicates the currently active view. You can jump from view to view in a drawing or directly to a desired view in a different drawing just by clicking its preview image.

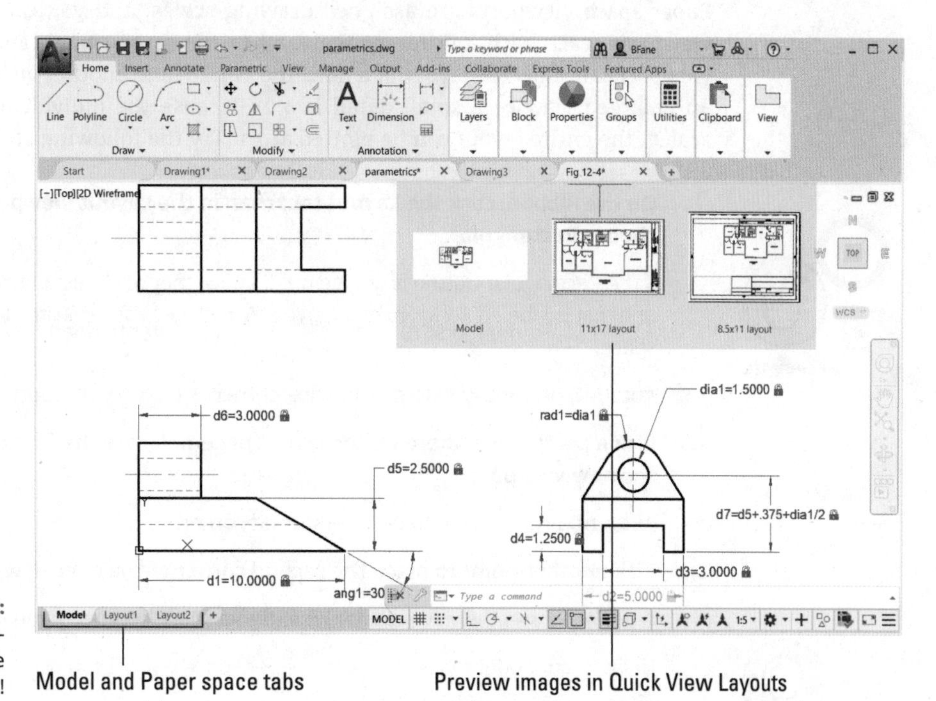

FIGURE 12-2: View those layouts, and make it quick!

TIP

You can rename a layout tab (but not the Model tab) by double-clicking the layout name and typing a new one.

Any Old Viewport in a Layout

TECHNICAL STUFF

The viewports I talk about in this chapter are paper space viewports. You can also create viewports in model space, but they're a completely different animal. Model space viewports are also known as *tiled* viewports because (like bathroom tiles) they can't overlap. You can use tiled viewports to take a close look at widely separated areas of the screen or different viewports of a 3D model. I cover tiled model space viewports in Chapter 22 when I discuss 3D modeling. The potentially confusing aspect is that AutoCAD uses the same command name, and even the same dialog box, for creating the two different types of viewports. In this chapter, ensure that you're in paper space when you create viewports.

Up and down the detail viewport scales

Paper space viewports are assigned drawing scales, and you can have multiple viewports, each with a different scale, all in the same layout. For example, one viewport can show the layout of a machine at 1:20, and another viewport can show an enlarged view of a small detail at 1:2. Because the individual viewports are scaled, the entire layout can be plotted at 1:1. Try the following steps:

1. **On the Ribbon, click the Layout tab; then in the Layout Viewports panel, choose Rectangular.**

   REMEMBER

   If the Rectangular button is grayed out (I know, they're *all* rectangular — it's the one that is labeled *Rectangular*), you're still in model space. Switch to paper space.

   AutoCAD prompts you to pick the first corner of the new viewport.

2. **Pick a point somewhere on the layout page to locate the first corner of the new viewport.**

   AutoCAD prompts you to pick the second corner.

3. **Pick another point to place the second corner of the new viewport.**

   AutoCAD draws the viewport, and the model space geometry appears inside it.

   TIP

   Unlike tiled model space viewports, paper space layouts can overlap each other or have spaces between them.

4. **Zoom and pan until the new viewport shows the desired region of model space.**

   When you zoom and pan in paper space, you simply zoom and pan the entire layout. If you zoom and pan in model space, it has no effect on the paper space viewport.

TIP

Okay, here's the clever part. While you're in paper space, double-click inside the viewport. You have now reached through the hole in the time-space continuum and are working in the parallel universe of model space. Any panning and zooming you do here *does* affect what's visible in the layout viewport. While you're in the parallel universe, you may want to look for your missing left socks. When things are to your liking, return to the current layout by double-clicking anywhere in paper space outside the viewport.

5. **Click the viewport boundary to which you want to apply a scale.**

   The Viewport Scale button appears toward the right end of the status bar.

6. **Find the scale that you want to apply to the active viewport, and then select it from the list.**

   The display zooms in or out automatically to adjust to the chosen viewport scale.

7. **Lock the viewport when the scale is correct.**

   Click the padlock icon, which appears to the left of the Viewport Scale button near the left end of the status bar, to toggle viewport scale locking on or off.

WARNING

Okay, here's the warning I warned you about earlier in the chapter. When you don't lock the viewport scale, others (not you, of course) can mess up the viewport scale factor in all innocence, thinking they're just zooming in for a closer look.

TIP

You can zoom directly to an exact scale factor while reaching into model space through a viewport. Start the Zoom command, and then enter the desired scale factor immediately followed by the letters **XP** ("times paper," not case sensitive). For example, **Z 2XP** will zoom in 2:1 relative to paper scale, and **Z .5XP** will zoom out 1:2. Now return to paper space and lock the viewport.

TIP

When a viewport is locked, you can still double-click in it to enter model space and create or edit objects that are visible in the viewport. Panning and zooming do not affect what appears in the viewport but do affect the entire layout, as though you had momentarily slipped back into paper space.

REMEMBER

Viewport boundaries have most of the properties and capabilities of other drawing objects. You can stretch, move, copy, delete, and so on, even when the viewport is locked. Grip editing, described in Chapter 11, can be particularly useful.

Earlier in this chapter, I say that viewport boundaries should normally be placed on their own non-plotting layer. That is generally true for the main viewport that shows all or most of the objects in model space. Standard practice, however, is for details at different scales to display their viewport boundaries, as shown in Figure 12-3. The solution is to create two layers for viewports, one that plots and one that doesn't.

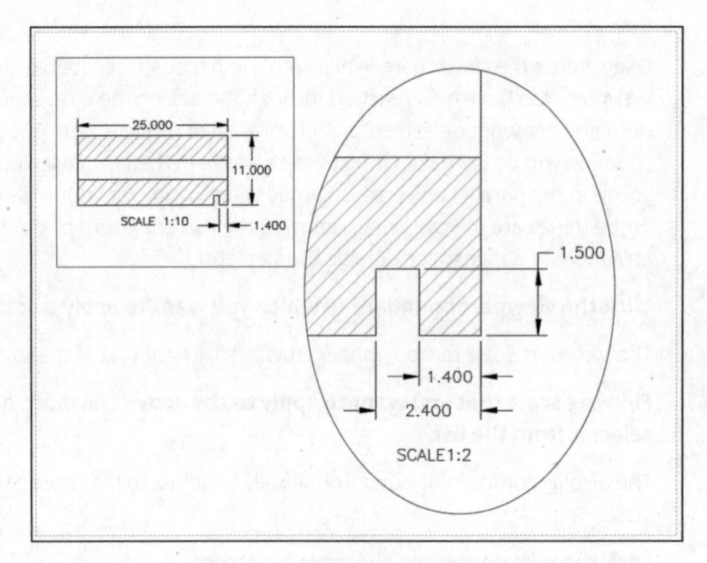

FIGURE 12-3: Show me the viewport!

Keeping track of where you're at

When you start working in layouts, it may not always be crystal-clear whether you're in model space or paper space. The status bar button helps because it says *PAPER* when you're in paper space or *MODEL* when you're in model space. You can tell the layout spaces apart in a few other ways:

- **Check the cursor.** If you're in paper space, you can move the cursor over the entire drawing area. If you're in model space by reaching through a layout viewport, you can move the cursor only in the active viewport; when you try to move the crosshair cursor outside the viewport, it turns into the Windows arrow cursor.
- **Select some model geometry.** Try clicking some objects that you know are in model space. If you can select and highlight them, you're in model space. If nothing happens when you click them, they're inaccessible because you're in paper space.
- **Check the UCS icon.** The User Coordinate System (UCS) icon is the symbol at the lower-left corner of the drawing area. The model space icon shows two lines at right angles to each other, with the letters indicating the direction of the X-axis and the Y-axis. The paper space icon is triangular — its closed, three-sided shape represents a flat plane. If you don't see this symbol, type **UCSICON** and press Enter, and then type either **ON** (to display the UCS icon in the lower-left corner of the display) or **OR** (to display the icon at the drawing's origin — its 0,0 coordinates). I explain more about user coordinate systems in Chapter 8.

Practice Makes Perfect

Best practice when creating new drawings is to use paper space layouts with viewports. This greatly simplifies the drawing scale problem and related plotting issues. In particular, note the following:

- **Model space:**
  - Here you draw everything full size.
  - You must scale plots (Chapter 16), notes (Chapter 13), dimensions (Chapter 14), crosshatching (Chapter 15), and non-continuous line types (Chapter 9) separately.
- **Paper space:**
  - Here you draw everything that isn't part of the drawing or model, such as the drawing border, title block, and detail views.
  - You configure layouts with appropriate plotter, printer, and sheet size specifications, and create appropriately scaled viewports.
  - Layouts are plotted at full size (1:1).

Clever Paper Space Tricks

The following list shows a few useful tips and tricks that will help you become a paper space guru:

- **Create additional layouts as necessary.** You can do this in several ways, including using the New button in the Layout tab of the Ribbon menu or simply clicking the + (plus) sign to the right of the rightmost layout tab.
- **Delete unused layouts.** Simply right-click the undesired layout tab, and then click Delete in the context menu that appears.

TIP

- **Viewports don't have to be rectangular.** The Ribbon's Layout tab includes three options in the Layout Viewports tab, but two will be hidden under the last one used. The options are Rectangular, Polygon (you pick a series of points that lasso the desired region), and Object, which lets you choose any existing closed shape and convert it to a viewport. That's how I created the elliptical detail in Figure 12-3.

» **Freeze layers in individual viewports.** You don't always want to show everything in every viewport. If you enter model space by double-clicking inside a viewport, Layer Properties Manager (see Chapter 9) displays extra columns that let you freeze selected layers in the current viewport. Conversely, the drop-down list below the Layers panel of the Home tab includes a command to freeze the layer of the selected object in all other viewports except the current one.

» **Snap into model space.** Even when you are in paper space, it's still possible to snap onto objects in model space that are visible in a viewport. This technique was used mostly before self-scaling annotations came along. Dimensions could be applied from paper space without having to worry about scaling them. The problem is that it's usually more helpful to have the dimensions in model space when you are working there.

» **Use templates.** Ah, now here's the biggie. I refer to template files in several chapters, but a quick refresher wouldn't hurt. A template file serves as the starting point for a new drawing and can contain anything that a normal drawing file can contain. This includes *everything* discussed in this chapter: layouts, viewports, plotting configurations, borders, title blocks, and so on. John Walker, one of the founders of Autodesk, is rumored to have once said that an AutoCAD user should never have to do anything twice. I don't think he said it again.

But wait! There's more! A bit of exploring will reveal that the commands for creating and configuring layouts and viewports all have a From Template option. You can browse through the current drawing and through *any* existing drawing or template file *anywhere* to inhale the complete layout setup specifications. *Complete* means everything; for example, when you're setting up a layout from a template and the source includes viewports, a border, and a title block, it all comes over to the new layout, including the creation of any new layer, text, and dimension specifications and block definitions required.

WARNING

If the current drawing contains any specifications with the same names as the source, the host drawing wins and the incoming specifications are ignored. The existing specifications will be applied to the incoming objects.

3

If Drawings Could Talk

IN THIS PART . . .

Place text, numbered and bulleted lists, and tables in your drawings.

Include text labels — dimensions — in your drawings.

Add hatches to your drawings.

Print — or plot, as AutoCAD aficionados call it — your drawings.

- Employing text styles to control text appearance
- Placing single-line and multiline text
- Using fields and background masks with text
- Making numbered and bulleted lists
- Working with columns and paragraphs
- Editing text contents and properties
- Creating tables
- Drawing multileaders

Chapter 13
Text with Character

It has been said that "a word is worth one-one-thousandths of a picture." On the other hand, the opposite is often true as well. Adding a few words to your drawing can save you from having to draw thousands of lines and arcs. It's a lot easier to write `Simpson A35 framing clip` next to a simple, schematic representation of the clip than to draw one in excruciating detail and hope that the contractor can figure out what it is.

Most CAD drawings include text in the form of explanatory notes, object labels, and titles. This chapter shows you how to add general drawing text and *leaders* (descriptive notes with arrows that point to specific drawing objects) to your drawings. In this chapter, I show you how to take advantage of AutoCAD's annotative text objects and text styles, how to find specific text, and how to check for speling erors. Chapter 14 covers text that's connected with dimensions.

In most cases, adding text, dimensions, and other descriptive symbols is a task that you should do later in the drafting process, after you've drawn at least some of the geometry. In CAD drawings, text and other annotations are usually intended to complement the geometry, not to stand alone. You generally need to have the geometry in place before you annotate it. Many drafters find that it's most efficient to draw as much geometry as possible first, and then add text annotations and dimensions to all the geometry at the same time. In this way, you develop a rhythm with the text and dimensioning commands instead of bouncing back and forth between drawing geometry and adding annotations.

As I mention in Chapter 4, AutoCAD's *annotative objects* present a streamlined way to add notes, dimensions, and other annotations to drawings. This chapter introduces annotative text, and subsequent chapters cover annotative dimensions and hatches. See the "Annotatively yours" sidebar later in this chapter for some background.

Getting Ready to Write

In AutoCAD, adding text to a drawing is similar to adding it to a word processing document. Here are the basic steps, which I explain in more detail in the sections that follow:

1. **Select an existing AutoCAD text style, or create a new style that includes the font and other text characteristics you want to use.**

   Just like a word processor, AutoCAD uses *styles* — collections of formatting properties — to control the appearance of drawing text. I explain text styles in the next section.

2. **Make an appropriate layer current.**

   To make your AutoCAD drawing efficient and easy to edit for you and others, create text on its own layer. Most drafting offices already have a set of CAD standards that establish specific layer names for text and other object types.

3. **Run *one* of these commands to draw text:**

   - *TEXT:* Places single lines text at a time
   - *mText:* Places paragraphs of text, also called *multiline* text

4. **Specify the text alignment points, justification, and (if necessary) height.**
5. **Type the text.**
6. **(Optional) For annotative text, assign additional annotation scales to the text you just typed, if desired.**

In the next few sections of this chapter, I review the particularities of AutoCAD text styles, the two kinds of AutoCAD text, and ways to control height and justification.

ANNOTATIVELY YOURS

One helpful aspect of AutoCAD is that it offers multiple ways to accomplish drafting tasks. This usually comes about because successive releases of AutoCAD add new or improved methods of performing existing tasks. One less-helpful aspect of AutoCAD is that it offers multiple ways to accomplish drafting tasks. You're almost always better off adopting the newer methods, but in this book I give you a taste of the older methods because literally billions of existing drawings were created using the older methods — and you're bound to encounter them sooner rather than later in your career. Text and dimensions are prime examples. I cover text in this chapter, and dimensions in Chapter 14.

AutoCAD supports these three methods for placing text at the correct size in drawings:

- **Adding text and dimensions in model space through multiplying the drawing scale factor times the desired plotted text height.** For example, suppose that you have drawn a floor plan in full size in model space, following mandatory AutoCAD practice. Now you want to plot it at a scale of `1/4" = 1'-0"` (corresponding to a drawing scale factor of 1:48) so that it fits on the desired sheet of paper. You want your notes to appear 1/8-inch high when the drawing is plotted to scale, so you have to create text that's 48 times 1/8-inch, or 6 inches, high. And that works fine — until you want to plot the drawing at a different scale or create a detail at a different scale. This method was the only one available on older drawings.

  This calculation is particularly cumbersome if you're working with feet, inches, and fractions, but it's a bit easier to complete when you're using decimals of inches or metric units.

- **Adding text and dimensions to a paper space layout.** I cover paper space layouts in Chapter 12. Because paper space is plotted at 1:1, you create annotations at their actual sizes when plotted. The scale calculations are simple ("Hey, boss, what's 1/8-inch times 1?" "I don't know; my calculator's broken."). It's sometimes beneficial to have drawing annotations in the same space as the drawing geometry, and therefore place text in model space. A potential problem is that you might have to create duplicate notes in each layout or view because you can create multiple layouts in paper space and multiple views in a layout.

(continued)

(continued)

- **Adding text and dimensions in model space by using annotative text and dimension styles.** This method is similar to the one described in the first bullet, except that AutoCAD automatically completes the size calculations for you based on a desired plotted size and specified annotation scale that you specify ahead of time.

Using annotative objects makes the most sense because they can cover all plot scales automatically, and they don't have to be duplicated on multiple layouts because they're in model space. One tricky concept to grasp in AutoCAD is the necessity of scaling text and dimensions to apparently ludicrous sizes (6-inch-high text, for example) so that they plot correctly; annotative objects do the scale calculations for you. Better yet, details at different scales are handled almost automatically.

Annotative scaling can be applied to objects like single-line and multiline text (covered in the sections "Using the Same Old Line" and "Saying More in Multiline Text," respectively, later in this chapter), leaders (covered in the later section "Take Me to Your Leader"), dimensions (covered in Chapter 14), hatches (see Chapter 15), and even blocks and attributes (see Chapter 17).

Creating Simply Stylish Text

AutoCAD assigns properties to text based on *text styles*. Text styles in AutoCAD are similar to the paragraph styles in a word processor: They contain font settings and others that determine the look and feel of text. An AutoCAD text style includes

- The name of the **font**, or basic shape style of the characters
- A **text height**, which you can set to a specific value or leave at 0 for later flexibility
- **Special effects** that depend on the font, such as *italic*, bold, and ***bold italic***
- ***Truly* special effects,** such as vertical, backwards, and upside down, that have highly specialized applications

Before you add text to a drawing, use the Text Style dialog box to select an existing style or create a new one with settings that are appropriate to your purpose.

TIP

You can retroactively assign the annotative property to text styles in old drawings by opening the Text Style dialog box in the drawing, choosing the individual text style(s) you want to update, selecting the Annotative check box, and then saving the drawing.

Most drawings require only a few text styles. You can create one style for all notes, object labels, and annotations, and then create another text style for special titles. You may also want to create a unique text style for dimensions. See Chapter 14 for more on dimension text. A title block may require one or two additional fonts, especially if you want to mimic the font used in a company, client, or project logo.

As with layers, your office should have standards for text styles. If so, you'll make everyone happy by following those standards. One of the best ways to make sure your use of text styles is efficient and consistent is to organize them into a template drawing file with which you start each new drawing. When your office is well organized, it should already have one or more template drawings with company-approved styles defined in them. See Chapter 4 for information about creating and using templates. Another technique is to copy existing text styles from one drawing to another by using the DesignCenter palette, but this strategy isn't always desirable because it doesn't enforce conformity.

Font follies

When you create a text style in AutoCAD, you can choose from a huge number of fonts. AutoCAD can use two different kinds of fonts: native SHX (compiled shape) fonts that come with AutoCAD and all of the TrueType fonts that come with Windows:

- **SHX:** In the Text Style dialog box, SHX font names appear with a drafting compass to the left of the name and display the `.shx` file extension. SHX fonts usually provide better performance because they're optimized for AutoCAD's use but are not filled and can look rough. SHX fonts are the only fonts you will find in very old drawings created before TrueType came along.
- **TrueType:** In the Text Style dialog box, TrueType font names appear with a TT symbol to the left of the name and no file extension. TrueType fonts give you more and fancier font options, but they can slow down AutoCAD when you zoom, pan, or select and snap to objects. TrueType fonts also can cause greater complications when you exchange drawings with other AutoCAD users, because their computers might not carry the same set of fonts.

AutoCAD defaults to using Arial, a TrueType font. SHX and TrueType fonts are available also with specialized mathematical, astronomical, and astrological characters.

Avoid complicated fonts. They can slow down AutoCAD, and they're usually more difficult to read than simpler fonts. Remember that it's CAD; clarity is needed, not fancy graphical designs or reproductions of medieval manuscripts! Unless, of course, that is exactly what you are doing. The best advice is, however, to stick with whatever font is standard for your office, your company, or your clients.

WARNING

Whenever possible, avoid fonts not supplied with AutoCAD or Windows. AutoCAD doesn't embed font files in drawings, and so must refer to font files installed locally by AutoCAD (for SHX fonts) or Windows (for TrueType fonts). If you use a custom font of either type, exchanging drawings with other people becomes more complicated. Among other things, there may be copyright issues if you send a copy of a purchased font to someone else. It's far less hassle to avoid custom fonts altogether. See Chapter 20 for additional information about how to deal with fonts when you send and receive drawings.

Get in style

The steps in this section describe how to create a new text style for text in drawings. When you want to experiment with an existing drawing that contains a variety of text styles, download the AutoCAD sample drawings from `www.autodesk.com/autocad-samples`. Look for the `architectural_-_annotation_scaling_and_multileaders.dwg` file.

Commonly, you create a text style once in a drawing template so it can be used by all future drawings. After a text style is created but before you add text to a drawing, you set the text style that you want to use with the Text Style drop-down list on the Text panel of the Annotate tab on the Ribbon.

WARNING

The Text panel on the Annotate tab provides lots of text-related functions, whereas the Text button on the Annotation panel of the Home tab only places text.

1. **Start the STyle command: On the Ribbon's Home tab, click the Annotation panel's label to open the slideout and then click the Text Style button.**

   The Text Style dialog box appears, as shown in Figure 13-1.

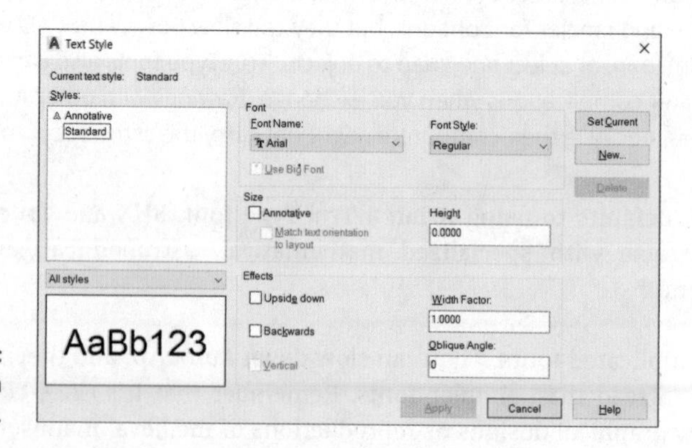

FIGURE 13-1: Text — with style.

2. **From the Styles list, select each style in turn to examine the properties of the text styles that have been created in this drawing.**

   Note the font name and look at the Preview panel in the lower-left corner of the dialog box to get a feel for what the different fonts look like.

3. **Click the New button to create a new style.**

REMEMBER

   Pay attention to the style that's current when you click New. If it's an annotative style, the new style is also annotative; if the current style isn't annotative, the new style isn't, either. It's easy enough to select or deselect the Annotative check box, but you may overlook this action at first.

   The New Text Style dialog box appears, with a text box for you to type a name.

4. **Type a name for the new text style and then click OK.**

   The new text style is added to the Styles list and becomes the current style.

5. **Choose a font from the Font Name list.**

   The best all-purpose font for most drafting work is `Romans.shx`. When you want to use a different font, review the font suggestions and warnings in the preceding section.

   The font that you choose becomes the font that's assigned to the new text style.

6. **To create an annotative text style, select the Annotative check box. Deselect the check box for non-annotative text.**

TIP

   Annotative is almost always the best choice for new drawings because it automates the process of setting the text height.

   See the "Annotatively yours" sidebar, earlier in this chapter, to find out more.

7. **Adjust the remaining text style settings.**

   The text style shown earlier (refer to Figure 13-1) has the following setup:

   - *Paper Text Height* (just *Height* for non-annotative styles) = `0.0`
   - *Width Factor* = `1.0`
   - *Oblique Angle* = `0`
   - All check boxes other than Annotative are deselected.

   A text style height of `0.0` makes the style of *variable height,* which means that you have to specify the height separately for each text object as it's created. Assigning a *fixed* (that is, nonzero) height to a text style forces all text that uses the style to be the same height. Variable height styles are more flexible, but fixed height styles usually make it easier to draw text of consistent height. The decision to use styles of variable height versus fixed height is another aspect of text that depends on office practice; when you work with other AutoCAD users, ask around.

TIP

Most drawings require only two or three text sizes, such as a Normal (1/8-inch) for general notes, Larger (1/4-inch) for view and detail labels, and Large (1/2-inch) for section line identifiers. The best way is to create several styles with fixed heights.

REMEMBER

In certain cases, the fixed height value isn't the final text size. It may need to be corrected by the drawing scale factor, as covered in the section "Calculating non-annotative AutoCAD text height," later in this chapter.

WARNING

Dimensions also use text styles to format the appearance of the dimension text. When you create a text style that you think you might use for your dimensions, you *must* set the height to `0`. Otherwise, the settings that control dimension text won't work, and the dimension text is likely to be either enormous or microscopic. This paragraph should have an extra Warning icon because this is one of the most common mistakes made by new AutoCAD drafters. The best practice is to create a Dimensions text style.

8. **Click Apply, and then click Close.**

   The Text Style dialog box closes, and the text style you selected or created is now the current style for new text objects.

Taking Your Text to New Heights

REMEMBER

Drawing scale is the traditional way to describe a scale. It uses an equal sign or colon — for example, `1/4" = 1'-0"`, 1:20, or 2:1. The *scale factor* represents the inverse relationship with a single number, 48, 20, or 0.5, respectively. The *drawing scale factor* is the multiplier that converts the first number in the drawing scale into the second number.

WARNING

You *must* determine the scale factor of a drawing before you add non-annotative text to it. This scale factor most often relates to the size of paper on which the drawing will be printed.

Plotted text height

Most industries have plotted text height standards, which AutoCAD refers to as *paper* text height. A plotted text height of 1/8″ or 3mm is common for notes. Some companies use slightly smaller heights (for example, 3/32″ or 2.5mm) to squeeze more text into small spaces.

Calculating non-annotative AutoCAD text height

To calculate non-annotative text height, you need to know the drawing scale factor, the desired plotted text height, and the location of the multiplication button on your calculator. Follow these steps to calculate the text height:

1. **Determine the drawing scale factor.**

   Drawing scale is based on the size of the object you're drawing compared to the size of the sheet of paper on which it will be printed. For example, a typical house floor plan would have to be scaled down 1:96 (1/8"=1') to fit on an A-size sheet, whereas a tiny watch gear would probably have to be scaled up by 50:1 to appear as more than a dot in the middle of the sheet. As noted, the drawing scale factor is the inverse of the drawing scale: 96 for the house plan, and 1/20 or 0.05 for the watch gear.

   WARNING

   You shouldn't select a scale factor at random. Different industries have lists of "preferred" (which you should read as "mandatory") scale factors, and individual companies in turn may have a short list in the preferred values.

   TIP

   When you edit an existing drawing, it should have a bar scale or text note indicating the drawing scale. If not, and if the drawing dimensions are in model space, you can check the value of the DIMSCALE variable (the system variable that controls dimension scale), as described in Chapter 14, or work backward from existing text sizes.

2. **Determine the height at which notes should appear when you plot the drawing to scale.**

   See the earlier section "Plotted text height" for suggestions.

3. **Multiply the numbers resulting from Steps 1 and 2.**

   After you know the AutoCAD text height, you can use it to define the height of a text style or of an individual text object.

If you assign a nonzero height to a text style (refer to Step 7 in the earlier section "Get in style"), all text that you create in that style uses the fixed height. If you leave the text style's height set to 0 (zero), AutoCAD asks you for the text height every time you draw single-line text objects, which soon becomes a real nuisance.

TIP

This discussion of text height assumes that you're adding non-annotative text in model space. In addition to annotative text in model space, you have a third alternative: Add annotative or non-annotative text to a paper space layout — for example, when you draw text in a title block or add a set of sheet notes that doesn't directly relate to the model space geometry. When you create text in paper space, you specify the actual, plotted paper height instead of the scaled-up height. (I cover paper space in Chapter 12.)

Entering Text

AutoCAD offers two different kinds of text objects and two corresponding text-drawing commands:

» **Single-line text:** TEXT (DT) places one line of text at a time. Although you can press Enter to place more than one line of text, each line becomes a separate text object. Text doesn't wrap if you add or remove anything.

» **Paragraph text:** MTEXT (T) places paragraphs of text, with word-wrapping as necessary. AutoCAD groups the paragraph(s) as a single object. Other special formatting, such as numbered and bulleted lists and multiple columns, is possible. The mText command basically launches a compact word processor.

Although you may be inclined to ignore the older, single-line text option, you should know how to use both kinds of text. The TEXT command is quite a bit simpler to use than the mText command, and it's still useful for quickly entering short, single-line pieces of text such as object labels and one-line notes all over the drawing. It's also the command of choice for CAD comedians who want to document their one-liners!

Both the TEXT and mText commands offer a full array of text *justification* options — the way the text flows from the point or points that you pick in the drawing to locate the text. For most purposes, the defaults are left justification (for single-line text) and top-left justification (for paragraph text), and they work just fine. Occasionally, you may want to use a different type of justification, such as center, for labels or titles. Both commands provide options for setting the text justification.

Using the Same Old Line

Despite its limitations, the TEXT command is useful for labels and other short notes for which mText (multiline) would be overkill. The following steps show you how to add text to a drawing by using AutoCAD's TEXT command.

The TEXT command doesn't use a dialog box, fancy formatting toolbar, or contextual Ribbon tab, like the mText command's In-Place Text Editor. You set options by typing them in the command line or in the Dynamic Input tooltip.

Here's how to add text by using the TEXT command:

1. **Set an appropriate non-annotative text style current, as described in the section "Creating Simply Stylish Text," earlier in this chapter.**

   It's possible to set an already created text style current at the TEXT command prompt, but it's usually more straightforward to set the style before starting the command.

   TIP

   An alternative to making an existing style current by opening the Text Style dialog box is to click the Text Style drop-down list and choose the style there. Look for the Text Style drop-down list on the Annotation panel's slideout (on the Home tab) or on the Text panel of the Annotate tab.

   When you know the name of the text style that you want to use, begin typing it at the command line. The AutoCAD command line searches for text styles and reports it like "TextStyle: Simplex." When you see the one you want, just select it.

2. **(Optional) Use the Object Snap button on the status bar to enable or disable running object snaps.**

   You may or may not want to snap text to existing objects. For example, you'd want to use a CENter object snap to locate a letter or number precisely at the center of a circle. Make sure that you specify middle-center (MC) text justification to ensure that the object snap works perfectly.

3. **On the Home tab's Annotation panel, click the lower part of the big button labeled *Text,* and then choose Single Line from the drop-down menu to start the TEXT command.**

   Don't click the upper part of the Text button. That action starts the multiline text command, mText, which I cover in the later section, "Saying More in Multiline Text."

   When your text style is annotative and this annotative object is the first one you're creating in this drawing session, AutoCAD usually displays the Select Annotation Scale dialog box, which advises you that you are indeed creating an annotative object and asks you to set the scale at which you want the annotation to appear; click OK to continue. Later, I explain how and why to turn off this annoying dialog box.

   AutoCAD tells you the current text style and height settings and then prompts you to either select a starting point for the text or choose an option for changing the text justification or current text style first:

   ```
   Current text style: "Standard" Text height: 0.2000
     Annotative: No
   Specify start point of text or [Justify Style]:
   ```

4. **If you want a justification style different from the default (left), type J, press Enter, and choose another justification option.**

   Look up the term *create single-line text* in the online Help system if you need help with the justification options.

5. **Specify the insertion point for the first text character.**

   You can enter the point's coordinates from the keyboard, use the mouse to click a point onscreen, or press Enter to locate new text immediately below the most recent single-line text object that you created.

   AutoCAD prompts you for the text height:

   ```
   Specify height <0.2000>:
   ```

6. **Specify the height of the text.**

REMEMBER

   The text height prompt doesn't appear if you're using a text style with a *fixed* (that is, nonzero) height. See the "Creating Simply Stylish Text" section, earlier in this chapter, for information about fixed-versus-variable text heights. Now here's the nuisance: If you haven't specified a fixed height for the current text style, you'll be asked for the text height every time you create text, and you'll have to enter the correctly scaled value to get things to come out right.

   AutoCAD prompts you for the text rotation angle:

   ```
   Specify rotation angle of text <0>:
   ```

7. **Specify the text rotation angle by typing the rotation angle and pressing Enter or by rotating the line onscreen with the mouse.**

   For example, enter **90** to rotate the text so that it reads sideways. AutoCAD prompts you to type the text.

8. **Type the first line of text and press Enter.**

9. **(Optional) Type additional lines of text, pressing Enter at the end of each line.**

   Each line of text is neatly aligned below the previous line and is a separate, independent object and cannot be edited as a single paragraph, even though it looks like it. If you want a single paragraph, use the mText command.

10. **To stop entering text and return to the command line, press Enter at the start of a blank line.**

    AutoCAD adds to the drawing the new, single-line text object — or objects, if you typed more than one line.

The TEXT command remembers its last-used settings (style, justification, height, and rotation angle), so you don't need to reapply them every time.

TIP

If your text will occupy more than one line, it is better to use mText, covered later in the chapter.

To edit single-line text, double-click it to open the In-Place Text Editor.

The in-place editing box highlights the selected text object, enabling you to edit the contents of the text string. When you want to edit other text properties such as text height, select the text, right-click, and choose Properties to display the Properties palette. Use this palette to change properties as needed.

TIP

If the Quick Properties button on the status bar is enabled, clicking a single-line text object opens the Quick Properties panel, allowing you to change some (but not all) of the same properties as you can on the Properties palette.

Saying More in Multiline Text

When you just can't shoehorn your creative genius into one or more one-line pieces of text, AutoCAD's multiline text object gives you room to go on and on and on. The following sections show you how to create multiline text with the mText command.

Making it with mText

The first part of the mText command prompts you for various points and options. Read the prompts carefully to avoid confusion.

Here's how to use the mText command:

1. **Set an appropriate text style current, and (optional) turn off running object snaps, as described in Steps 1 and 2 in the "Using the Same Old Line" section, earlier in this chapter.**

   Make sure to also set an appropriate layer current before creating the text object.

2. **On the Home tab's Annotation panel, click the upper part of the split button, labeled Text, to start the mText command.**

   The command line displays the current text style and height settings, and then prompts you to select the first corner of a rectangle that determines the word-wrapping width of the text object:

   ```
   Current text style: "Standard" Text height: 0.2000
     Annotative: No
   Specify first corner:
   ```

3. **Pick a point in the drawing.**

   The command line prompts you for the opposite corner of a rectangle that will determine the word-wrapping width; it also gives you the option to change settings first:

   ```
   Specify opposite corner or [Height Justify Line spacing
     Rotation Style Width Columns]:
   ```

4. **To change the default text height, type H and press Enter.**

   The command line prompts you for a new default text height if the current text style has a height of 0.0:

   ```
   Specify height <0.2000>:
   ```

5. **If applicable, type an appropriate text height.**

   See the "Taking Your Text to New Heights" section, earlier in this chapter, for information. If you're adding text in model space, I highly recommend that you use annotative text.

   The prompt for the opposite corner of the mText rectangle reappears. The command line displays:

   ```
   Specify opposite corner or [Height Justify Line spacing
     Rotation Style Width Columns]:
   ```

6. **If you want a different justification from the default (top left), type J, press Enter, and choose another justification option.**

   Enter *justify multiline text* in the Search box of the online Help system when you want an explanation of the other justification options.

7. **Pick another point in the drawing.**

TIP

   Don't worry about the *height* of the rectangle that you create by choosing the second point; the width of the rectangle is all that matters. AutoCAD adjusts the height of the text rectangle to accommodate the number of lines of word-wrapped text. Don't worry too much about the width, either, because you can adjust it later.

   The rectangle changes its size and the In-Place Text Editor window appears with the tab and indent ruler above it (as shown in Figure 13-2), and the previously hidden Text Editor contextual tab appears in blue on the Ribbon.

TIP

When you create multiline text in either AutoCAD or AutoCAD LT, text objects default to Dynamic Column mode. As I indicate earlier, the initial selection rectangle primarily sets the width. By default, if you enter enough text to fill the rectangle, mText defaults to splitting the text into two columns, as in a newspaper. If you don't like this arrangement, you can dynamically stretch the text editor window to become wider or longer.

If you never want to use columns in the current drawing, then click Columns in the Insert panel of the Text Editor tab, which appears only when you're creating or editing multiline text. This action sets the MTEXTCOLUMN system variable to 0. It affects only the current drawing, so if you never want columns, set it up in your template drawings. (I cover templates in Chapter 4.)

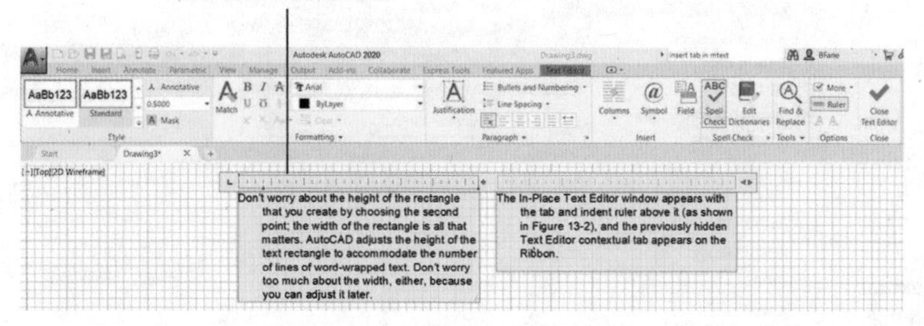

FIGURE 13-2: Adding immortal multiline text.

8. **Verify the text font and height.**

   The text font and height should be correct if you properly completed Steps 1, 4, and 5. If you didn't, you can change these settings in the Font drop-down list and the Text Height text box on the Text Editor tab (or on the classic Text Formatting toolbar).

9. **Type text into the text area of the In-Place Text Editor.**

   AutoCAD automatically word-wraps and extends the text box downward as you add more text. If you want to force a line break at a particular location, press Enter.

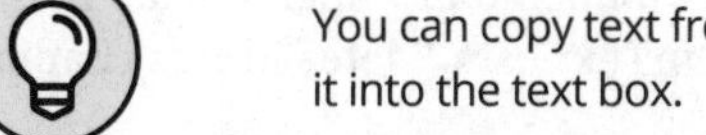

TIP

You can copy text from other applications, such as Word documents, and paste it into the text box.

10. **If you want to set other formatting options, select some text, right-click, and make an appropriate choice from the menu, as shown in Figure 13-3.**

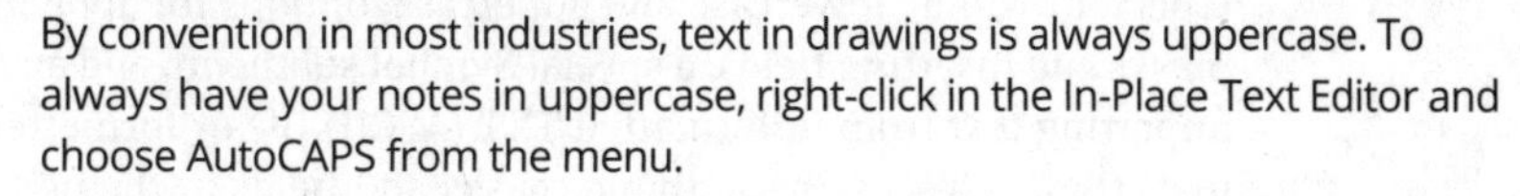

TIP

By convention in most industries, text in drawings is always uppercase. To always have your notes in uppercase, right-click in the In-Place Text Editor and choose AutoCAPS from the menu.

11. **Click Close Text Editor on the Ribbon or OK on the classic Text Formatting toolbar.**

    The In-Place Text Editor window closes, and AutoCAD adds your text to the drawing.

TIP

You can close the text editor much more easily by simply clicking outside its window. But if you like clicking buttons instead, AutoCAD has amply provided for you.

12. **If you set an annotative text style current in Step 1, assign an annotation scale to the multiline text object.**

    See the "Turning on Annotative Objects" section, later in this chapter.

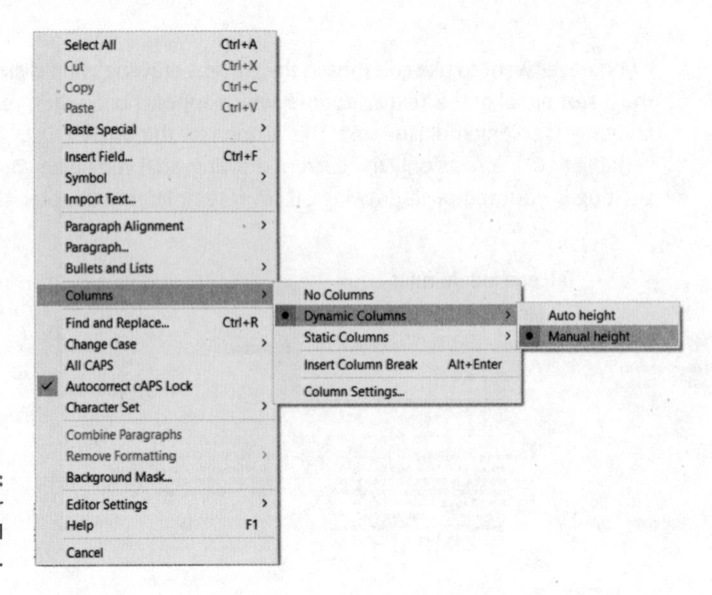

FIGURE 13-3: Right-click your way to textual excellence.

You can change the layout of text easily. Simply double-click the text and then use grip editing to change the width or height of the text box. If you make the box too short to hold all the text, AutoCAD automatically splits it into side-by-side newspaper-style columns as needed.

As you can tell by looking at the Text Editor tab (or the Text Formatting toolbar) and multiline text right-click menu, the mText command gives you plenty of other options. You can show or hide the toolbar, the ruler, and the Options buttons, and you can give the In-Place Text Editor an opaque background. Other tool buttons give you access to columns and numbered or bulleted lists. Both covered in the section "Doing a number on your mText lists," later in this chapter.

Between them, the Text Editor tab (or the Text Formatting toolbar) and the right-click menu also include a Stack/Unstack feature for fractions, a Find and Replace utility, tools for toggling lowercase and uppercase, options for applying colored background masks and inserting fields, a special Symbol submenu, and an Import Text option for importing text from unformatted TXT (ASCII) file or formatted RTF (Rich Text Format) files. (I discuss background masks and fields in the next section.) If you think you can use any of these other features, choose Resources ⇨ Commands ⇨ mText in AutoCAD's online Help browser to find out more about them.

mText dons a mask

When you turn on background masking, AutoCAD hides the portions of any objects that lie under the multiline text. This feature is used primarily when adding text to an area that has already been crosshatched. I cover hatching in Chapter 15. Use these steps to turn on, and control, this feature:

1. **Right-click in the In-Place Text Editor and choose Background Mask from the menu.**

   The Background Mask dialog box appears.

2. **Mark the Use Background Mask check box so that this option is turned on.**
3. **Either click Use Drawing Background Color to make the mask the same color as the drawing area's background color or choose a color from the drop-down list to make the text appear in a solid rectangle of the specified color.**
4. **Click OK to return to the In-Place Text Editor.**

TIP

If you've turned on background masking but it isn't having the desired effect, use the TEXTTOFRONT command to move the text on top of other objects. Click the Bring to Front drop-down button on the Home tab's Modify panel slideout, and choose Bring Text to Front.

Insert Field

The Insert Field option in the In-Line Text Editor's right-click menu creates a text field that updates automatically every time you open, save, plot, or regenerate the drawing. Fields can contain data such as the current date, filename, and author's name. Fields draw information from the operating system settings, Drawing Properties dialog box, sheet sets feature (not covered in this book), AutoCAD system variables (see Chapter 23), and specific drawing object properties, such as the circumference of a circle or the area enclosed by a closed polyline. Follow these steps to add a field while you're creating multiline text:

1. **Right-click in the In-Place Text Editor and choose Insert Field from the menu.**

   The Field dialog box appears.

2. **Choose a Field Name in the left column.**
3. **Choose a Format in the right column or, for date fields, type a format in the Date Format box.**
4. **Click OK.**

   AutoCAD adds the field to the mText object that you're creating or editing. Field text is distinguished from regular text with a gray background and is updated automatically under certain conditions, such as while saving and printing drawings.

Doing a number on your mText lists

The mText command supports bulleted and numbered lists. This feature is especially useful for creating general drawing notes, as shown in Figure 13-4.

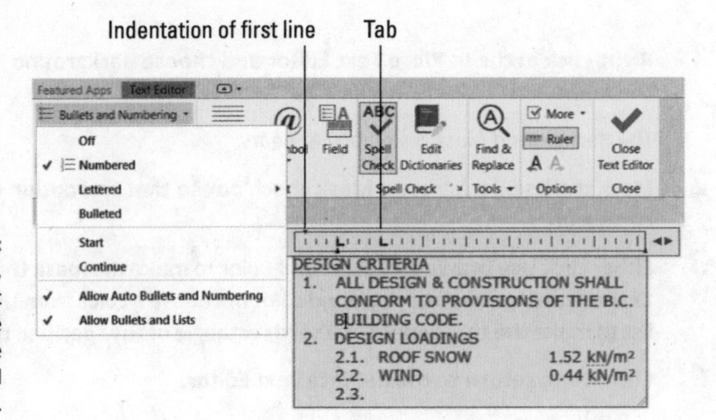

FIGURE 13-4: Tabs, indents, and automatic numbering are set to create numbered lists.

AutoCAD automates the process of creating numbered lists almost completely. Follow these steps:

1. **Follow Steps 1–8 in the earlier section, "Making it with mText," to open the In-Place Text Editor.**
2. **Type a title (for example, DESIGN CRITERIA).**

   If you want the title underlined, click Underline on the Formatting panel before you type the title; click Underline again to turn it off. Press Enter to go to the next line, and press Enter again to leave a little more space.
3. **On the Paragraph panel of the Text Editor tab, click the Bullets and Numbering drop-down button; verify that Allow Auto-List, Use Tab Delimiter Only, and Allow Bullets and Lists are selected; then click Numbered, as shown in Figure 13-4.**

   The number `1` followed by a period appears on the current line, and the cursor jumps to the tab stop that's visible on the ruler, at the top of the In-Place Text Editor window. Enabling the Numbered option places numerals followed by periods in front of items in a list. The Bulleted option places bullet characters in front of items in a list. Allow Auto-List enables automatic numbering; every time you press Enter to move to a new line, AutoCAD increments the number.
4. **Type the text corresponding to the current number or bullet.**

   As AutoCAD wraps the text, the second and subsequent lines align with the tab stop, and the text is indented automatically.
5. **Press Enter at the end of the paragraph to move to the next line.**

   As with creating numbered lists in your favorite word processor, AutoCAD automatically inserts the next number at the beginning of the new paragraph, with everything perfectly aligned.

TIP

To create nested numbered or bulleted items (as shown in Figure 13-4), simply press Tab at the start of the line. If you change your mind, you can bump up a nested text item up one level by selecting the item in the In-Place Text Editor and pressing Shift+Tab.

6. **Repeat Steps 4 and 5 for each subsequent numbered or bulleted item.**

TIP

For legibility, you sometimes add spaces between the notes. If you press Enter twice to add a blank line, AutoCAD — like every good word processor — thinks you're finished with the list and turns off numbering. AutoCAD is smart, so you need to be smarter. If you put the cursor at the end of the first note and press Enter, you get a blank line. The problem is, the blank line is now numbered, and the intended Note 2 is now Note 3. Just press the Backspace key. The number on the blank line disappears, and Note 2 is back to being Note 2. When you delete a numbered item, the remaining numbers adjust automatically.

If you don't like the horizontal spacing of the numbers or the alignment of subsequent lines, you can adjust them easily by manipulating the tab and indent markers in the In-Place Text Editor's ruler.

7. **In the ruler, drag the upper slider (the triangle pointing down) to the right a short distance. Drag the lower slider (the triangle pointing up) a slightly greater distance to the right.**

   The upper slider controls the indentation of the first line in each paragraph. The lower slider controls the indentation of the second and subsequent lines. An indent of one to two of the short, vertical tick marks usually works well for the first line. An indent of two to four tick marks works well for the second and subsequent lines.

8. **Click in the ruler just above the lower slider.**

   A small *L* appears above the lower slider. The *L* shows the left-justified tab stop.

   Make sure that the corner of the *L* aligns horizontally with the point of the lower slider triangle. If not, click and drag the *L* until it aligns.

TIP

If you prefer to manually enter tab and indent distances (rather than adjust them with the cursor), open the Paragraph dialog box by either

» Clicking the little arrow at the right end of the Paragraph panel label (in AutoCAD-ese, it's a Panel dialog box launcher)

» Right-clicking inside the In-Place Text Editor and choosing Paragraph

Whichever way you do it, if you select text first, the tab and indent changes apply to the selected text. If you don't select text first, the changes apply to new text from that point forward in the multiline text object.

Line up in columns — now!

The text functionality of AutoCAD keeps becoming more word processor-like. A few releases back, it could handle simple indents, and then came numbered and bulleted lists. A more recent addition to multiline text is columns.

Columns come in two flavors:

- **Static:** You specify the number of columns into which you want text to flow. Columns are always the same height, and text is allowed to overflow the final column if there's too much of it.
- **Dynamic:** As you might expect, these columns are friendlier and more flexible, and they're the life of any party. You can individually adjust column heights, and new columns are added automatically to accommodate the text.

Selecting either column type also offers you the Column Settings dialog box, where you can specify values numerically rather than drag grips. Figure 13-5 shows a block of multiline text imported as an RTF file from a word processor and then formatted in dynamic columns using the Manual Height option.

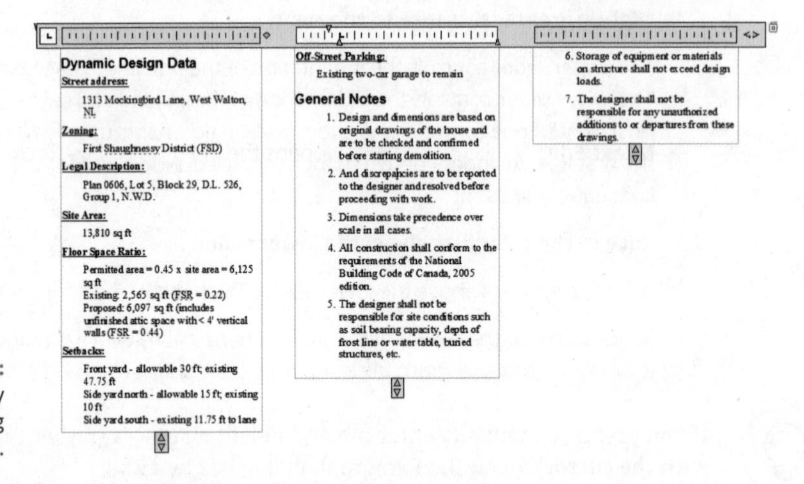

FIGURE 13-5: Dynamically columnizing the text.

Setting up columns is straightforward, as these steps explain:

1. **Open a drawing that contains a large, multiline text object, or create a large, multiline text object in a new drawing.**

TIP

If you already have drawing specifications or general notes in a word processing document, or even in a text file, you can right-click inside the mText command's In-Place Text Editor and choose Import Text. The Select File dialog box opens, giving you the choice of Rich Text Format (RTF) or ASCII text (TXT) files.

2. **If the In-Place Text Editor isn't already open, either double-click the text; or select it, right-click, and choose Mtext Edit.**
3. **Click Columns in the Insert panel and choose either Dynamic Columns or Static Columns.**

   If you choose Dynamic Columns, select either Auto Height or Manual Height. Selecting Manual Height puts grips on each column so that you can adjust their height individually (refer to Figure 13-5). Auto Height displays a single grip so that the heights of all columns remain the same, but new columns are still added as required by the amount of text.

   If you choose Static Columns, select the number of columns you want from the menu. Clicking 2, for example, creates two columns regardless of the length of the text. You may end up with overflowing text or empty columns.
4. **Click Close Text Editor on the Ribbon (or OK on the Text Formatting toolbar) when you're satisfied with the column arrangement.**

   You can revert to a noncolumnar arrangement by clicking the Columns button in the Insert panel and choosing No Columns.

Modifying mText

After you create a multiline text object, you can edit it in the same way as a single-line text object: Select the object, right-click, and choose Mtext Edit or Properties.

- **Mtext Edit:** Selecting this option opens the In-Place Text Editor window so that you can change the text contents and formatting.
- **Properties:** Selecting this option opens the Properties palette, where you can change the overall properties of the text object.
- **Quick Properties:** Enable this setting on the status bar so that when you select a multiline text object, the Quick Properties palette opens, in which you can modify a subset of the mtext object's properties.

The easiest way to change the word-wrapping width of a paragraph text object is to *grip-edit* it. Select the text object, click a corner grip, release the mouse button, move the cursor, and click again. Chapter 11 describes grip-editing in detail.

Just like any good word processor or text editor, AutoCAD includes both a spel chekcqing and a find-and-replace tool for text. To check the spelling of selected annotation objects or the entire drawing, click the Annotate tab on the Ribbon and choose Check Spelling from the Text panel to display the Check Spelling dialog box. On the same panel, clicking Find Text (or typing **FIND** and pressing Enter) displays the Find and Replace dialog box. Handily, both Find and Replace and Spell

Check are also accessible from the Text Editor tab's Spell Check and Tools panels, in case you want to replace text or check spelling in a single multiline text object. Look up *SPell* or *FIND* in the online Help system when you need more information on either command.

Turning On Annotative Objects

Annotative objects were introduced in AutoCAD 2008, ending 25 years of calculation agony, but a quick show of hands in classes at Autodesk University and AUGI CAD Camp reveals that only 20 percent of students, on average, have ever tried using annotative objects — and most of the students abandoned them to return to the old way, in the belief that they weren't working. I provide an explanation in a moment.

Make sure that you're in model space, and follow these steps to enable the use of annotative scaling for new or modified annotation objects.

If you don't see on the status bar the buttons you need to access, click the hamburger button (three horizontal bars) at the end of the status bar and then click any status bar items that you want to turn on or off.

1. **Set the Annotation controls.**

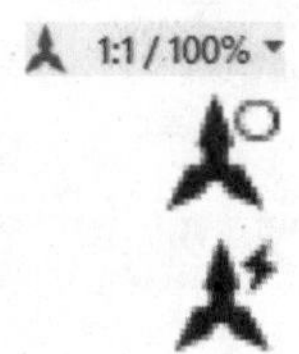

   Near the right end of the AutoCAD status bar are three buttons with variants of a triangular symbol. They represent the end view of an engineer's triangular scale; an old-timer in your office should be able to show you one and how they worked. If the person has a rubber band in the other hand, be ready to duck. Click the Annotation Visibility (middle) and Automatically Add Scales to Annotative Objects (right-hand) buttons, until they're grayed out and don't show a small, yellow splash in them. If the Annotation Scale (left-hand) button doesn't display the scale *1:1*, click it and select 1:1 from the drop-down list.

2. **Select an annotative text style.**

   Use the STyle command as described in the earlier section "Get in style," and set the Annotative text style current. Set its paper text height to 0.125.

3. **Create a text object.**

   Use the TEXT command and create a line of text that says *Scale 1:1*.

   The first time you place annotative text, you're asked to select a drawing scale, and it defaults to the current scale, which is probably what you want anyway. Eventually, it becomes a nuisance, so the check box lets you turn it off.

4. **Change the drawing scale.**

   Click the Annotation Scale (left-hand) button and choose 1:2 from the drop-down list. This step sets the current drawing scale to 1:2. Hey, the first piece of text disappears! Not to worry — that's what it's supposed to do.

5. **Create a text object.**

   Use the TEXT command again to create a line of text that says *Scale 1:2*, and another one that says *Both scales*. Note that they're twice as tall as the now-invisible Scale 1:1 text, even though you didn't specify a height.

6. **Change the drawing scale.**

   Click the Annotation Scale (left-hand) button and choose 1:1 from the drop-down list. Interesting! The first piece of text reappears, but the other two disappear!

7. **Add another scale to an object.**

   Click the Annotation Scale (left-hand) button and choose 1:2 from the drop-down list. Then click the *Both scales* text and right-click. Choose Annotative Object Scale and then choose Add/Delete Scales. When the dialog box appears, click the Add button in the upper-right corner and then choose 1:1 from the list. Click OK, and then click OK again to return to the drawing screen.

8. **Change the drawing scale.**

   Click the Annotation Scale (left-hand) button and choose 1:1 from the drop-down list. Interesting! The *Scale 1:1* text reappears, *Scale 1:2* disappears, and the *Both scales* size changes to match the height of the *Scale 1:1* text!

9. **Change the drawing scale again.**

   Click the Annotation Scale (left-hand) button and choose 1:5 from the drop-down list. All three text items disappear!

Here's what I want you to remember about annotative objects:

- Upon creation, annotative objects automatically size themselves to suit the current drawing scale; they're visible only when their annotation scales match the current drawing scale; and they can have more than one scale attached, so they're visible in more than one drawing scale setting.

  Better yet, they exhibit this same behavior in paper space viewports. It thus becomes trivially easy to set up drawings with details at other scales. I cover this topic in more detail in Chapter 15.

- If an annotative object has multiple scales attached and you edit its location at one scale, its location at the other scales isn't affected.

» When viewing an object with multiple scales, you aren't looking at multiple objects but rather at a single object with multiple representations. To see what I mean about different representations, change the text string of the *Both scales* text object that you created earlier to a different value. The new value automatically appears in all scale representations for that text object.

Now I reveal the dirty, dark secret of why most people give up on annotative objects. It all goes back to the annotative controls:

» When the Annotation Visibility (middle) button is turned on, all the annotative objects at all scales are always visible, regardless of the value of the current drawing scale, even if none of the objects has a scale that matches the current drawing scale. You should only turn this option on when you want to move the different representations of your annotative objects around onscreen.

» If you turn on the Automatically Add Scales to Annotative Objects (right-hand) button and then change the drawing scale, every annotation object retroactively gets the current scale added to it. This is not something you want to do except in the situation of when you create a drawing to print at a specific scale and then decide to change the scale to suit a different printer or paper size.

By themselves, both options can give you some headaches, but if both options are enabled at the same time and you switch the current scale, it's not obvious that you might have accidentally added the current scale to all the annotative objects in your drawing.

Most of the time, for most people, there are way too many scales listed in the Add Scales to Object dialog box. AutoCAD has a handy-dandy Edit Drawing Scales dialog box that lets you remove those imperial scales from the current drawing if you never work in feet and inches. And vice versa, for the metrically challenged. To run through the scales, choose Scale List on the Annotation Scaling panel of the Ribbon's Annotate tab. If you make a mistake, pressing the Reset button restores all default scales. You can remove those extra scales for *all* drawings in the Default Scale List dialog box, accessible from the User Preferences tab of the Options dialog box.

Figure 13-6 shows two views of the same drawing. The hatching and the 1.400 dimension were created once but have resized themselves to appear correctly in the two views at two different scales.

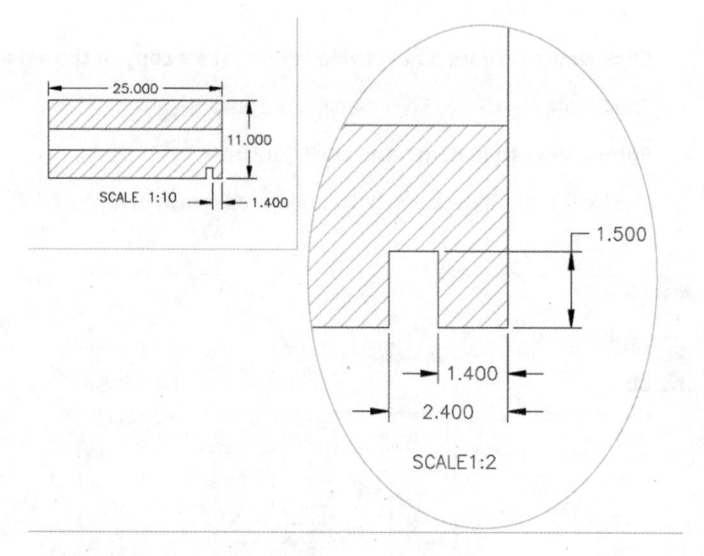

FIGURE 13-6: Annotative objects resize automatically as the annotation scale changes.

Gather Round the Tables

You don't know the meaning of the word *tedious* unless you've tried to create a column-and-row data table in older versions of AutoCAD with the Line and TEXT commands. AutoCAD's table object, along with the TABLESTYLE and TABLE commands for creating it, make the job almost fun.

REMEMBER

Table objects in AutoCAD are *not* annotative, so you have just two methods of adding them to drawings: Create them in model space, scaling them up by the drawing scale factor (see Chapter 4 for more) or — this one seems more sensible — create them in paper space, defining them by their actual plotted (paper) dimensions.

Tables have style, too

You control the appearance of tables — both text and cells — with *table styles* just as you control the appearance of text with text styles. Use the TABLESTYLE command to create and modify table styles. Follow these steps to create a table:

1. **On the Home tab, click the Annotation panel's label to open its slideout, and then choose Table Style.**

   The Table Style dialog box appears.

2. **In the Styles list, select the existing table style whose settings you want to use as the starting point for the settings of the new style.**

   For example, select the default table style named Standard.

3. **Click New to create a new table style that's a copy of the existing style.**

   The Create New Table Style dialog box appears.

4. **Enter a new style name and click Continue.**

   The New Table Style dialog box appears, as shown in Figure 13-7.

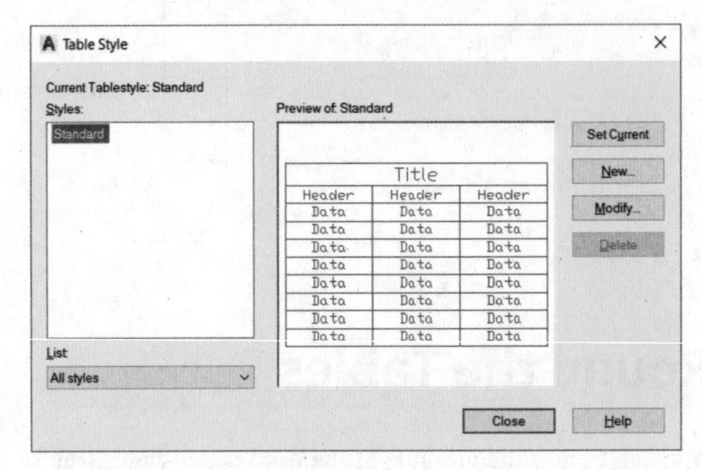

FIGURE 13-7: Setting the table.

5. **In the Cell Styles area, with Data showing in the list box, specify settings for the data alignment, margins, text, and borders.**

   The settings you're likely to change are Text Style, Text Height, and perhaps Text Color. All three are on the Text tab. Grid Color is on the Borders tab. If you leave colors set to ByBlock, the text and grid lines inherit the color that's current when you create the table. That color will be the current layer's color.

6. **In the Cell Styles area, open the drop-down list and repeat Step 5 for the Headers (that is, the column headings) and the Title styles.**

7. **Click OK to close the New Table Style dialog box.**

   The Table Style dialog box reappears.

8. **Click Close.**

   The new table style becomes the current table style that AutoCAD uses for future tables in this drawing, and the Table Style dialog box closes. Now you're ready to create a table, as described in the next section.

TIP

You can access the Manage Cell Styles dialog box directly from the Cell Styles drop-down list of the New Table Style dialog box. The Table Cell Format dialog box (on the General tab, in the Format row, click the ellipsis button) provides a number of additional options for formatting cells by data type.

AutoCAD stores table styles in the DWG file, so a style that you create in one drawing isn't easily available in others. You can copy a table style between drawings with DesignCenter. Use the steps for copying layers between drawings that I outline in Chapter 9, but substitute `Tablestyles` for `Layers`. It's better to define table styles in drawing template files, which I discuss in Chapter 4.

Creating and editing tables

After you create a suitable table style, adding a table to the drawing is easy, with the TABLE command. Here's how:

1. **Set an appropriate layer current.**

   Assuming that you leave the current color, linetype, and lineweight set to ByLayer, as I recommend in Chapter 9, the current layer's properties control the properties of all table parts that were set to ByBlock when in the assigned table style. Refer to Step 5 in the preceding section, "Tables have style, too."

2. **On the Home tab's Annotation panel, click Table.**

   The Insert Table dialog box appears.

3. **Choose a table style from the Table Style drop-down list.**
4. **Choose an insertion behavior:**

   - *Specify Insertion Point:* Pick the location of the table's upper-left corner; use the lower-left corner instead if you set Table Direction to Up in the table style. When you use this method (the easiest), you specify the default column width and number of rows in the Insert Table dialog box.

   - *Specify Window:* Pick the upper-left corner and then the lower-right corner. When you use this method, AutoCAD automatically scales the column widths and determines how many rows to include.

5. **Specify column and row settings.**

   - If you chose Specify Insertion Point in Step 4, you must set the number of columns and data rows that the table should be created with along column width and row height.

   - If you chose Specify Window in Step 4, AutoCAD sets the column width and the number of data rows to Auto. AutoCAD then figures them out, basing those values on the overall size of the table that you specify in Steps 7 and 8. While AutoCAD calculates the column width and number of data rows to use for the table automatically, you must still specify the number of columns and row height values to create the table.

6. **Click OK.**

   AutoCAD prompts you to specify the insertion point of the table.

7. **Click a point or type coordinates:**
   - If you chose Specify Insertion Point in Step 4, AutoCAD draws the table gridlines, places the cursor in the title cell, and displays the Text Editor tab on the Ribbon.
   - If you chose Specify Window in Step 4, specify the diagonally opposite corner of the table.

   AutoCAD draws the table based on the table size you indicated and chooses the column width and number of rows.

8. **Type a title for the table.**
9. **Type values in each cell, using the arrow keys or Tab key to move among cells.**

   The cell right-click menu offers many other options, including copying contents from one cell to another, merging cells, inserting rows and columns, changing the formatting, and inserting a *block* which is a graphical symbol; see Chapter 17 for information about blocks.

TIP

You can insert a field into a table cell. For example, you might create part of a title block, with fields serving as the date and drawn by data.

10. **Click Close Text Editor on the Ribbon or click OK on the Text Formatting toolbar.**

    Figure 13-8 shows a completed table, along with the Insert Table dialog box.

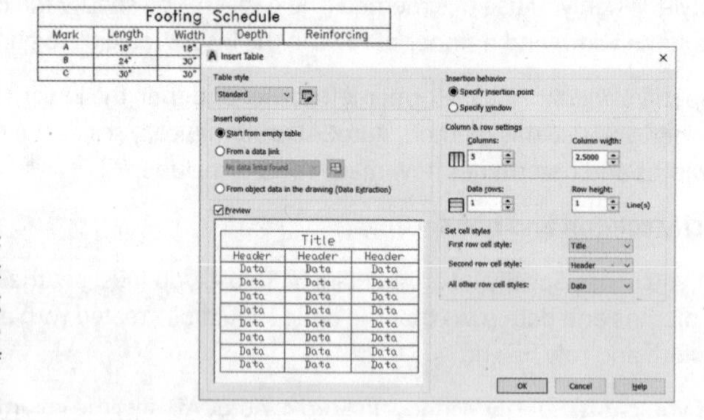

FIGURE 13-8: The Insert Table dialog box and one result of using it.

Edit cell values by double-clicking in a cell. To change the column width or row height, click the table grid and then click and move the blue grips. To change the width of one column without altering the overall width of the table, hold down the Ctrl key while you move the grip. If you want to change other aspects of a table or individual cells in it, select the table or cell and use the Quick Properties palette or the Properties palette to make changes.

The DATAEXTRACTION and DATALINK commands are powerful tools in AutoCAD. You can use them to link all or part of an Excel spreadsheet into a table in a drawing; or interrogate a drawing to produce a great variety of information from it, such as sizes, areas, perimeters, and quantities of drawing objects, and then create a linked table of the data; or write the data to an Excel spreadsheet — or all three options at once. A detailed description of these functionalities is beyond the scope of this book, but several online references are available that cover them. These two can get you started:

`www.cadalyst.com/cad/autocad/tell-me-all-you-know-attribute-extraction-part-1-learning-curve-autocad-tutorial-6271`

`www.cadalyst.com/cad/autocad/tell-me-all-you-know-and-even-more-advanced-attribute-extraction-part-2-learning-curve-a`

Take Me to Your Leader

Don't worry — I'm not a space alien, popular opinion to the contrary. No, I'm talking about notes with arrows pointing to drawing objects that you need to embellish with some verbiage.

Multiline leaders (*mleaders,* for short) are a vast improvement over the old-style leaders, available in releases before AutoCAD 2008. In fact, they're so good that they should be running the United Nations! MLeaDers, unlike the old-style leaders, are single objects. They can also point in multiple directions at a time; just don't ask them which way is the bus station. Finally, multileaders — just like text, dimensions, hatching, and other objects you use to document drawings — can be annotative.

Electing a leader

You can draw multileader objects that consist of leader lines and multiline text at the same time by using the MLeaDer command; follow these steps:

1. **Set a multileader style that's appropriate for your needs.**

   Choose an existing style from the Multileader Style drop-down list on the Home tab's Annotation panel slideout, or create a new style by clicking the Multileader Style button on the Annotation slideout. Visit the online Help system for tips on creating new multileader styles.

OLD LEADER STYLES

Two older-style leaders are stored in AutoCAD's attic. Autodesk hates to throw anything away because someone will complain about needing it for a *vital* custom menu macro that was created 20 years ago. These semiretired leaders are no longer accessible on menus or toolbars, but they're still lurking up there, waiting to . . . er, lead you astray. If you've become attached to typing commands, you may be inclined to type **LEADer** to create those pointy-headed thingies. If you do, be aware that the LEADER command runs the positively ancient command for creating notes-with-arrows. There are no options, so you see only straight leader lines and infinitely long text strings. LE (the alias of the qLEader command) runs the less elderly command for creating notes with arrows; this one has a Settings dialog box where you can set many options, and it asks you to specify a width for the mText note. However, unlike multileaders, both LEADer and qLEader create the leader lines and the text as separate objects.

2. **Choose Multileader from the Annotation panel.**

   The command line prompts you to select the location of the pointy end of the leader arrowhead:

   ```
   Specify leader arrowhead location or [pre enter Text/leader
     Landing first/Content first/Options] <Options>:
   ```

   The initial default method is to locate the arrowhead, followed by the *leader landing* (that is, the short horizontal line between the leader line and the text), at which point AutoCAD displays the In-Place Text Editor and you enter the text of your note. If you'd rather place the leader landing first, type **L** and press Enter. If you'd rather type the content first, type **C** and press Enter.

   New in AutoCAD 2023 is the capability to toggle which prompt the MLeaDer command asks you first:

   - **Content first** causes MLeader to first ask you, "Specify first corner of text." You enter text, close the mtext editor by clicking the check mark at the end of the text editor panel, and then position the leader line.
   - **leader arrowHead first** causes MLeader to first ask you, "Specify leader arrowhead location." You position the leader line and then enter the text.
   - **pre enter Text** lets you enter mText anywhere in the drawing, which you relocate after you exit the editor, and then last of all place the leader.
   - **select Mtext** prompts you to select mText, which is then turned into an mleader.

If you want to draw spline-curved rather than straight leader lines, specify the number of pick points for the leader lines and press Enter to display multileader options at the command line or the Dynamic Input tooltip. Refer to the online Help for more information on the command options.

3. **Pick a location in the drawing that you want the leader to point to.**

   If you use an object snap mode, such as NEArest or MIDpoint, to pick a point on an object, AutoCAD associates the leader with the object. If you later move the object, AutoCAD updates the leader so that it points to the new location.

   The command line prompts you for the next point.

4. **Pick a second point.**

   AutoCAD draws a shaft from the arrowhead to this point.

   WARNING

   Don't pick a second point that's too close to the arrowhead point. If AutoCAD doesn't have enough room to draw the arrowhead, it omits the arrowhead.

   By default, AutoCAD lets you pick two points for the leader line: The first point locates the arrowhead, and the second point locates the start of the short horizontal leader landing. If you need more points than that, restart the command and choose Options, Maxpoints and set a new value. After you pick the point for the leader landing, the In-Place Text Editor opens.

5. **Type your text.**

   You're now in the same In-Place Text Editor I describe earlier in this chapter, with the same Text Editor tab and options.

6. **Click Close.**

   Your text is added to the drawing, next to the leader.

Figure 13-9 shows several different leaders with notes.

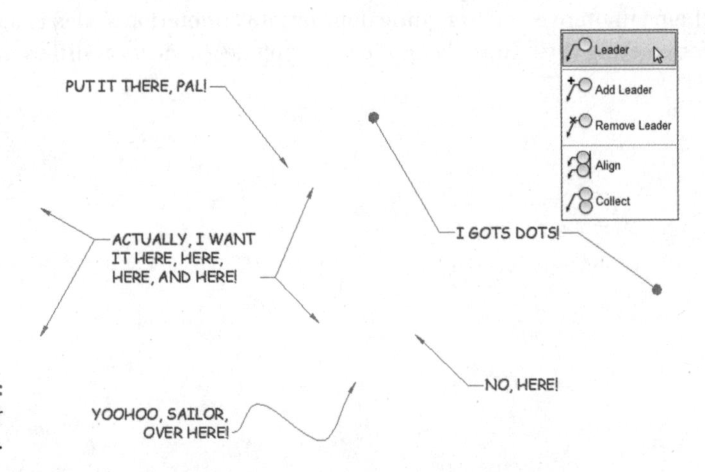

FIGURE 13-9: No loss for (multi)leaders.

You can use grips to manipulate a multileader's individual components, just as if it were several objects instead of a single object. Hold down the Ctrl key and click a multileader to select different pieces. You can even use grips to adjust the width of multileader text as if it were an everyday mText object.

Multi options for multileaders

In addition to using the MLeaDer command itself, multileaders come with a slew of formatting, drawing, and editing commands:

- **MLeaderEdit:** Use this command to add or remove leader lines from multileader objects. You'll find each of those functions on the Multileader drop-down list (refer to Figure 13-9).
- **MLeaderStyle:** Multileaders, like text and dimensions, are formatted according to named styles. To display the Multileader Style Manager dialog box, click Multileader Style on the Home tab's Annotation slideout.
- **MLeaderAlign:** A tedious chore in earlier AutoCAD versions was making all the leaders line up by using a construction line and then using object snaps to move common points on the leaders to the construction line. To line up multileaders, choose Align from the Multileader drop-down list.
- **MLeaderCollect:** Multileaders can contain blocks as well as text. I cover blocks in Chapter 17. Choose Collect from the Multileader drop-down list to gather a group of leaders containing blocks; they'll rearrange themselves as a single multileader containing the multiple blocks.

For more information on any of these commands, look them up by name in the Command Reference section of the online Help system.

Multileaders and multileader styles can be annotative or non-annotative. Making them annotative and assigning appropriate annotation scales is a huge timesaver for creating them initially and for creating detail views at different scales.

» Choosing a dimensioning method

» Creating and modifying your own dimension styles

» Adjusting dimension sizes to suit the drawing plot scale

» Placing and modifying annotative dimensions, including details at other scales

» Modifying dimensions

Chapter 14 Entering New Dimensions

In drafting, whether CAD or manual, *dimensions* clearly indicate the size of things. They are special text labels with attached lines that point to what is being dimensioned. Although it's theoretically possible to draw all the pieces of a dimension by using AutoCAD commands such as Line and mText, dimensioning is such a common drafting task that AutoCAD provides specific commands to do the job efficiently. These dimensioning commands place all parts of a dimension into a convenient, easy-to-edit package, much like a block. In fact, AutoCAD actually produces each dimension as something it calls an *anonymous block*.

Even better, as you edit an object, such as stretch a line, AutoCAD automatically updates the measurement displayed in the dimension text label to indicate the object's new size. And perhaps best of all, AutoCAD's annotative dimensions adjust their text height and arrowhead size automatically to suit the annotation scale on the Model tab or the viewport scale in a layout. I explain the general principles of annotative objects in Chapters 13 and 15. In this chapter, I take a close look at annotative dimensions.

Adding Dimensions to a Drawing

I start off with an exercise to introduce AutoCAD's dimensioning function by creating linear dimensions that show the horizontal or vertical distance between two points:

1. **Start a new drawing, using the `acad.dwt` template file.**

   This step creates a drawing that uses imperial units, even if your default installation uses metric. It saves lazy writers from having to duplicate instructions for metric users.

2. **Use the Line command to draw a non-orthogonal line.**

   A *non-orthogonal* line is a segment that's neither horizontal nor vertical. Make the line about 6 units long, at an angle of about 30 degrees upward to the right.

3. **Set a layer that's appropriate for dimensions as current.**

   Okay, you started from a blank template, so it doesn't have specific layers, but I included this step as a gentle reminder. As I discuss in Chapter 9, you normally set dedicated layers for visible edges, hidden edges, text, dimensions, section lines, hatching, and so on.

TIP

   AutoCAD 2022 added a significant enhancement to dimensioning. A single command, DIM, handles all dimension types. The older method of using specific commands for designated dimension types is still available, and I discuss it later in "Dimensioning the Legacy Way." As of AutoCAD 2016, you can define an override layer on which all dimensions automatically end up. See the "And the Correct Layer Is . . ." section at the end of this chapter.

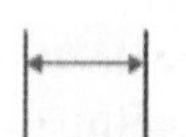

4. **Start the DIM command by clicking the big DIM button on the Annotate tab's Dimensions panel, or type** DIM **and press Enter.**

   The DIM command has nothing to do with brightness. It can replace five older dimension commands, resulting in the same dimensions but with a lot less effort: DImLinear, DimALigned, DimANgle, DimDIameter, and DimRAdius. I talk about these individual dimensioning commands in "Dimensioning the Legacy Way."

5. **Select the object you want dimensioned, in this case the line.**

   You can pick an entity, and AutoCAD make a best guess at what kind of dimension it should receive. Or you can specify the dimension type. See the different kinds of dimensions AutoCAD produces later in "A Field Guide to Dimensions."

```
Select objects or specify first extension line origin or
   [Angular Baseline Continue Ordinate align Distribute
   Layer Undo]:
```

6. **At the next prompt, move the cursor around the line, and then pick a point to locate one of the extension lines.**

   As you move the cursor, notice that the dimension type changes. For a line, AutoCAD can dimension it with an aligned dimension (it appears parallel to the line), a horizontal dimension, or a vertical one.

   ```
   Select line to specify extension lines origin:
   ```

7. **Press Enter to end the command.**

   AutoCAD automatically starts the DIM command again, so that you can carry on dimensioning objects in the drawing. Pressing Enter ends the command and returns you to the command prompt.

The DIM command has an additional set of characteristics to enhance your dimensioning experience:

- Unlike the older individual commands, you do not need to differentiate between picking exact start and end points as opposed to selecting an object. If you select an object such that your pick point is close to an object snap point, DIM will snap to it. But if you select an object in such a way that it doesn't find a snap point, DIM automatically selects the picked object.
- Running object snaps won't snap to the end of an extension line that is closest to the object being dimensioned. Instead, DIM snaps to the end of the dimension line itself and hence creates a continuing dimension.
- After you have placed a dimension, you don't need to restart the DIM command; it reverts to asking you to begin defining the next dimension. In this way, you can place the vast majority of dimensions in a typical drawing with just one run of the DIM command.
- While the DIM command is running, the command line shows several options that can be invoked, of which my favorite is Undo.
- To terminate the DIM command, press ESC or Enter.

Dimensioning the Legacy Way

Prior to release 2016, AutoCAD had one command for every dimensioning type. There was DimLinear for linear dimensions, DimAligned for aligned ones, and DimAngular for dimensioning angles. If you are using a version of AutoCAD older than 2016, or even if you want to apply a specify dimension type and not have

AutoCAD guess at what you want, you need to use the old system. Here's how it works:

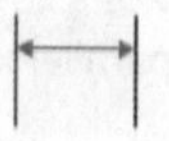

1. **Start the DimLInear command by clicking the Linear button at the bottom of the Annotate tab's Dimensions panel, or type** DLI **and press Enter.**

   AutoCAD prompts you:

   ```
   Specify first extension line origin or <select object>:
   ```

2. **To specify the origin of the first extension line, snap to the lower-left endpoint of the line by using an ENDpoint object snap.**

   If you don't have ENDpoint as one of your current running object snaps, specify a single endpoint object snap by holding down the Shift key, right-clicking, and choosing ENDpoint from the menu that appears. See Chapter 8 for more about object snaps.

   AutoCAD prompts you:

   ```
   Specify second extension line origin:
   ```

   AutoCAD makes a link between dimensions and the objects being dimensioned. Therefore, to ensure that the values reported by dimensions are accurate and that later editing works properly, you *must* use object snaps or object selection when applying dimensions.

3. **To specify the origin of the second extension line, snap to the other endpoint of the line by using the ENDpoint object snap again.**

   DimLInear draws a *horizontal* dimension (the length of the displacement in the left-to-right direction) if you move the cursor above or below the line. It draws a *vertical* dimension (the length of the displacement in the up-and-down direction) if you move the cursor to the left or right of the line.

   AutoCAD prompts you:

   ```
   Specify dimension line location or [Mtext Text Angle
     Horizontal Vertical Rotated]:
   ```

4. **Move the mouse to generate the type of dimension you want — horizontal or vertical — and then click wherever you want to place the dimension line.**

   AutoCAD draws the dimension.

WHY USE DIMENSIONS IN CAD?

You may think that the precision of CAD would have rendered text dimensions obsolete. After all, you comply with all my suggestions of using AutoCAD's precision techniques when you draw and edit, and you're careful to draw each object at its true size, right? The contractor or machinist can use AutoCAD to query distances and angles in the DWG file, right? Sorry, but no (to the last question, anyway). Here are a few reasons why the traditional dimensioning CAD inherited from manual drafting is likely to be around for a long while:

- **Some people need to — or want to — use paper drawings when they build something.** Even with mobile devices (such as AutoCAD Mobile running on Apple iPad or Google Android tablets), we're still some time away from the day when contractors haul computers around on their tool belts, never mind mousing around drawings while hanging from a scaffold.
- **In many industries, paper drawings still rule legally.** Your company may supply both plotted drawings and DWG files to clients, but your contracts probably specify that the plotted drawings govern in the case of any discrepancy. The contracts probably also warn against relying on any distances measured by recipients of drawings — whether using measuring commands in electronic drawing files or scale rulers on plotted drawings. The text of dimensions is *supposed* to supply all the dimensional information that's needed to construct the project.
- **Dimensions sometimes carry information in addition to basic lengths and angles.** For example, dimension text can indicate allowable manufacturing tolerances, or show that particular distances are *typical* of similar ones elsewhere on the drawing, or can simultaneously show imperial and metric values.

When you're specifying the location of the dimension line, you usually *don't* want to object-snap to existing objects. Rather, you want the dimension line and its text to sit in a relatively empty part of the drawing rather than have it bump into existing objects. If necessary, temporarily turn off running object snaps (for example, click the OSNAP button on the status bar) to avoid snapping the dimension line to an existing object.

If you want to be able to align subsequent dimension lines easily, turn on Snap mode and set a suitable snap spacing (more easily done than said!) before you pick the point that determines the location of the dimension line. See Chapter 8 for more information about Snap mode. To automatically space several existing dimensions equally apart, you can use the DIMSPACE command.

5. **Repeat Steps 1–4 to create another linear dimension of the opposite orientation (vertical or horizontal).**
6. **Click the line to select it.**
7. **Click one of the grips at an end of the line and drag it around.**

   The dimensions automatically update, live and in real time, to reflect the current values as you move the mouse.

You probably don't want dimensions to display four decimal places, or maybe you want to use a different font for dimension text, or use both imperial and metric units, or show manufacturing tolerances. Not a problem! AutoCAD controls the look of dimensions by means of dimension styles, just as it controls the look of text and tables with styles. In fact, AutoCAD uses text styles also to control the appearance of text in dimensions. I cover dimension styles in greater detail later in this chapter, but suffice it to say that AutoCAD has over 80 variables that can be used to warp dimensions into just about any perversion that your industry or company can imagine.

A Field Guide to Dimensions

AutoCAD provides several types of dimensions and commands with which to draw them; most commands are shown in Figure 14-1. These commands are found on the Dimensions panel of the Annotate tab on the Ribbon. If you can't find the button you want, it's probably hidden in a drop-down list, under the smaller Linear button on this panel. This panel remembers the last button you used, so at any given time, it may be displaying any one of the following buttons.

You can also find these basic dimension commands in the upper-right corner of the Annotation panel of the Home tab on the Ribbon. Remember that the last one you used stays on top, so the one you want now may be underneath the one that is displayed.

In the following list, the three characters (in parentheses) is the shortened command name (command alias), which you can enter instead:

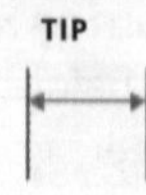

» **Linear — DimLInear (DLI):** Indicates the linear extent of an object or the linear distance between two points. Most linear dimensions are either horizontal or vertical, but you can draw dimensions that are rotated to other angles, too.

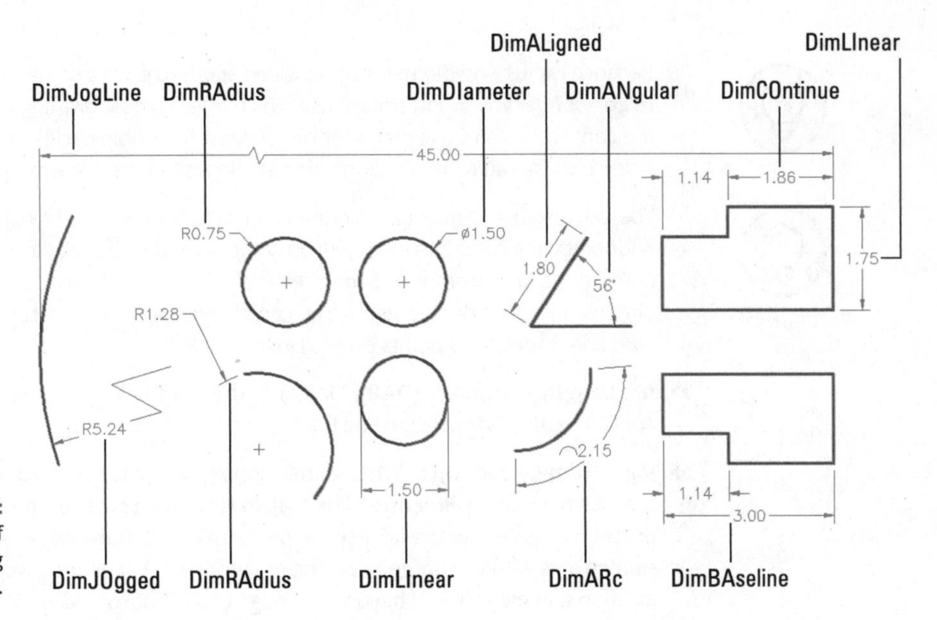

FIGURE 14-1: Examples of dimensioning commands.

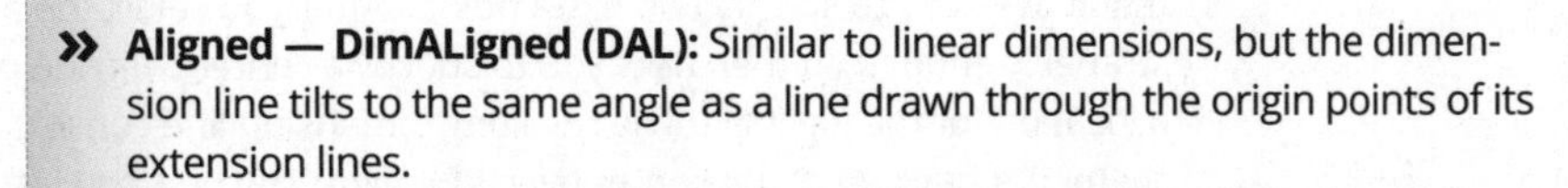

» **Aligned — DimALigned (DAL):** Similar to linear dimensions, but the dimension line tilts to the same angle as a line drawn through the origin points of its extension lines.

TIP

You don't always have to pick points when placing dimensions with these commands. Watch the command line; it may ask you to

```
Specify first extension line origin or <select object>:
```

If you press Enter instead of picking the first point, AutoCAD prompts you to select an object. And when you do, it automatically selects each end of the object for you. (The DIM command reverses things: it first prompts you to pick an object.)

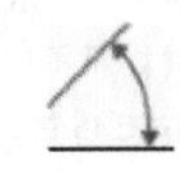

» **Angle — DimANgle (DAN):** Indicates the angular measurement between two lines, the two endpoints of an arc, or two points on a circle. The dimension line appears as an arc that indicates the sweep of the measured angle.

» **Radial — DimDIameter (DDI), DimRAdius (DRA):** Indicates the diameter of a circle or an arc, or a radius of a circle or an arc. You can position the dimension text inside or outside the curve; refer to Figure 14-1. When you position the text outside the curve, AutoCAD draws (by default) a little cross at the center of the circle or arc. AutoCAD automatically adds the diameter and radius symbols to the appropriate dimension type.

TIP

Because radius and diameter dimensions seem to do the same thing, you might wonder when you should use which one. In most drafting disciplines, the convention is to use diameter dimensions for whole circles (for example, a hole) and use radius dimensions for part circles or arcs (for example, a fillet).

TIP

You can also use DimLinear to dimension a circle or an arc. Trying to pick the endpoints of a circle can keep you busy for several hours, but if you press Enter at the first prompt and then select a circle or an arc, it applies a linear dimension across the diameter. You can also use a QUAd object snap to pick two directly opposite points on the circle.

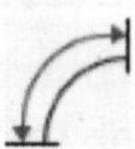

- **Arc Length — DimARc (DAR):** Measures the length along an arc, and then applies a curved dimension line that's concentric with the arc.

- **Jog — DimJOgged (DJO):** When dimensioning arcs, it is standard drafting practice to create a dimension line that starts at the center of the arc and proceeds outward to the arc itself. A problem arises, however, when dimensioning arcs with very large radii. The center may be at an inconvenient location and may be well beyond the edge of the final drawing. DimJOgged neatly solves this problem for you. It works almost exactly like DimARc except that it asks you to specify two extra points. After you select the arc (or circle, strangely enough), it then asks you to specify a center override point and then a location for the jog itself. The resulting dimension line consists of three segments, starting at the center override point: The first and last point to the real center; the middle segment connects the first and last points by putting a jog in the line.

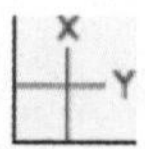

- **Ordinate — DimORdinate (DOR):** Applies X or Y coordinate values from the origin point. Certain types of mechanical parts can have a great many holes and other features. The drawing would become impossibly cluttered if you were to apply traditional dimensions to show the X and Y locations of every detail. Common practice in this situation is to use ordinate dimensioning. Each ordinate dimension appears as a single extension line and a number showing the X or Y distance from the origin. The original is 0,0, unless a user coordinate system (UCS) is current, which is a handy way to locate the starting point for ordinate measurements. I cover UCSs in Chapter 21. Normal practice is to specify the lower-left corner of the part as the origin. The X and Y values are usually aligned neatly in a row and column along the edge of the part. There are no dimension lines or arrowheads.
- **Center lines — DIMCENTER:** Creates center lines or center marks on circles and arcs without applying a dimension. The definition of lines versus marks matches the current dimension style, as described later in this chapter.

Self-centered

AutoCAD 2017 introduced associative center lines. This involves two commands:

- CENTERLINES: Prompts you to select two lines. AutoCAD then draws a centerline midway between the two lines using a long-dash–short-dash line style.
- CENTERMARK: Prompts you to select a circle or an arc. AutoCAD then draws two centerlines that cross at the exact center of the circle or arc.

Now comes the magic part. Centerlines created using these two commands are fully associative to the lines, circles, or arcs that were used to define them. When you move an arc or a circle or change its radius, or when you move either the line or either end of either line, the centermarks and centerlines obediently follow and remain properly centered.

Associative centerlines are automatically placed on a layer named "Center lines" with the Center2 linetype override. If this layer doesn't exist, AutoCAD creates it automatically with a white/black color and the Continuous linetype. If this layer already exists in your template file, AutoCAD uses the color you specified, but ignores the linetype assigned to the layer and applies the Center2 linetype override anyway.

Quick, dimension!

Three dimensioning commands can help you place multiple dimensions very quickly. They're found near the middle of the Dimensions panel of the Annotate tab on the Ribbon.

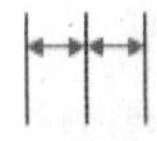

- **Continue — DimCOntinue (DCO):** Having placed one linear dimension in the drawing, you can now carry on placing a series of dimensions end-to-end. AutoCAD picks the end of the previous dimension as the start of the next one automatically, so you need only locate the extension of the next one.

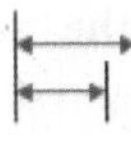

- **Base — DimBaseline (DBA):** This one works much like DimCOntinue except that AutoCAD selects the start of the previous one as the start of the next one. You thus end up with a series of stacked dimensions, all measuring from a common starting point.

- **Quick Dimensions — QDIM:** Interestingly, this is one of the few dimensioning commands that doesn't start with DIM*xxx*. When you invoke it, it invites you to select one or more objects, and then it dimensions them all at once. The best bet usually is to use one or more window selections (pick from left to right), each of which completely surrounds several objects. Now whenever you press

Enter, AutoCAD automatically applies DimLInear dimensions wherever it can, and you might be surprised at some of those places. It can place dimensions in several formats, including continue, baseline, staggered, ordinate, and several others. For convenience in repeat operations, it remembers the last mode you used.

TIP

If you want to be superefficient, memorize the three-letter command aliases for the dimension commands you use most often, of which DIM should top the list. Refer to Figure 14-1 to see the aliases of the most common commands.

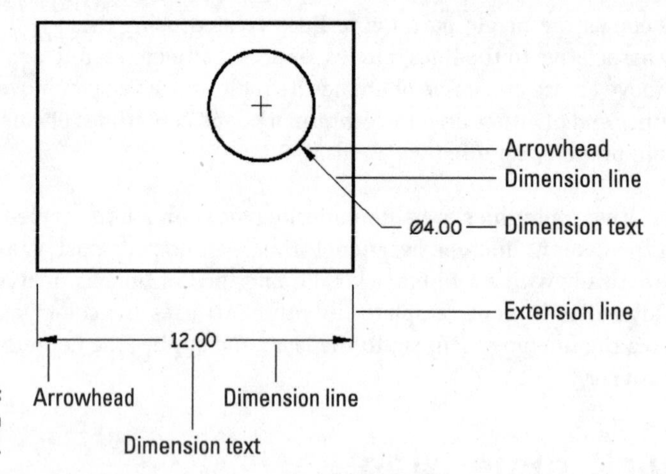

FIGURE 14-2: The parts of a dimension.

REMEMBER

The AutoCAD dimensioning commands prompt you with useful information at the command line or at dynamic input prompts. Read the prompts during every step of the command, especially when you're trying a dimensioning command for the first time. When all else fails . . .

Where, oh where, do my dimensions go?

In this section, I explain where to *put* dimensions, not where to *place* them. Huh? Right. (By now you should know to place dimensions outside the part, with smaller ones closer to the part and longer ones farther away, to avoid dimension lines crossing each other.) In this section, I show you the location in the drawing where your dimensions will live. You can choose one of three different methods for dimensioning your drawings:

» **You can place dimensions in model space using non-annotative dimensions.** This method, the oldest, for many years was the only possible way. I mention it here because you're bound to encounter many of the millions and millions of existing drawings that were created this way.

- *Advantage:* Dimensions are available in model space as you edit the drawing.
- *Disadvantage:* As with text and hatching, dimension objects have to be scaled to suit the drawing scale. Placing dimensions on details at other scales can get truly ugly. I cover scale factors in Chapters 13 and 15 as well as later in this chapter.

» **You can place dimensions in paper space.** This was introduced as the second-generation method in an attempt to overcome the scaling issue. You can reach from paper space through a layout viewport and snap onto the model space objects. Dimensions remain associative and usually change when the model space objects change. Paper space layouts are discussed in Chapter 12.

- *Advantage:* You don't need to set a dimension scale factor to suit the drawing scale. The dimension scale factor is always 1:1 because paper space layouts are usually plotted at full size, and the viewport scale sets the model space scale factor.
- *Disadvantages:* The dimensions aren't available when editing in model space, so you're flying blind. In addition, the associative connections between the paper space dimensions and the model space objects are fragile and easily broken. There's a repair command, but it's a nuisance to have to use it.

» **You can place dimensions in model space using annotative dimensions.** This is the newest, latest, greatest, and best way to place dimensions. Annotative dimensions automatically size themselves to suit the drawing scale that you choose. I discuss annotative text in Chapter 13, annotative hatching in Chapter 15, and annotative dimensions later in this chapter.

- *Advantages:* Dimensions are available in model space while you're drawing and editing, you don't need to mess with dimension scale factors, and applying dimensions to drawing details that are at a different scale from the main drawing becomes trivial.
- *Disadvantage:* Annotative dimensions have been slow to catch on because people are stuck in the older ways ("That's how we've always done it!") and because they don't understand a couple of subtleties about them that I reveal in Chapter 13. If you're the first in your office to adopt annotative dimensions, everyone else will hail you as the Great AutoCAD Guru and will pester you to help them learn how to use them.

The Latest Styles in Dimensioning

In this section, I discuss how to set the factors that control the format and appearance of dimensions, but I start with a brief anatomy lesson. AutoCAD uses the names shown in Figure 14-2, and described in the following list, to refer to the parts of each dimension:

- **Dimension text:** The numeric value that indicates the true distance, radius, diameter, or angle between definition points (also known as defpoints). Dimension text can include other information in addition to, or instead of, the number. For example, you can add a suffix, such as `TYP.`, to indicate that a dimension is typical of several similar configurations; add manufacturing tolerances; and show dual dimensioning in alternate units.
- **Dimension lines:** In linear dimensions, indicate the true distance between points. Angular dimensions have curved dimension lines with the center of the curve at the vertex point of the objects being dimensioned. For radius and diameter dimensions, the dimension line simply points at or through the center of the object being dimensioned. Refer to Figure 14-1 for examples of these dimension types.
- **Arrowheads:** Appear at the end or ends of the dimension lines and emphasize the extent of the dimensioned length. AutoCAD's default arrowhead style is the closed, filled type shown in Figure 14-2, but you can choose 20 other symbols, such as dots or tick marks, to indicate the ends of the dimension lines. Don't get ticked off about it, but AutoCAD calls the line ending an *arrowhead* even when, as in the case of a tick mark, it doesn't look like an arrow.
- **Extension lines:** Extend outward from the defpoints that you select to the dimension lines, usually by snapping to points on an object. By drafting convention, a small gap usually exists between the defpoint and the beginning of the extension line. Also by convention, extension lines usually extend just past the dimension lines; refer to Figure 14-2 for examples. AutoCAD makes dimensions look tidier by assigning fixed gap sizes and projection lengths for the extension lines, and if you need to dimension to circles or center lines, you can assign dash-dot linetypes to either or both extension lines.
- **Definition points (or defpoints):** When you create any kind of dimension, defpoints are placed on a special layer named (what else?) Defpoints. The program creates the layer automatically the first time you issue a dimension command. These tiny points are usually invisible because the objects being dimensioned sit on top of them, but you can see them by selecting a dimension to turn on its grips. The grips appear on the definition points. Because you wouldn't want these points to appear when you plot drawings, nothing created on the Defpoints layer ever plots.

USING SYSTEM VARIABLES WITH DIMENSIONS

In the early days, AutoCAD controlled the appearance of dimensions by using dimensioning system variables. For example, to switch dimension text from showing decimal values to showing fractions, you would have to change the setting of DIMLUNIT from 2 to 5. Similarly, DIMDEC sets the number of decimal places (such as 0.001) or specifies the fraction, such as 1/256 of an inch. I discuss system variables in Chapter 23.

The problem was that every time you wanted to switch between decimals and fractions, you had to change the value of these two variables. To set a system variable, or *sysvar*, you enter its name, followed by a suitable value as determined from the Help system. The only reason I mention it is that you're bound to run into older drawings that were done this way.

The good news is that Autodesk later added named dimension styles. You now set up styles for dimensions, such as making all dimension values shown with decimals or fractions, or a style with specific manufacturing tolerances, and so on. You can create as many dimension styles as you need; all it takes is a few mouse clicks to switch between them quickly.

By default, no matter which type of dimension you create, AutoCAD groups all parts of each dimension (extension lines, dimension lines, arrowheads, and text) into a special *associative dimension* object. *Associative* has two meanings:

- **The different parts of the dimension function as a single object.** When you click any part of the dimension, AutoCAD selects all its parts.
- **The dimension is connected to the points on the object that you specified when you drew the dimension.** If you change the size of the object (for example, stretch a line), the dimension updates appropriately, as shown in Figure 14-3. The lines and arrows move, and the text changes to reflect the line's new size.

TECHNICAL STUFF

The associative dimensions I'm talking about here first appeared in AutoCAD 2002. Before that, AutoCAD had a more primitive kind of dimensioning. Dimensions were single objects, and they updated when you stretched an object *if* you were careful to include the dimension itself in the crossing selection for the Stretch command. Here's where things can become confusing: AutoCAD used to call these old-style, single-object dimensions *associative*, but now calls them *non-associative*. And what used to be called *non-associative dimensions* before AutoCAD 2002 are now called *exploded dimensions*. But when you explode a dimension, you get four lines, two arrowheads, and a piece of text. Sometimes, you can't tell the players even with a program. For more information about how to determine which kind of dimension AutoCAD draws, see the "Controlling and editing dimension associativity" section later in this chapter. I mention the old style here only because you will probably encounter it in old drawings.

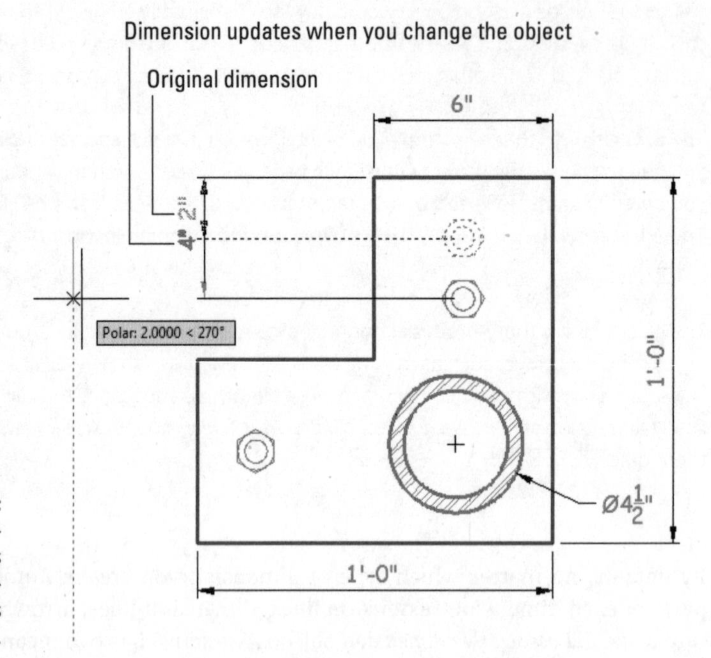

FIGURE 14-3: Changing objects automatically updates dimensions.

REMEMBER

Some people will try to warn you that AutoCAD dimensioning is a big, complicated, difficult subject. Don't be alarmed, however: The basic principles are quite simple. The problem is that every industry has its own dimensioning conventions, habits, and quirks. As usual, AutoCAD tries to support them all — and, in so doing, makes things cumbersome for everyone.

The good news is that you should have to adjust things to suit your specific industry or company only once, and then all dimensions will suit the specified standard. The really good news is that it usually takes only a bit of fine-tuning of the default settings to cover most of your dimensioning needs.

Dimension styles are saved in the current drawing. The really, really good news is that you can save this drawing as a *template* file (I cover templates in Chapter 4) so that all new drawings created from this template will have the dimension styles predefined. Or if you're working in an office, then someone may already have set up suitable dimension styles. Or you can use DesignCenter to copy styles from one drawing to another.

REMEMBER

You add dimensions to a drawing *after* you've drawn at least some of the geometry; otherwise, you won't have much to dimension! (Well, duh.) Your dimensioning and overall drafting efficiency improve if you add dimensions in batches: First draw all geometry and then place dimensions. Don't draw a line, place a dimension, draw another line, place another dimension. . . .

Creating dimension styles

A *dimension style* (or *dimstyle*, for short) is a collection of drawing settings, called *dimension variables* (or *dimvars*, for short), which are a special class of the system variables that I describe in Chapter 23.

If you *do* need to create your own dimension styles or you want to tweak existing ones, use the Dimstyle command. You can invoke it by clicking the small, diagonal arrow in the lower-right corner of the Dimensions panel of the Annotate tab on the Ribbon. The Dimension Style Manager dialog box opens, as shown in Figure 14-4.

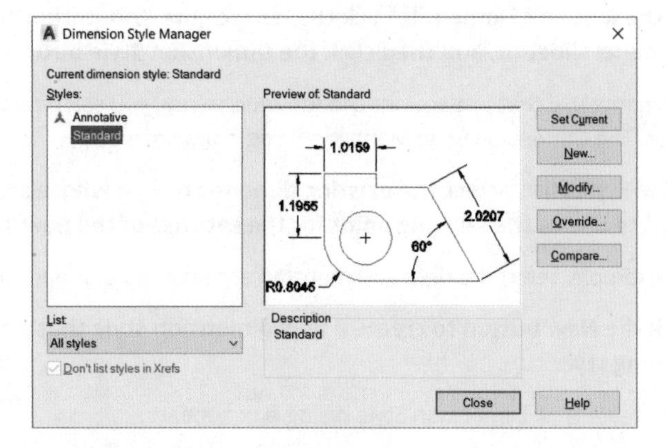

FIGURE 14-4: Yet another manager, this one for dimension styles.

Every drawing comes with a default dimension style named Standard when you start with an imperial (feet-and-inches) template, or ISO-25 with a metric template, and matching annotative styles. (I cover annotative dimensions later in this chapter.) Although you can use and modify the Standard or ISO-25 style, I suggest that you leave them as they are, and create your own dimension style(s) for the settings appropriate to your work. This approach ensures that you can always use the default styles as a reference. More important, it avoids a potential conflict that can change the way your dimensions look when the current drawing is inserted into another drawing: AutoCAD refers to styles by name, so if you have two styles of the same name, one will override the other. Chapter 18 describes this potential conflict.

When you install AutoCAD, it checks with the operating system (OS) to see what country you selected when the OS was installed, and then sets its measurement system accordingly. If you're in the United States or one of the few other countries that still use imperial units, the default dimension style is Standard; most of the rest of the world defaults to ISO-25. Canada, on the other hand, is nominally metric, but most people still use imperial. The system variable MEASUREINIT controls the default action. When set to 0 (zero), AutoCAD uses imperial units; when set to 1, metric units are used. Among other things, this setting also affects text styles (see Chapter 13), hatching (see Chapter 15), and noncontinuous line types (see Chapter 9).

Starting a new drawing from an ISO template, such as `acadiso.dwt`, forces everything to metric, and starting a new drawing from a non-ISO template (for example, `acad.dwt`) forces everything to imperial, regardless of the setting of MEASUREINIT. The system variable MEASUREMENT overrides the default for a specific drawing, but it affects only text, hatching, and noncontinuous line types.

Follow these steps to create your own dimension style(s):

1. **On the Ribbon's Home tab, click the label of the Annotation panel to open the panel slideout, and then click the Dimension Style button.**

   Alternatively, if that just sounds like too much work, you can type **D** and press Enter. The Dimension Style Manager dialog box appears.

2. **In the Styles list, select the existing dimension style whose settings you want to use as the starting point for the settings of the new style.**

   For example, select the default dimension style named Standard or ISO-25.

3. **Click the New button to create a new dimension style that's a copy of the existing style.**

   The Create New Dimension Style dialog box appears.

4. **Enter a New Style Name and then select or deselect the Annotative check box. Click Continue.**

   Select the Annotative check box to create an annotative dimension style, or deselect it for a non-annotative style.

   The New Dimension Style dialog box appears. This dialog box is virtually identical to the Modify Dimension Style dialog box, which is displayed when you edit a dimension style.

5. **Modify dimension settings on any of the seven tabs in the New Dimension Style dialog box.**

   See the descriptions of these settings in the next section of this chapter.

6. **Click OK to close the New Dimension Style dialog box.**

   The Dimension Style Manager dialog box reappears.

7. **Click Close.**

   The Dimension Style Manager dialog box closes, and the new dimension style becomes the current dimension style that AutoCAD uses for future dimensions in this drawing.

8. **Draw dimensions to test the new dimension style.**

WARNING

Avoid using the Dimension Style's Modify button to change existing dimension styles you didn't create, unless you know for sure what they're used for. When you use Modify to change a dimension style setting, *all* existing dimensions using that style change to reflect the revised setting. Thus, a change to one minor dimension variable can affect a large number of existing dimensions! To play it safe, rather than using the Modify button to change an existing dimension style, click the New button to create a new style, which automatically copies an existing one; then you can change the settings in the copy.

TECHNICAL STUFF

A further variation on the already convoluted dimension styles picture is that you can create dimension *substyles* (also called *style families*), which are variations of a main style that affect only a particular type of dimension, such as radial or angular. If you open the Dimension Style Manager dialog box and see names of dimension types indented beneath the main dimension style names, be aware that you're dealing with substyles.

Adjusting style settings

After you click New or Modify in the Dimension Style Manager dialog box, AutoCAD displays a tabbed New Dimension Style dialog box or Modify Dimension Style dialog box (the two dialog boxes are identical except for their title bars) with a mind-boggling — and potentially drawing-boggling, if you're not careful — array of settings.

Fortunately, the dimension preview that appears on all tabs — as well as on the main Dimension Style Manager dialog box — immediately shows the results of most setting changes. With the dimension preview and some trial-and-error setting changes, you can usually home in on an acceptable group of settings. For more information, use the Help feature in the dialog box: Just hover the mouse pointer over the setting that you want to know more about.

TIP

If you find the preview image hard to read in the Dimension Style Manager, New Dimension Style, or Modify Dimension Style dialog boxes, click and drag the right edge of the dialog box to increase the size of the preview image.

Before you start messing with dimension style settings, know what you want your dimensions to look like when they're plotted. If you're not sure how it's done in your industry, ask others in your office or profession, or look at a plotted drawing that someone in the know represents as being a good example. A general rule, that I've found to be helpful in virtually all aspects of life, is to stick with the defaults unless you know specifically what you want to change and why.

The following few sections introduce you to the more important tabs in the New/Modify Dimension Style dialog boxes and highlight the most useful settings. Note that whenever you specify a distance or length setting, you should enter size you want it to be when the drawing is *plotted*. I discuss plotting size and scale factors in Chapters 13 (text) and 15 (hatching) and later in this chapter.

When you have everything set the way you want it, click OK to close the Modify Dimension Style dialog box, and then click Close to exit Dimension Style Manager. The new style is now current.

Following lines and arrows

The settings on the Lines tab and the Symbols and Arrows tab control the basic look and feel of all parts of your dimensions, except text.

Symbolically speaking

The settings on the Symbols and Arrows tab control the shape and appearance of arrowheads and other symbols.

A useful setting is Center Marks. Depending on which radio button you select, placing a radius or diameter dimension also identifies the center by placing a small center mark, or by placing center lines that extend just beyond the circle or arc, or none. The default is the tick mark, but I prefer the line because normal drafting practice is usually to show center lines.

Tabbing to text

Use the Text tab to control how dimension text looks, which includes the text style and height to use (see Chapter 13) and where to place the text with respect to the dimension and extension lines. In particular, note the Text Style drop-down list, which shows the text styles available in the drawing. Click the three-dot Browse button at the right end of the list to open the Text Style dialog box, and edit or create a suitable text style if one doesn't already exist in your current drawing. The default Text Height in imperial units (0.180) is too large for most situations; set it to 1/8″, 0.125′, 3mm, or another height that makes sense.

You should define the text style for dimensions with a height of 0 (zero) in the Text Style dialog box. See Chapter 13 for more information about variable-height and fixed-height text styles. If you specify a fixed-height text style for a dimension style, the text style's height overrides the Text Height setting in the New/Modify Dimension Style dialog boxes. Use a zero-height style to avoid the problem. A zero-height text style can be a real nuisance when placing text, which means that it's almost mandatory to have at least two text styles defined: one for text (fixed height) and one for dimensions (zero height).

Getting fit

The Fit tab includes a bunch of options that control when and where AutoCAD shoves the dimension text when it doesn't quite fit between the extension lines. The default settings leave AutoCAD in "maximum attempt at being helpful" mode. That is, AutoCAD moves the text, dimension lines, and arrows around automatically so that nothing overlaps. On rare occasions, AutoCAD's guesses might be less than perfect. It's usually easier to adjust the text placement by grip-editing the placed dimension, as I describe in the section "Editing dimension geometry," later in this chapter, instead of messing with dimension style settings.

Even at its most helpful, AutoCAD sometimes makes a bad first guess about how you want dimension text and arrows arranged. If you're having problems getting the look you want, flip the arrows to the other side of the dimension lines by selecting the dimension and choosing Flip Arrow from the multifunction grip on the arrow.

Most important, the Fit tab includes the Annotative check box. Using annotative dimensions, as I recommend in this chapter, makes dimensioning go a lot more smoothly!

The Use Overall Scale Of setting corresponds to the DIMSCALE system variable, and you'll hear long-time AutoCAD drafters refer to it as such. In Chapters 13 (text) and 15 (hatching), I refer to a scale factor by which text height and hatch patterns need to be multiplied so that they plot properly. This principle also applies to dimensions: When you're using old-style non-annotative dimensions, this is where that number goes. It resizes text height, arrowhead sizes, and gaps accordingly. When Scale Dimensions to Layout (for paper space layout dimensioning) is selected, DIMSCALE is automatically set to 0 (zero). When Annotative is selected, DIMSCALE is ignored, and a suitable scale factor is applied to each dimension when it's created.

Using primary units

The Primary Units tab gives you highly detailed control over how AutoCAD formats the characters in the dimension text string. You usually set the unit format and precision, and maybe specify a suffix for unitless numbers, when it's not clear from the drawing which units you're using. You may also change the Zero Suppression settings, depending on whether you want dimension text to read `0.5000`, `.5000`, or `0.5`. "Zero Suppression!" also makes a great rallying cry for organizing your fellow AutoCAD drafters.

AutoCAD 2010 introduced an interesting tweak to dimension text: dimension *subunits.* If the main unit of measure on a drawing is meters, rather than have a bunch of smaller distances dimensioned as, say, 0.450, you could create a centimeter subunit so that any dimension of less than 1 meter would be shown in centimeters. I was always taught a strict drafting standard specifying that all dimensions on a drawing must be in the same units. In other words, in a drawing with meters as the dimension unit, 0.45 would be correct and 45cm would be incorrect. NASA once crashed an expensive probe on Mars because some people were using imperial units and others were using metric. Check your own office standards before incorporating this feature, which you can find on the Primary Units tab of the New/Modify Dimension Style dialog box.

Other style settings

When your work requires that you show dimensions in two different systems of measure, such as inches and millimeters, use the Alternate Units tab to turn on and control alternate units. Alternate Units display both dimensions at a time. Use 25.4 as the Multiplier for Alt Units to display millimeters alongside inches. When your work requires listing construction or manufacturing tolerances, such as `3.5 +/-0.01`, use the Tolerances tab to configure the tolerance format.

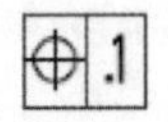

AutoCAD includes a separate TOLerance command that draws special symbols called *geometric tolerances.* If you need these symbols, you probably know it and you have my sympathies; if you've never heard of them, just ignore them. Search for the term *Geometric Tolerance dialog box* in the AutoCAD Help system for more information.

Changing styles

To switch to a different style, click in the dimension style window in the Dimensions panel of the Annotate tab on the Ribbon, and select a style from the drop-down list. Beginning with AutoCAD 2014, if you know the name of the style that you want to use, just begin typing it at the command line. The AutoCAD command line searches for dimension styles; when you see the one you want, just click it.

Scaling Dimensions for Output

You need to adjust the size of text and dimensions that are applied in model space to suit the final plotting scale of the drawing. By far, the best way to do this is to use annotative dimensions.

Follow these steps to use the Annotative dimension style and apply an annotative scale to a dimension:

1. **Start a new, blank Imperial drawing.**

   Use the `acad.dwt` template.

2. **Draw a horizontal line about 20 units long.**
3. **Apply a linear dimension (DimLInear, or DLI) to the line.**

   Pretty hard to read, isn't it?

4. **Switch to the Annotative dimension style.**

   Find the Dimension Style drop-down list that reads `Standard` in the upper-right corner of the Dimension panel of the Annotate tab on the Ribbon. Click it and then choose Annotative.

5. **Change the current drawing annotation scale.**

   Select the 1:5 scale from the drop-down list, under the Annotation Scale button in the lower-right corner of the application window.

6. **Use the DIMLIN command again and place a second dimension for the line.**

   Ah, that's better, and it didn't require any esoteric calculations. You now have two dimensions that measure the length of the line, but the text and arrowheads appear at different sizes now.

SCALING DIMENSIONS THE HARD WAY

You used to have to draw the complete object full size, including all the fine details and zooming in as necessary. Then you'd copy those elements that made up the fine detail and use the SCale command to resize them to suit the detail scale. Finally, you'd apply dimensions to the scaled one that had two suitable scale factors: Set the dimension scale factor to suit the overall drawing scale (say 1:2), and then set the DIMLFAC (DIMension Linear FACtor) to suit the detail scale.

The problem is that your detail is at a different scale. DIMLFAC (DIMension Linear FACtor) is a fudge factor applied to dimensions that corrects the measured value, so a 1" line in a 5:1 detail measures as 1 inch, not 5 inches. Ouch.

To avoid confusing results, which is what has turned off most users from using annotative annotations, make sure that the Automatically Add Scales button near the right end of the status bar is turned off. When Automatically Add Scales is enabled, AutoCAD updates all annotation objects that support annotative behavior when you change the current annotative scale. Having all scales attached to every annotation object is not ideal as it is much harder to control where the new scale representations appear in your drawing, especially when they might not all be needed. I discuss this topic more in Chapter 13.

Dimensioning details at different scales is always the most difficult type of dimensioning, unless you use annotative dimensions.

Here's the easy way to create a multiscale drawing, such as the one in Figure 14-5:

1. **Draw the object at full size in model space, including the small notch detail.**

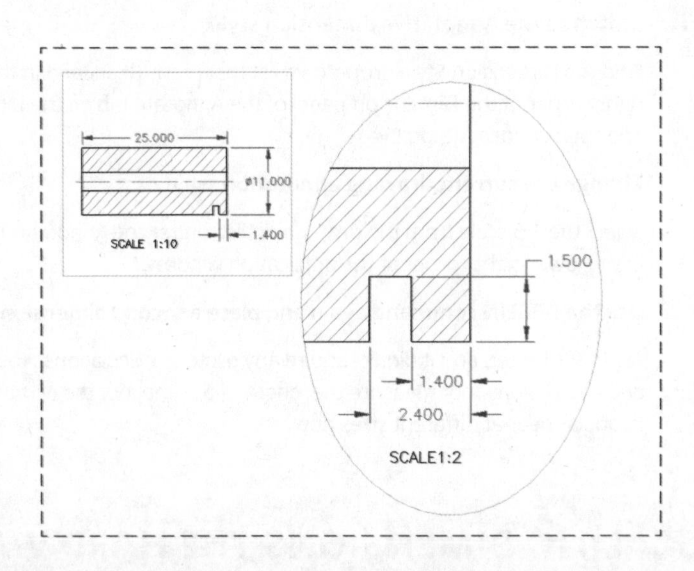

FIGURE 14-5: A drawing with a detail at another scale.

2. **Select the 1:10 scale from the drop-down list under the Annotation Scale button in the lower-right corner of the AutoCAD window.**
3. **Apply the three dimensions that show in the Scale 1:10 view in Figure 14-5, using an annotative dimension style.**

   The figure shows paper space, but you're working in model space.
4. **Apply an annotative hatch and then draw the center line.**

   Turn to Chapter 15 if you're not sure how to apply an annotative hatch.

5. **Edit the properties of the hatch pattern and the 1.400 dimension to add 1:2 scale factors.**

   Chapter 15 discusses editing hatch patterns.

6. **Make sure that the Automatically Add Annotative Scales button is turned off, and change the Annotation Scale to 1:2.**

   Three existing dimensions disappear, and the hatch and the 1.400 dimension resize themselves.

7. **Add the 1.500 and 2.400 dimensions.**

8. **Switch to the paper space Layout1 tab. Click the viewport boundary and then grip-edit it approximately to the size and location shown.**

   Your model space drawing is probably not properly located, and the hatch and dimensions don't show.

9. **Click the viewport boundary again, click the Viewport Scale button (which might contain an AutoCAD-generated best-fit value like `0.694694`), and select 1:10 from the scale list.**

   The viewport zooms accordingly, and the hatch and 1:10 dimensions appear. If necessary, double-click in the viewport to enter model space, and then pan accordingly.

WARNING

   Don't zoom or else you lose the scale setting!

10. **Double-click outside the viewport to return to the paper space layout.**

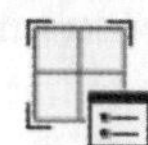

11. **Create a second viewport by using the VPORTS command, or simply copy it and then modify the existing one.**

12. **Repeat Steps 8 and 9 on the new viewport, but use a Viewport Scale of 1:5.**

TIP

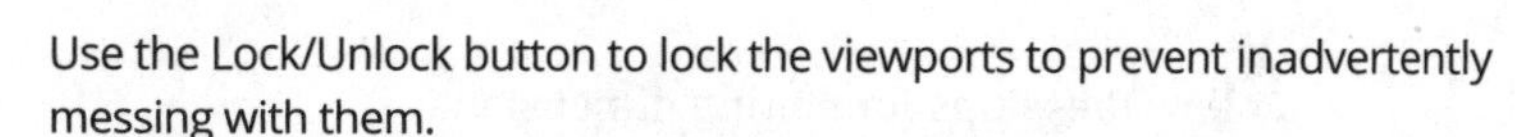

   Use the Lock/Unlock button to lock the viewports to prevent inadvertently messing with them.

There you have it — a multiview, multiscale detail drawing without duplicating any geometry or annotations. To see the real magic of it, go to the model space tab and observe that there's still only one model of the part and no detail view at the other scale. Use the Stretch command to play with the depth and location of the notch. Go back to the Layout1 tab to see how everything has automatically updated in both views. Magic!

Editing Dimensions

After you draw dimensions, you can edit the position of the various parts of each dimension and change the contents of the dimension text. AutoCAD groups all parts of a dimension into a single object.

Editing dimension geometry

The easiest way to change the location of a part of a dimension is with grip editing. I describe editing with grips in Chapter 11. Just click a dimension, click one of the grips (which generally appear at the text, the ends of the dimension lines, and the defpoints), and then maneuver away. You'll discover that certain grips control certain directions of movement.

In AutoCAD 2012, dimensions joined the group of objects that feature multifunction grips. Hover the cursor over the text grip of a linear dimension, and notice the menu that appears, with numerous options for adjusting the location of the text. Hover over an arrow grip, and you can stretch the extension line(s) longer or shorter, start a continuous or baseline dimension from that end of the dimension, or flip the arrow. You can do these things by selecting a dimension and changing items in the Properties palette, but the multifunction grips are more immediate.

When you want to change the look of a component of a specific dimension, use the Properties palette. For example, you can substitute a different arrowhead (Arrow 1 and Arrow 2) or suppress an extension line (Ext line 1 and Ext line 2). See Chapter 11 for more on the Properties palette. All dimension settings in the New/Modify Dimension Style dialog boxes (see "Adjusting style settings," earlier in this chapter) are available in the Properties palette after you select one or more dimensions.

Follow these tips for editing dimensions:

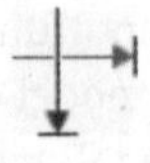

- **Use smart breaks.** In manual drafting, it's considered bad form to cross object lines (that is, real geometry) with dimension lines or extension lines, or to have anything cross a dimension line. Extension lines can be crossed. Dimension Break (DIMBREAK) prompts you to select a dimension and then an object to break the dimension line. The wild aspect is that it's a smart break. If you later change where they cross, the break follows accordingly — the break even heals itself when you remove the crossing elements. Better yet, if you change things *again* so that they cross again, the break reappears!

- **Use the DIMSPACE command.** The Dimension Space (DIMSPACE) command applies a specified separation between existing linear or angular dimensions. If you don't use the DimBAseline command as the dimensions are created, spacing dimensions equally afterwards will require tedious manipulation with Snap Mode and the Move command.
- **You don't always have to draw everything at full size.** A fundamental mantra in AutoCAD is that you should always draw everything at full size. On occasion, however, it isn't always practical. For example, you may design a power-transmission shaft for a large machine. The shaft is 4 inches in diameter and 12 feet long. It has a variety of splines, keyways, and bearing shoulders on each end, but the 11-foot section in the middle is simply a straight cylinder. If you draw it at full size and scale the plot to fit a suitable paper size, you'll never see the details at each end.

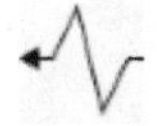

  Common practice would be to draw the interesting end details, break out and remove the boring center section, and then bring the ends closer together. Now you can create a reasonable plot. The problem is that any dimension that crosses over the break, such as the overall length, doesn't show the correct value. The solution is to override the value, and then to use the DIMJOGLINE command to insert a jog in the dimension line to indicate that the dimension line isn't the true length. I explain how a little later in the section "Editing dimension text."

REMEMBER

  In spite of the name similarities, don't confuse DIMJOGLINE with the DimJOgged command, covered earlier in this chapter. DIMJOGLINE is for linear dimensions, and DimJOgged is for radial dimensions.

- **Right-click for useful options.** If you select one or more dimensions and right-click, the menu displays a number of useful options for overriding dimension settings or assigning a different style.
- **Sometimes, it's better to create a new style.** When you change a setting in the Properties palette, you're *overriding* the default style setting for that dimension. If you need to make the same change to a bunch of dimensions, it's usually better to create a new dimension style and assign that style to them. You can use the Properties palette or the right-click menu to change the dimension style that's assigned to one or more dimensions.

WARNING

  If you manually change a dimvar setting, the setting is applied to the current dimension style as an override, and all subsequent dimensions that are placed by using this style have the overridden appearance. This can cause much the same problem as overriding object properties instead of using different layers; if you edit a dimension style, all existing dimensions that use it update, including the ones that you hadn't expected because they seem to be different from the ones you want.

» **Use a mask for the dimension text.** You can use the Properties palette to turn on AutoCAD's background mask feature (described in Chapter 13) for the text of individual dimensions: Select the dimensions, display the Text area in the Properties palette, and find the Fill Color item. Click in the list box, scroll down, and select Background to use the drawing background color, which usually gives the best results. To ensure that dimension text lies on top of other objects, use the TEXTTOFRONT (Chapter 13) or DRAWORDER (Chapter 18) commands.

» **Don't explode dimensions.** The AutoCAD eXplode command on the Home tab's Modify panel blows a dimension to smithereens — or at least into a bunch of line, polyline, and multiline text objects. *Resist the temptation.* Exploding a dimension makes it much harder to edit cleanly and eliminates AutoCAD's ability to update the dimension text measurement automatically.

Editing dimension text

In most cases, you shouldn't have to edit text in dimensions. Assuming that you draw the geometry accurately and pick the dimension points precisely, AutoCAD always displays the right measurement. When you change the size of the associated object, AutoCAD updates the dimension text. However, you may occasionally want to *override* the dimension text (that is, replace it with a different measurement), *add* a prefix or a suffix to the true measurement, or change its display accuracy.

AutoCAD creates dimension text as multiline text, so dimension text has the same editing options as mText. You have several ways to launch the dimension text editor, but the easiest is to simply double-click the dimension text.

AutoCAD displays the true dimension length as text in the actual dimension, and updates the text when you change the size or location of the object. You can override the true length by replacing it with a specific length or another text string. You can preserve the true length by adding a prefix or suffix by inserting <> (that is, the left and right angle-bracket characters) as placeholders for the dimension value. For example, if you enter **About <>** and the actual distance is 12.000, AutoCAD displays `About 12.000` for the dimension text. If you stretch the object later so that the actual distance changes to 14.528, AutoCAD changes the dimension text automatically to read `About 14.528`. Now you can appreciate the importance of drawing and editing geometry precisely!

WARNING

Avoid the temptation to override the default dimension text by replacing the angled brackets with a numeric value unless you plan to use the DIMBREAK command on the dimension. Doing so eliminates AutoCAD's ability to keep dimension measurements current; even worse, you get no visual cue that the default distance has been overridden unless you edit the dimension text. If you're overriding dimension text a lot, then it's probably a sign that the creator of the drawing didn't pay enough attention to using precision techniques when drawing and editing. I don't point fingers, but you probably know whom to talk to.

Controlling and editing dimension associativity

When you add dimensions by selecting objects or by using Object Snap modes to pick points on the objects, AutoCAD normally creates associative dimensions, which are connected to the objects and move with them. This is the case in new drawings that were originally created in any release of AutoCAD starting with 2002.

WARNING

When you have to work on drawings created or last edited in versions older than AutoCAD 2002, you must set the DIMASSOC system variable to 2 before later releases will create associative dimensions. An easy way to make this change for the current drawing is to open the Options dialog box (click the Application button and choose Options from the bottom of the Application menu), click the User Preferences tab, and turn on the Make New Dimensions Associative setting. Be aware that this setting affects only new dimensions that you draw from now on. To make existing non-associative dimensions associative, use the DIMREASSOCiate command described in the following list. Search for the term *DIMASSOC system variable* in the AutoCAD Help system for more information.

TECHNICAL STUFF

You aren't likely to need any of these three commands often, but when you do, look up the command name in the online Help system:

- **DIMREASSOCiate:** If you have dimensions that aren't currently associative (probably because they were created in older versions of AutoCAD) or they're associated with the wrong objects, you can use the DIMREASSOCiate command to associate them with points on the objects of your choice.
- **DimDisAssociate:** You can use the DimDisAssociate (DDA) command to sever the connection between a dimension and its associated object.
- **DIMREGEN:** In a few special circumstances, AutoCAD doesn't automatically update geometry-driven associative dimensions. In those cases, the DIMREGEN (no command alias) command fixes things.

And the Correct Layer Is . . .

Starting with AutoCAD 2016, you can set up AutoCAD so that dimensions are automatically placed on the correct layer. Follow these steps:

1. **Start the LAyer command.**

   Click the Layer Properties button on the Layers panel.

2. **Create a new layer.**

   Give it a suitable name, such as Dims, and assign it a desired color. I discuss layers in Chapter 9.

3. **Click the Annotate tab of the Ribbon menu.**
4. **In the Dim Layer Override drop-down list in the middle of the Dimensions tab, select your dimension layer.**

   That's it! Whenever you add a dimension to this drawing, it will automatically land on your dimension layer, regardless of which layer is active.

WARNING

DO NOT use this layer selection drop-down list when you want to change the current layer for other drawing objects. This list sets only the dimension layer.

TIP

Create and set your default dimension layer in your template drawing. Any new drawings that are created from your template will automatically have your dimension layer predefined. I discuss templates in Chapter 4.

- » **Applying annotative hatch patterns**
- » **Choosing hatch boundaries**
- » **Using predefined and user-defined hatch patterns**
- » **Editing hatches**

Chapter 15
Down the Hatch!

When you need to fill closed areas of drawings with special patterns of lines (*crosshatches*, or simply *hatches*) or fill them with solid colors, this is your chapter. If you were hoping to hatch a plot or plot a hatch, see Chapter 16. If you want to hatch an egg, look at *Raising Chickens For Dummies*, by Kimberly Willis and Robert T. Ludlow (Wiley).

A *hatch* in AutoCAD is an object that fills an area. Its appearance is dictated by the pattern or fill color assigned to it. It is associated with the objects that bound the area — typically, lines, polylines, and arcs make up the boundary, much like borders of a country. When you move or stretch the boundary, AutoCAD normally updates the hatch to match the moved or resized area.

Drafters often use hatches to represent the types of material from which the objects they are designing are made or filled, such as metal, insulation, or concrete. In other cases, hatch patterns help emphasize and clarify particular elements in drawings — for example, the location of walls in a building plan or swampy areas on a map so that you know where to avoid building the road.

Figure 15-1 shows an example of using hatches to specify a concrete footing. The four patterns indicate materials to be used; from top to bottom, they are concrete, gravel, sand, and soil. In mechanical design, hatches are used to show which faces are cut in cross sections, as well as to indicate the material.

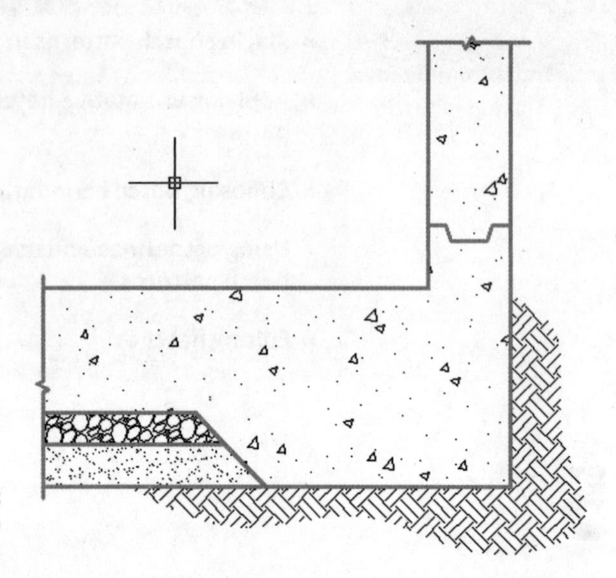

FIGURE 15-1: A big batch o' hatch.

Creating Hatches

This section gives you a jump-start on the basic process AutoCAD uses to create hatches and shows you how easy it is. I also cover many of the options included on the Hatch Creation contextual tab on the Ribbon, shown in Figure 15-2, and how to edit existing hatched areas.

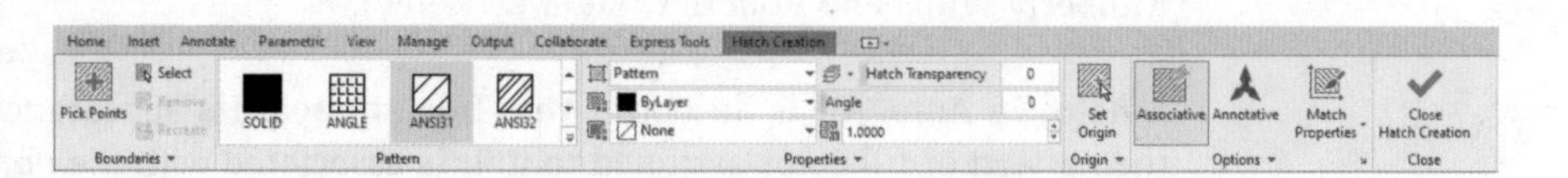

FIGURE 15-2: The Hatch Creation contextual tab on the Ribbon.

Follow these steps to hatch an enclosed area by using the pick-points method of showing the area to be hatched:

1. **Start a new drawing, using the `acad.dwt` template for imperial units or `acadISO.dwt` for metric units.**

   Draw a circle with a radius of 5 units (imperial) or 50 units (metric). Draw a second circle inside the first with a radius of 2 units (imperial) or 20 units (metric).

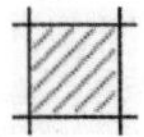

2. **Start the Hatch command by typing H and pressing Enter or by clicking the Hatch button in the Draw panel of the Home tab on the Ribbon.**

   The Hatch Creation contextual tab on the Ribbon appears, as shown in Figure 15-2. Ignore it for now.

3. **Move the cursor on the screen.**

   As the cursor moves within any enclosed area in the drawing, a preview of the pattern shows you how the final hatch will look after you pick a point inside that particular area. Starting from outside the larger circle, move the cursor into the space between circles, and then into the inner circle, and then back into the space between circles, noting how the quick preview shows the area to be hatched.

4. **Change some hatch options.**

   The quick preview updates also as you select different patterns and change values in the Hatch Ribbon tab. Try different patterns. The three little arrows along the right side of the Pattern panel scroll through the different patterns that are available. My favorite pattern is Escher.

5. **Pick a point in the region between the circles and then press Enter or the spacebar.**

6. **Click the Close Hatch Editor button to exit the command.**

   Congratulations — in only a few seconds, you've done something that would have taken an hour or more in the days of pencil and paper!

7. **Change the diameter of the outer circle:**

   a. *Click the outer circle and then click and drag one of its four outer grips to change the diameter of the circle.* The hatch pattern updates to match. This behavior, which is due to AutoCAD associating the pattern to its boundaries (in this case the two circles), is known as *associative hatching*.

   b. *Click the inner circle, and then click and drag its center grip to move it outside the larger circle.* The hole in the pattern of the larger circle fills in and the smaller circle gets filled with hatching.

   c. *Click the smaller circle and then click and drag its center grip to move it back inside the larger circle.* The smaller circle loses its hatch and reverts to cutting a hole in the hatch of the larger circle.

When working with hatches, keep these tips in mind:

- **Place the hatches on a dedicated layer or layers.** You can then easily turn hatches off, which is useful for making drawings clearer to read and printing draft plots more quickly. See "Hatching Its Own Layer."

- **Always use the Continuous line type for hatch layers.** Most hatch patterns display patterns of lines generated by the hatch process. If you place hatches on a layer with a noncontinuous line type, the Hatch command tries to create each of its own noncontinuous line segments from noncontinuous lines and the hatch process looks a little bizarre.
- **Modify each hatch separately.** By default, when you select two or more separate, closed areas, the areas are hatched as a single object. Later, when you edit one hatched area, all areas in the original set update. A good example of this is the cross-section of a complex mechanical part. When you want to apply a lot of hatches at once, but be able to modify each hatched area on its own, choose Create Separate Hatches on the Options slideout panel.
- **Set the Draw Order** to specify whether the hatch objects are in front of or behind the hatch boundary or other drawing objects, as shown in Figure 15-3. By default, they're sent behind their boundary, which is typically what you want. If a hatch is a different color from its boundary and if it's in front of its boundary, the ends of the hatch lines produce an unwanted dotted effect along the boundary.

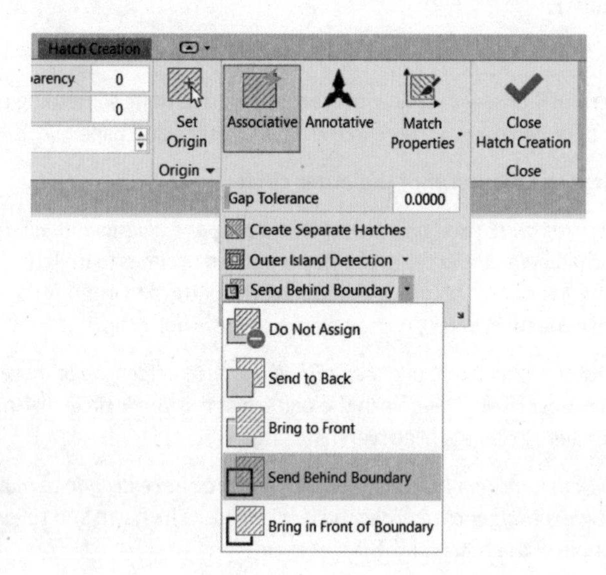

FIGURE 15-3: Hatch options for every occasion.

Hatching Its Own Layer

Starting with AutoCAD 2017, you can set up drawings so that hatches are placed on a layer of their own, automatically. Follow these steps:

1. **Start the LAyer command.**

   To do so, click the Layer Properties button on the Layers panel of the Home tab.

2. **Create a new layer.**

   Give it a suitable name, such as Hatch, and assign it a color. I discuss layers in Chapter 9.

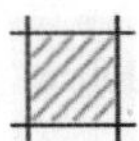

3. **Start the Hatch command.**

   Type **H** and press Enter or click the Hatch button on the Draw panel of the Home tab.

4. **Select your Hatch layer.**

   Click the down arrow at the bottom of the Properties tab to see the slideout panel. In the Hatch Layer Override drop-down list that appears, change the layer from Use Current to the new layer you created.

5. **Close the Hatch Creation tab by clicking the Close Hatch Creation button.**

   That's it! From now on, whenever you create a hatch pattern in this drawing it will automatically land on the Hatch layer, regardless of which layer is currently active. Remember to add this layer to your template drawings; you can do this by using the HpLayer system variable.

TIP

Create and set your default Hatch layer in your template drawing. Then any new drawings you create from your template will automatically have the Hatch layer predefined. I discuss templates in Chapter 4.

Using the Hatches Tab

Now that you know how easily you can hatch areas, you can explore the main options in the contextual Hatch Creation tab of the Ribbon. Any time you start the Hatch command, you see the Hatch Creation tab. (Refer to Figure 15-2.) If you double-click an existing hatch, you see the Hatch Editor tab instead, but the two tabs are virtually identical.

Working from left to right, the Hatch Creation tab contains these options:

- **Boundaries:** The easiest, most intuitive, and most common way to define the boundary of a hatch is to simply pick inside a region. AutoCAD searches for objects nearest the pick point that would enclose an area, and then highlights them in blue. If you pick multiple regions in one run of the Hatch command, each region is hatched, and the result is one big hatch object, even when parts look disconnected. If you edit the properties of any region later, then connected hatches are updated. This strategy is commonly used when hatching the cut faces of a cross section; all faces common to a single item automatically maintain matching hatch properties.

  Choose Create Separate Hatches on the Options slideout panel when you want to modify each hatched area independently.

  You can also select an area to hatch by picking objects that could define the boundary, but you must select enough objects to define a fully enclosed region.

  Starting with AutoCAD 2014, when you're picking in multiple bounded areas or selecting individual objects, the Hatch command includes an Undo option in case you accidentally pick the wrong area or object.

- **Island Detection:** In the section "Creating Hatches," you can see how the hatch pattern behaves when you select the region between the circles; it doesn't hatch the inner circle. This is *island detection*. If you select specific objects for the boundary, island detection is turned off. If island detection is off and you select only the outer circle, the hatches run right over the inner circle. To learn more about island detection, consult the Help system.

WARNING

No matter which selection method you choose, the boundary must usually be airtight. Boundary objects can overlap, but they can't have leaks — not even microscopic ones. Technically, a fuzz factor can be set in AutoCAD to allow for tiny leaks. However, I don't tell you how to set a fuzz factor because it defeats the purpose of drawing with precision.

TIP

Whenever you see the error message `A Closed Boundary Could Not Be Determined`, you need to adjust lines and other boundary objects so that they define a fully airtight boundary. The Hatch command displays red circles at gaps in the not-quite-enclosed area you want to hatch. Even when it doesn't fix the gaps, it shows you where you should fix them. Sometimes, you can use the Fillet command with a 0 (zero) fillet radius or the Join command to force two lines to meet exactly,. Another possibility is to use grip editing to align one endpoint precisely with another with the assistance of Object Snap (see Chapter 11).

- **Pattern:** The scroll arrows at the right edge of the Pattern panel give access to the 82 predefined hatch patterns, including nine solid and gradient fills that ship with AutoCAD. You can define two or three colors for the gradient fills, making them suitable for simulating curved surfaces or sunsets in the desert.

Hatch patterns are defined by external files named `acad.pat` or `acadlt.pat` in imperial units, and `acadiso.pat` or `acadltiso.pat` in metric. Each file includes the definitions for all patterns. You can create your own hatch patterns; the Customization Guide in the online Help system explains how. You can buy libraries of custom hatch patterns. Any pattern not defined in `acad.pat` or `acadiso.pat` is referred to as a *custom pattern*, but you must be careful when using them. Because they're external files, they must be available to AutoCAD whenever anyone opens a drawing containing them. If you send the drawing to someone else, you must also send the pattern definition file(s), which can have copyright issues if you've purchased them.

Autodesk created a set of hatch patterns whose names begin with the characters *AR-* (that's a hyphen at the end) for use in architectural drawings. Unlike non-AR patterns, they *do* represent real objects such as brick and roof shakes. The AR patterns were designed with a final hatch scale of 1.0 in mind, but in some cases you have to adjust up or down to achieve a suitable scale.

» **Properties:** The Properties panel contains buttons for nearly a dozen properties that can be assigned to hatches. I talk about the most important ones here. They are identified by their default values — your results may differ:
  - *Pattern:* See the Pattern panel.
  - *By Layer:* This button refers to the color of the hatch lines and should nearly always be set to ByLayer, which is what the current general color override should be.
  - *None:* This button refers to the background color, which can be useful when using hatching to portray actual objects. For example, a brick pattern can have black hatch lines with a red background so that the hatched region looks like red bricks, as shown in Figure 15-4.
  - *Hatch Transparency:* The options for this button range from 0 percent (opaque) to 90 percent (nearly transparent). Sometimes a faded hatch patterns helps reduce its visual effect.
  - *Angle:* Hatch patterns can be rotated at any angle you want. For example, standard practice in mechanical design is to hatch cut faces in cross sections of assemblies, usually with pattern ANSI31, and then to adjust the angle of each pattern to help distinguish one from its adjoining one. Note that some patterns (the ANSI series, in particular) usually have an initial 45-degree rotation built in.
  - *1,0000:* Ah, here's the big one (pun intended), scale. I cover it in more detail in the "Scaling Hatches" section later in this chapter.

- **Set Origin:** AutoCAD typically creates hatched areas by having them radiate outward from the origin (0,0 of the drawing, by default), effectively turning hatches on and off as they encounter boundaries. This behavior can, however, cause problems with certain real-object patterns. You wouldn't want a brick wall to start with 3/17 of a brick, so the Origin option lets you snap to an exact point, such as the lower-left corner of a wall. See Figure 15-4. You may also want to check out Chapter 9's reference to the need for a brick layer.

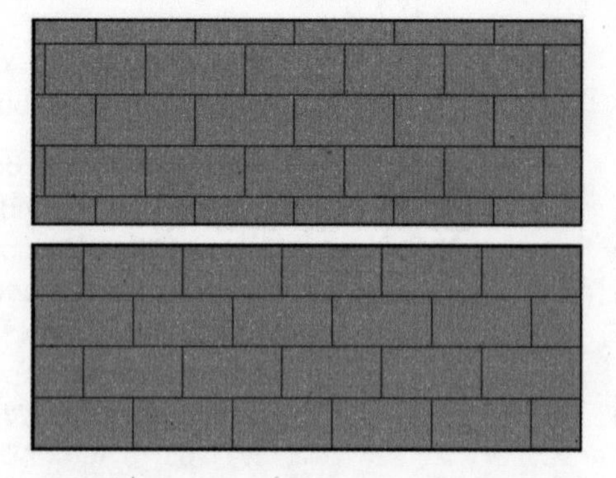

FIGURE 15-4: Poorly laid red bricks (top); properly laid red bricks (bottom).

- **Options:** This one has many suboptions, but these three are the most important:
  - *Associative:* By default, hatches are associative to their boundaries. One knows about the other, which explains what happens when you change and move boundary objects after hatching is applied.
  - *Annotative:* Like text (described in Chapter 13) and dimensions (described in Chapter 14), hatches must be scaled properly to match the drawing scale. I cover this topic in more detail in the later section "Scaling Hatches."
  - *Match properties:* This function lets you select an existing hatch object and then copy its properties onto another hatch object, just like the Format Painter in many Windows applications. I talk about Format Painter under "Adding style."
- **Close Hatch Creation:** Surprise! This option closes the Hatch tab.

Scaling Hatches

Hatches are annotation objects, just like text (see Chapter 13) and dimensions (see Chapter 14). Their size is independent of the geometry in the drawing. They impart important information and so must be clearly visible on the screen and on the printed page.

Like text and dimensions, hatches need to be scaled to suit the final drawing scale. Start with the common ANSI31 pattern. As defined, it produces parallel lines spaced 1/8" apart. So far, so good — but what if you're applying the cross section to a large part, such as the 10′ boom on a backhoe? Remember that you always draw at full size and then scale to suit the plot.

If you were to apply hatches at the nominal 1:1 scale and then plot at, say, 1:10 to fit the big boom on a small sheet of paper, the 1/8" hatch spacing becomes a tiny 0.0125-inch spacing, which is plotted effectively as a solid fill. The information imparted by the pattern is lost. So, as with text and dimensions in this drawing, you apply to the hatches a scale factor, say 10. The hatch lines in the drawing file then are bumped up to 1.25 inches apart, which scales down to the correct 1/8" apart when plotted. I discuss scale factors also in Chapter 4.

Scaling the easy way

The easy way to scale a hatch pattern is to select the drawing scale from the Scale List button near the right end of the status bar, the same as you do for text (see Chapter 13) and dimensions (see Chapter 14), and then to turn on the Annotative option in the Options panel of the Hatch tab on the Ribbon. Now whenever you create hatches, they scale themselves correctly to suit the current plot scale.

WARNING

The setting of the Annotative button is good for only the current editing session, and it must be reset if you add more hatches later. The AutoCAD Help facility states that this setting is saved in the drawing, but it isn't.

In Chapter 14 (which describes dimensions), I strongly recommend turning off the Automatically Add Scales option when placing annotative text and dimensions, because you normally don't want everything to show at every scale, especially when creating details at other scales. Hatches, on the other hand, normally *do* show in every view at every scale, scaled accordingly. You might be tempted to turn this option back on for hatches, but if you do, the first time you change the drawing scale, the new scale is added to all existing annotative objects. The best practice is to edit the hatches and manually add scales, the same as I discuss in Chapter 13 for text and Chapter 14 for dimensions.

REMEMBER

Don't confuse *annotative* and *associative*, even though both are polysyllabic words that start with *a* and sit next to one another in the Options panel on the Hatch Creation tab. *Annotative* hatch objects scale themselves automatically to suit the drawing scale. *Associative* hatch objects (enabled by default) update to the new area when you change the hatch boundary. A hatch can be either, both, or neither.

SCALING THE HARD WAY

The hard way (the old way) to scale hatches is to calculate the scale factor before placing the hatches and then enter it into the appropriate window in the Properties panel on the Hatch tab of the Ribbon, in the hope that you don't need to change the plotting scale later. You don't even want to think about setting the scale for a hatched detail at a different scale in the same drawing.

Annotative versus non-annotative

Figure 15-5 illustrates two versions of the same drawing, one half dressed up with annotative hatch patterns and the other with non-annotative hatch patterns. As shown by the annotation scales displayed on the drawings' status bars, annotative hatches change their scales automatically, whereas the size of the non-annotative hatches remains unchanged.

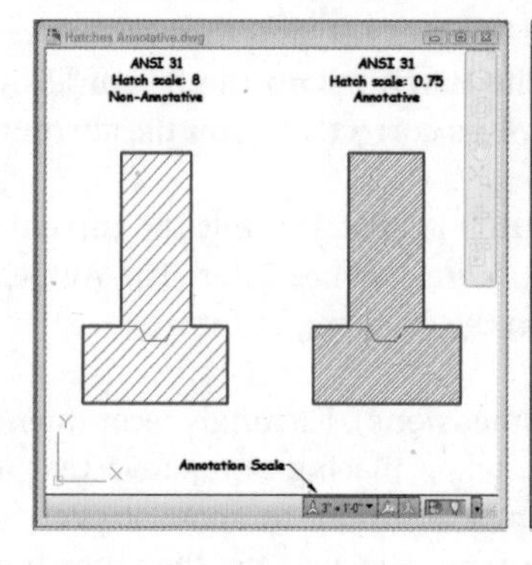

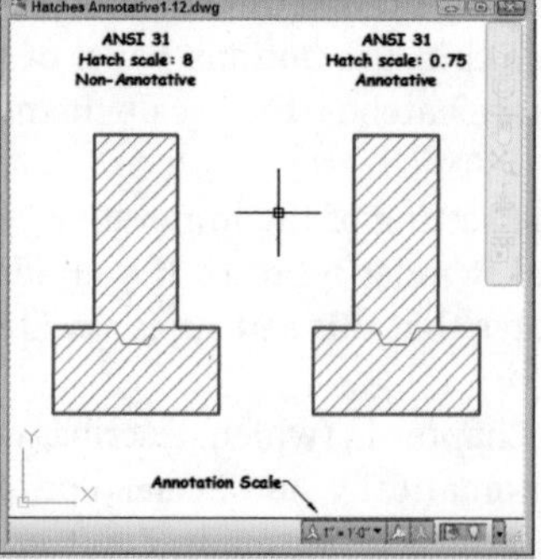

FIGURE 15-5: Hatches annotative (and not).

Chapter 14 shows another example: The hatches in both views are the same object, but because the viewports have different scales, the hatches adjust accordingly.

Before annotative hatches first appeared in AutoCAD 2008, the only way to create the effect in both drawings was to create separate layers (one for each hatch scale), hatch the object twice, and then freeze or thaw layers as needed.

Pushing the Boundaries of Hatch

In the remainder of this chapter, you discover how to refine hatch techniques. I describe how to copy existing hatches, take advantage of additional options in the Hatch and Gradient dialog box (which offers a bit more control than the Hatch Creation contextual tab), and choose more complicated hatch boundaries.

Adding style

A minor problem is that you can't create named styles for hatches, as you can for text and dimensions. The good news is that AutoCAD provides three workarounds to help offset this issue.

Plagiarism 101 — matching hatches: Suppose that a drawing has several different hatches with incorrect specifications. Fix them like this: Select a problem hatch, click the Match Properties button on the Hatch Editor tab, and then select a hatch with the correct properties. AutoCAD copies the properties from the second hatch to the first one, the reverse of what you might expect.

TIP

Select more than one problem hatch at a time to correct multiple hatches at once.

Plagiarism 201 — cloning hatches: Suppose that a drawing has several different styles of hatches applied, and now you want to hatch another area. You start the Hatch command, and the Hatch Creation contextual tab appears. Oops — it shows the specifications of the last hatch that was applied, which is quite different from what you want to do now. Click the Match Properties button in the Options panel of the Hatch Creation tab, or click the Inherit Properties button in the Hatch and Gradient dialog box. Pick an existing hatch, and — presto! — all settings are updated to match the selected hatch. You can use the cloned settings as is, or you can modify them.

Plagiarism 301 — creating hatches: Here's a technique that even many experienced users miss. It's remarkably easy to customize AutoCAD so that a single mouse click can produce any hatch style you want. Here's how:

1. **Apply a hatch with the properties you want.**

   Make sure that the existing hatch is on the correct layer and that the layer has all the correct properties.

REMEMBER

2. **Display the Tool Palette window.**

   Select the Tool Palettes tool in the Palettes panel on the View tab of the Ribbon, or enter the TOOLPALETTES command. (I discuss tool palettes in Chapter 2.)

3. **Access the Hatches and Fills palette by clicking its tab.**

   If you can't see this tab, click the overlapping edges of the tabs at the bottom of the palette window and select Hatches and Fills from the list that appears.

4. **Create a new Hatch tool.**

   Click once to select the existing hatch object, pause, and then select it again. Drag it to the tool palette and drop it. Everything about the hatch is copied to the palette: pattern name, scale, colors, layer, and so on.

WARNING

   Don't double-click too quickly, and don't click the blue grip, or else you'll be editing the existing object.

5. **Use the new Hatch tool.**

   Click the Hatch tool, and then click inside a closed boundary in the drawing. Bingo — instant hatches to your specification.

   You can also click and drag the desired pattern from the tool palette into the desired boundary.

   If you start a new drawing, create a boundary, and then use the Tool Palette to hatch it, you may be amazed to find that the new hatches in the new drawing are on the correct layer. If the correct layer doesn't exist in the new drawing, AutoCAD creates it to match the original specifications automatically — a process known as *standardization through customization*.

Hatches from scratch

You can use predefined, custom, and user-defined hatch patterns. Most of the time, you'll choose predefined hatch patterns, unless some generous soul did the hard work of writing a custom pattern for you. On the other hand, I haven't employed a user-defined pattern in over 25 years of using AutoCAD because these hatches consist solely of continuous lines: All you can define are the spacing, the rotation angle, and whether the lines are cross-hatched. All this already exists in the set of predefined patterns.

An alternative to using the Ribbon to define patterns is the Hatch and Gradient dialog box, as shown in Figure 15-6. You don't see the hatch object updating as you change settings (as you do with the Ribbon), but the dialog box gives you more control over what you end up with. To display the Hatch and Gradient dialog box, click the dialog box launcher (the tiny arrow at the right end of the Options panel on the Hatch Creation tab of the Ribbon).

TIP

By default, the right third of the Hatch and Gradient dialog box is hidden; to see additional hatch options at the right side of the dialog box (as shown in Figure 15-6), click the More Options arrow beside the Help button.

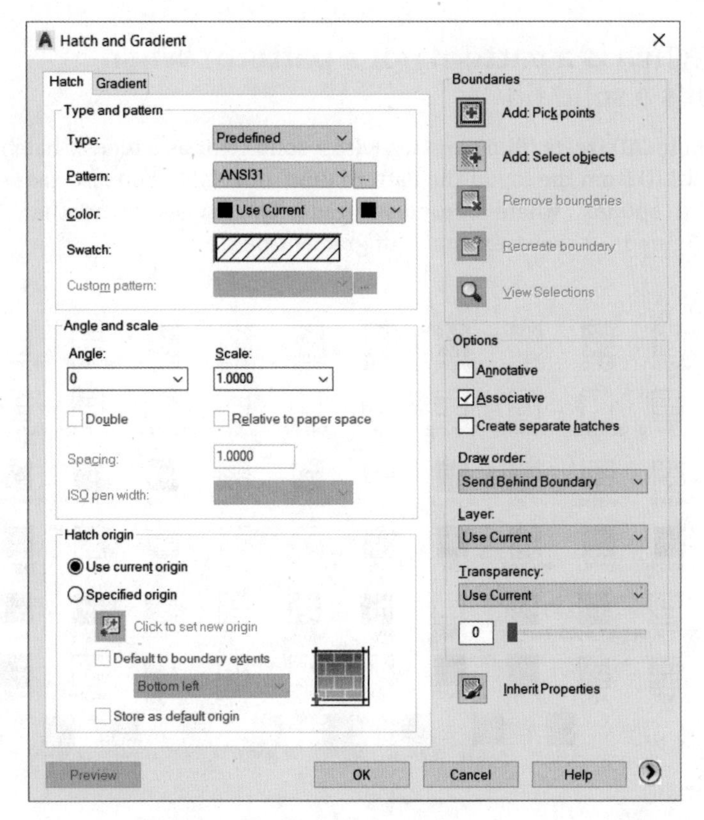

FIGURE 15-6: The Hatch tab in the expanded Hatch and Gradient dialog box.

Pick a pattern — any pattern: Predefined hatch patterns

Using predefined hatch patterns in the Hatch and Gradient dialog box requires two steps. First, select Predefined from the Type drop-down list at the top of the dialog box. Then specify the pattern you want in one of two ways:

- **Pattern drop-down list:** If you know the name of the hatch pattern, select it from the Pattern drop-down list. The list is alphabetical, except SOLID (that is, a solid fill), which appears at the beginning.
- **Pattern button:** If you don't know the pattern name or you prefer the visual approach, click the . . . button to display the Hatch Pattern Palette dialog box with its pattern previews and names. The dialog box groups patterns in tabs: ANSI, ISO, Other Predefined, and Custom.

Figure 15-7 shows all predefined hatch patterns included with AutoCAD, covering everything from dirt to Escher to stars.

When is a pattern not a pattern? When it's a solid fill

AutoCAD treats filling an area with a solid color as a type of hatch. Simply choose SOLID from the top of the Pattern drop-down list. You also see several gradient-fill options, where one color gradually changes to another. The bottom of Figure 15-7 shows examples of gradient fills.

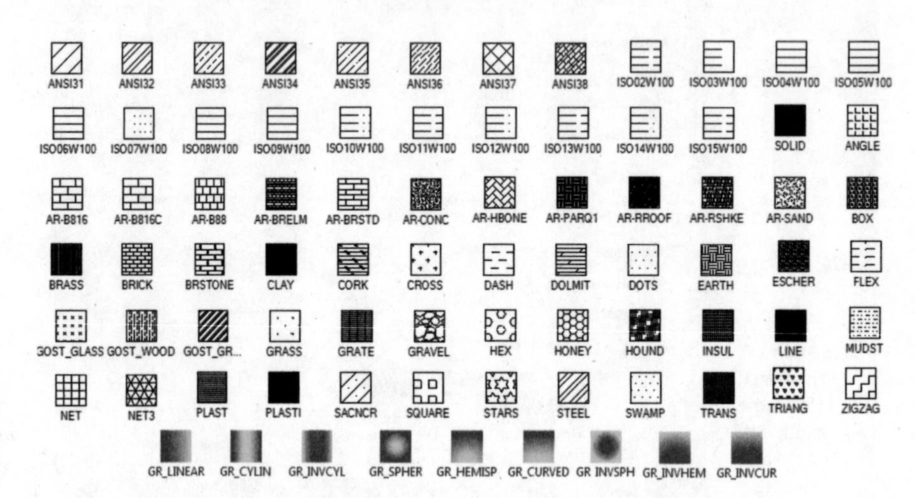

FIGURE 15-7: A veritable plethora of hatch patterns.

Like any other object, a solid hatch takes on the current layer's color when you maintain (as you should) the color as ByLayer. It takes on the current object color override when people who don't know what they're doing alter the drawing. Therefore, check whether the current object layer and color are set appropriately before you use the Solid hatches option. (See Chapter 9 for details.)

Here's looking through you, kid

The transparency object property in AutoCAD is probably most useful (in 2D, anyway) when applied to solid fills. You can use transparent solid fills to demarcate (distinguish) areas on architectural floor plans or aerial photographs of project sites, while letting text and symbols show clearly. In addition to ByLayer and Solid settings, make sure that the current object or layer transparency is set correctly, too.

Editing Hatch Objects

Editing an existing hatch pattern is easy after you're familiar with the Hatch Creation tab on the Ribbon. Follow these steps:

1. **Select the hatch object.**

   AutoCAD opens the Hatch Editor contextual tab on the Ribbon and displays the hatch object's current settings.

2. **Make any changes you want and watch the real-time preview as you do. When you're happy, click Close Hatch Editor to retain the changes.**

Alternatively, you can use the Properties palette or the Quick Properties palette (described in Chapter 9) to make most of the changes to existing hatch patterns. AutoCAD always displays the Hatch Editor tab when you click a hatch object, and it opens the Quick Properties palette when the QP button is turned on. The Properties palette is especially useful for changing several hatches at a time.

Simple grip-editing is also available. Select a hatch and then hover the cursor over the round, blue center grip. A contextual menu opens to give you quick access to changing the hatch origin, angle, and scale. Don't be misled by the Stretch option on the context menu, though: It only moves the hatch, and then it loses its associativity to its boundary.

Here are a few other hatch tips:

- **To make one hatch look like another (without even opening the Hatch Editor tab):** Click the Match Properties button in the Clipboard panel on the Home tab of the Ribbon.
- **To mirror drawing geometry that includes hatches without mirroring the hatch angle:** Set the MIRRHATCH system variable to 0 to retain the hatch pattern or to 1 to mirror the pattern. This tip can be quite useful when creating drawings of symmetrical objects. Model one half, apply the hatch, and then use the MIrror command (see Chapter 11) to get the other half. The hatch angle remains consistent on both sides of the mirror line if MIRRHATCH is 0.
- **To find the area of any hatch object:** Simply select the hatch object and then open the Properties palette. The area is listed in the Geometry section, near the bottom of the palette.

WARNING

Don't go overboard with hatches. Their purpose is to clarify, not to overwhelm, other geometry in the drawing. If your plots comprise a patchwork of hatch patterns, it's time to simplify.

REMEMBER

Always apply hatches last, if you can. The Hatch command is smart enough to recognize text and dimensions that fall within the hatched area, and so treats them like sterile eggs and automatically omits hatch zones around them. If you forget, or if you need to add a note later in a hatched area, apply background masking to the text, as discussed in Chapter 13.

IN THIS CHAPTER

» Plotting model space

» Plotting to scale

» Plotting paper space layouts

» Plotting with lineweights and colors

» Controlling plotters with plot styles

» Troubleshooting plots

Chapter 16
The Plot Thickens

Despite the infinitesimally small number of offices without a computer (or two) on every desk or in a pocket, many people still want or need to work with easily readable, dead-tree paper drawings or electronic drawings (can you spell P-D-F?). You may need to give hard-copy printouts or PDF files to less savvy colleagues who don't have AutoCAD, or to people on construction sites where relatively delicate computers wouldn't survive long. You may want to print hard copies to review during your bus ride home. You may even find that checking drawings the old-fashioned way, with a hard-copy printout and a red pencil, turns up errors that manage to remain hidden on the computer screen.

Hard copies may also survive longer as historic records. Some 10-year-old CAD files can no longer be opened because the company that produced the software no longer exists or the software can't run on newer operating systems. On the other hand, for the ultimate in hard copies, we can still read Babylonian clay tablets that are four thousand years old.

Whatever the reason, you'll want to print drawings at some point, and probably sooner rather than later to check your progress. Depending on where you are in a project, plotting can be the pop quiz, the midterm exam, or the final exam of your drawing-making semester. This chapter helps you ace the test.

You Say "Printing," I Say "Plotting"

Plotting originally meant creating hard-copy output on a device capable of printing large sheets, such as D-size or E-size (or A1 or A0 for the metrically inclined), which measure several feet (or a meter or more) on each side. The plotters mostly used pens to draw, robot-fashion, on large sheets of vellum or drafting paper. The sheets could then be run through *diazo machines,* which were copying machines that output less expensive blueline prints more quickly than plotters.

Printing, on the other hand, meant creating hard-copy output on ordinary printers that use ordinary-sized paper, such as A-size (letter size, 8½ x 11 inches) or B-size (tabloid or ledger size, 11 x 17 inches): A4 or A3, for you metric folk.

Nowadays, AutoCAD and most CAD users make no distinction between plotting and printing. AutoCAD veterans usually say "plotting," so if you want to be cool, you can do so, too.

Whatever you call it, printing an AutoCAD drawing is potentially more complicated than printing text documents or spreadsheets because CAD has a larger range of plotters and printers, drawing types, paper sizes, and output procedures than most other software applications. AutoCAD tries its best to help you tame the vast jungle of plotting permutations, but you'll probably find that you have to take some time to get the lay of the land and clear a path through the jungle of options to get to the hard copy you want. After you find the sweet spot, plotting becomes quite routine.

The Plot Quickens

Even if you already realize that you won't master AutoCAD plotting in five minutes flat, your boss, employee, spouse, construction foreman, or 11-year-old child may still want a quick check plot of your drawings.

Plotting success in 16 steps

Here's the quick, cut-to-the-chase section for plotting a simple drawing — a mere 16 steps! This isn't, however, as bad as it sounds. You see later how to save your printing setups, so you can reuse them with considerably less effort.

For the purpose of this section, I'm assuming that you plot from model space: that is, that clicking the Model tab on the status bar shows you the drawing you want to plot. I cover plotting paper space layout tabs in the later section "Plotting the Layout of the Land." This section doesn't deal with controlling plotted lineweights; see the "Plotting Lineweights and Colors" section, later in this chapter, for those details. The result should be a piece of paper that bears some resemblance to the drawing AutoCAD displays on your computer monitor.

Follow these steps to make a simple, not-to-scale, *monochrome* (black-and-white) plot of a drawing:

1. **Open the drawing in AutoCAD.**

Model

2. **Click the Model tab to ensure that you're plotting the model space contents.**

   If AutoCAD doesn't have display the Model and Layout tabs, click the Model button (not the MODEL/PAPER button) on the status bar.

   I explain model space and paper space in Chapter 12, and how to plot paper space layouts in the section "Plotting the Layout of the Land," later in this chapter.

3. **Zoom to the drawing's current extents.**

   Click the Zoom Extents button on the Navigation bar. If necessary, click the tiny down arrow below the Zoom button, and choose Zoom Extents from the menu. Or easier yet, type **Z E** (press the spacebar between the letters) and then press Enter.

   The *extents* of a drawing consist of the edges of a rectangular area just large enough to include all objects in the drawing.

4. **Click the Plot button on the Quick Access toolbar, or type the PLOT command in the command line.**

   The Plot dialog box appears, as shown in Figure 16-1.

   AutoCAD automatically appends whatever you're about to plot in the dialog box's title bar. For example, in Figure 16-1, the dialog box title is Plot - Model. If you are plotting from a paper space layout and have changed the layout name, the dialog box title is Plot - First Floor Plan (or whatever you renamed it). In this book, I call the dialog box (simply) "the Plot dialog box."

5. **In the Printer/Plotter area, select a device from the Name drop-down list.**

   If your computer has a printer attached to it, select *Default Windows System Printer pc3*. If a printer isn't connected, choose Microsoft Print to PDF, which creates a PDF file that can be viewed in the free Acrobat Reader software.

6. **In the Paper Size area, select a paper size that's loaded in the printer or plotter.**

   For most desktop printers, it's called Letter (8.5 x 11 inches, portrait), ANSI A-size (11 x 8.5 inches, landscape), or A4 metric (210mm x 297mm).

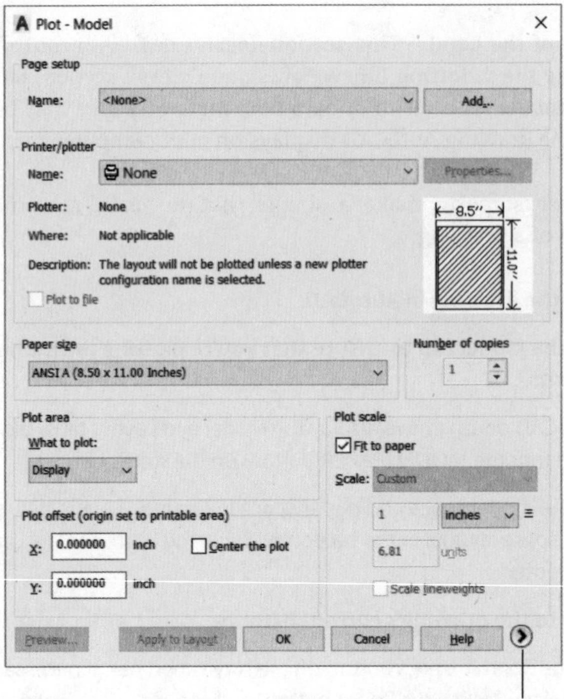

FIGURE 16-1: The Plot dialog box.

7. **In the Plot Area area (sponsored by the Department of Redundancy Department), select Extents from the What to Plot drop-down list.**

   This step plots the entire drawing. The alternatives are for plotting parts of drawings.

8. **In the Plot Offset (Origin Set to Printable Area) area, select the Center the Plot check box.**

   This step centers the drawing on the paper. Alternatively, you can specify an X or a Y offset (or both) of 0 (zero) or other amounts to position the plot at a specific location on the paper to, for example, avoid a title block.

9. **In the Plot Scale area, select the Fit to Paper check box.**

   This step ensures that the drawing, no matter how large (or small), is printed to fit the paper. For most real plotting, however, you have to plot to a specific scale. I cover plotting to a specific scale in the "Instead of fit, scale it" section, later in this chapter, for guidance.

10. **Click the More Options button (at the lower-right corner of the dialog box, next to the Help button).**

    The Plot dialog box reveals additional settings, as shown in Figure 16-2.

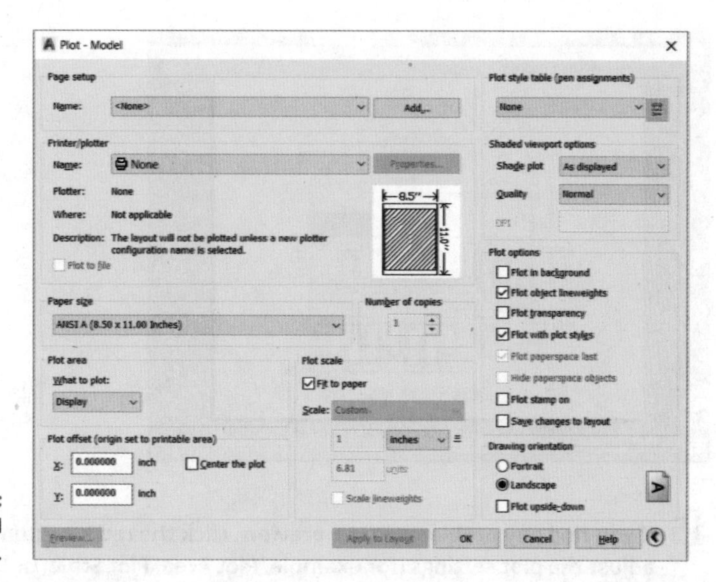

FIGURE 16-2: The expanded Plot dialog box.

TIP

Starting with AutoCAD 2016, the default is for the Plot dialog box to be expanded.

11. In the Plot Style Table (Pen Assignments) area, choose `monochrome.ctb` or `monochrome.stb` from the drop-down list.

This step converts all colors in the drawing to shades of gray. If AutoCAD asks whether to assign this plot style table to all layouts, click No for now.

12. In the Plot Options area, make sure that the Plot with Plot Styles check box is selected and that the Save Changes to Layout check box is not selected. (Refer to Figure 16-2.)

Leaving the Save Changes to Layout check box deselected tells AutoCAD that the changes you are making to the plot settings are *only for this plot.* AutoCAD reverts to the original plot settings the next time you plot the drawing.

TIP

After you are confident as to which settings you want changed, select this check box so that AutoCAD *does* save your plotting settings changes as the default. Alternatively, click the Apply to Layout button to make the current plot settings the defaults for future plotting of this tab (that is, the Model tab) in this drawing.

13. In the Drawing Orientation area, choose Portrait or Landscape.

The icon (the letter *A* on a sheet) in the lower-right corner may help you choose the orientation that best fits the paper. If you don't find it helpful, see the full preview mode described in Step 14.

14. Click the Preview button to ensure that the drawing appears on the paper at the correct orientation and size, as shown in Figure 16-3, and then right-click and choose Exit to return to the Plot dialog box.

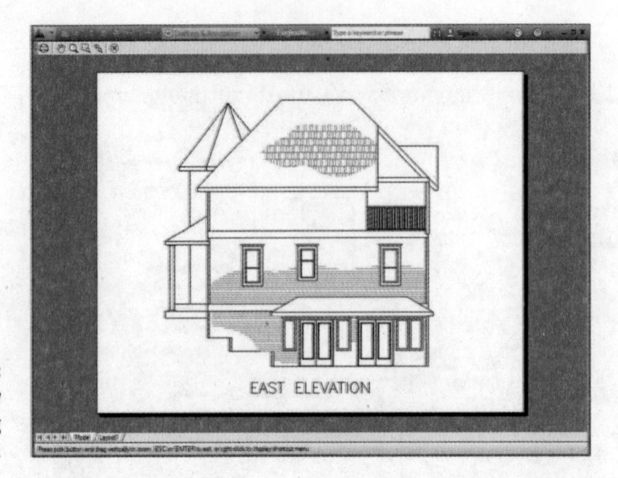

FIGURE 16-3: A preview of coming plot-tractions.

15. **If you find any problems in the preview, click the red X button and adjust the plot settings (for example, Plot Area, Plot Scale, or Drawing Orientation). Repeat the preview until the plot looks right on your screen before committing to paper.**

16. **In the Plot dialog box, click OK to create the plot. In the Preview window, click the Printer icon to start the plot.**

    When AutoCAD finishes generating and sending the plot, it displays the yellow Plot and Publish Job Complete balloon notification near the right end of the status bar. To hide these notifications, right-click the Plot/Publish Details Report Available icon near the right end of the status bar, and then deselect Enable Balloon Notification.

There — only 16 steps, as promised.

On the other hand, I never promised that this plot would be usable. For example, if you had tried to plot a large, complex, architectural drawing meant for D- or E-size media on an A-size sheet of paper, it isn't readable without a microscope. The bottom line, however, is that you produced a plot! Read the rest of this chapter for all the details about the numerous other plotting options available in AutoCAD. If you had trouble with this plot, jump ahead to the troubleshooting section "The Plot Sickens," later in this chapter.

Getting with the system

In your attempts to create hard copy, you might face a complication. AutoCAD has two distinct ways by which it communicates with plotters and printers: through the operating system, and on its own.

Operating systems, and most programs that run on them, use a special intermediary software, the *printer driver*, to reformat the document into data suitable for printing, and then send the data to the printer or plotter. When you connect a new printer to your computer or network, you set up Windows to recognize it by installing the printer's driver. AutoCAD, like other Windows programs, can work with the printers you've set up in Windows. AutoCAD calls these *system printers* because they're part of the Windows operating system.

But AutoCAD, unlike most other Windows programs, can't leave well enough alone. The Windows system printer drivers don't efficiently control certain output devices, especially certain larger plotters. For that reason, AutoCAD comes with its own set of *non-system drivers* (that is, drivers that aren't installed as part of the Windows system) for plotters from companies such as Hewlett-Packard, Xerox, Epson, and Canon-Océ. These drivers ignore Windows' own rules for communicating with these printers to get things done more accurately and a bit more quickly.

Most of the time, using system printers already configured with Windows is easiest, and they work well with many devices, such as laser and inkjet printers that print on smaller paper. However, when you have a large-format plotter, you *may* be able to get faster plotting, better plot quality, and more plot features by installing a non-system driver provided by the printer manufacturer. To find out more, search for your specific plotter using the AutoCAD online Help system.

Configuring your printer

To ensure that AutoCAD recognizes the devices you want to use for plotting, follow these steps:

1. **Launch AutoCAD and open an existing drawing or start a new, blank drawing.**
2. **Choose Options on the Application menu or type OP and press Enter to open the Options dialog box; then click the Plot and Publish tab.**
3. **Click the drop-down arrow to view the list just below the Use as Default Output Device option, as shown in Figure 16-4.**

   The list includes two kinds of device configurations, designated by two tiny, difficult-to-distinguish icons to the left of the device names:

   - *Windows:* A little printer icon with a sheet of paper coming out the top indicates a Windows system printer configuration.
   - *Non-system:* A little plotter-with-legs icon with a piece of paper coming out the front indicates a non-system (that is, AutoCAD-specific) configuration.

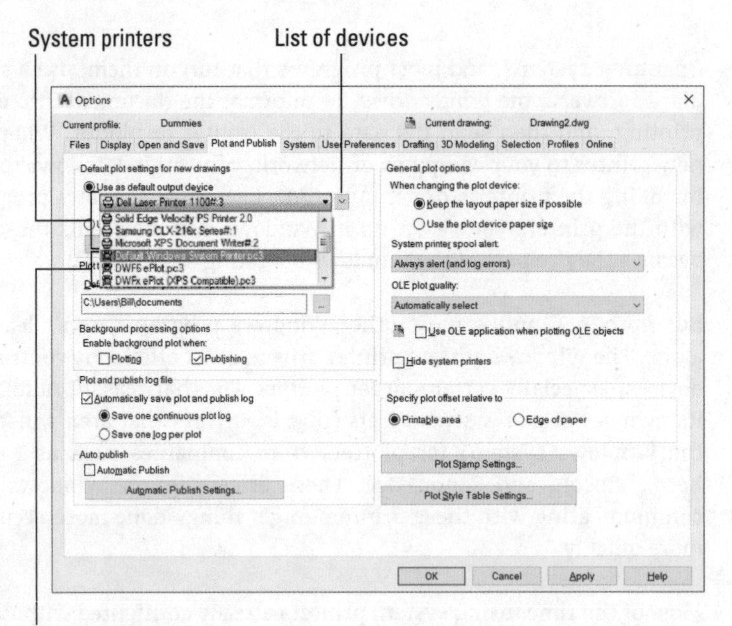

FIGURE 16-4: System and non-system printer configurations.

TIP

The non-system configuration names always end in `.pc3` because they're stored in AutoCAD Plotter Configuration version 3 files. So, if you can't distinguish the difference between the icons, look for the `.pc3` at the end of the name.

4. **Verify that the list includes the printers and plotters you want to have available in AutoCAD.**

   If they're not in the list, how you add one depends on your operating system:

   - *Windows 7:* Choose Start ➪ Devices and Printers ➪ Add a Printer, and then click Add a Local Printer.
   - *Windows 8:* Hold down the Windows key and then press W. In the Search field, enter **printers**. Under Settings, choose Add Printer.
   - *Windows 10 and 11:* Tap the Windows logo, enter **printers** in the Search field, choose Devices and Printers, and then choose Add a Printer.

   If your printer isn't on Windows' default list, cancel the Add a Printer Wizard and hunt down a driver disc that came with the printer. Better yet, download the current driver from the printer manufacturer's website.

5. **Choose the output device that you want to designate as the default for new drawings.**

6. **Click OK to close the dialog box and retain the changes you made in Step 5.**

To create configurations with non-system drivers, you use AutoCAD's Add-a-Plotter Wizard. On the Ribbon's Output tab, choose Plotter Manager in the Plot panel. AutoCAD displays an Explorer window containing a shortcut to the Add-a-Plotter Wizard. The wizard is similar to the Windows Add Printer Wizard; if you can handle adding an ordinary printer in Windows, you can probably handle adding a non-system plotter configuration to AutoCAD. When you complete the wizard's steps, AutoCAD saves the information in a PC3 (Plotter Configuration version 3) file. In some cases, such as when you add an HP DesignJet printer or certain Canon-Océ wide-format printers, the Add-a-Plotter Wizard advises you to exit and instead install the device as a Windows system printer. Many people find that the standard drivers work fine, but as I mention later in this chapter, custom drivers may include additional paper sizes as well as other handy settings.

Preview one, two

One key to efficient plotting is the liberal use of AutoCAD's plot preview feature. To maintain political balance, I recommend conservative use of other AutoCAD options elsewhere in this book.

The postage-stamp-size partial preview in the middle of the Plot dialog box is a quick reality check to ensure that the plot fits on the paper and is in the correct direction. If the plot area at the current scale is too large for the paper, AutoCAD displays thick, red warning lines along the side(s) of the sheet where the drawing will be truncated.

Click the Preview button to see a full preview in a separate window. You see exactly how the drawing lays on the paper, as well as how the various lineweights, colors, and other object plot properties appear. You can zoom and pan around the preview by using the right-click menu.

Zooming and panning are simply ways to get a better look at different areas of the plot preview. Neither action affects the size of the drawing when it's plotted.

Instead of fit, scale it

In most plotting situations, you plot to a specific scale dictated by your industry, rather than let AutoCAD choose an oddball scale (such as Fit) that maximizes the drawing's size on the paper. And if you plan to plot to scale a drawing in model space, you need to know its drawing scale factor. Chapter 4 describes setup concepts, and Chapter 13 provides some tips for determining the scale factor of a drawing that someone else created.

If the drawing was created at a standard scale, such as 1:50 or 1/4" = 1'-0", you simply choose the scale from the handy Scale drop-down list in the Plot dialog box. If the scale isn't on the list, something is probably wrong, because you should almost always be using standard scales. On the other hand, if you must use a custom scale, type the ratio between plotted distance and AutoCAD drawing distance into the two text boxes below the Scale drop-down list, as shown in Figure 16-5. Usually, the easiest way to express the ratio is to type **1** in the upper box and the drawing scale factor in the lower box. See Chapter 4 for more information.

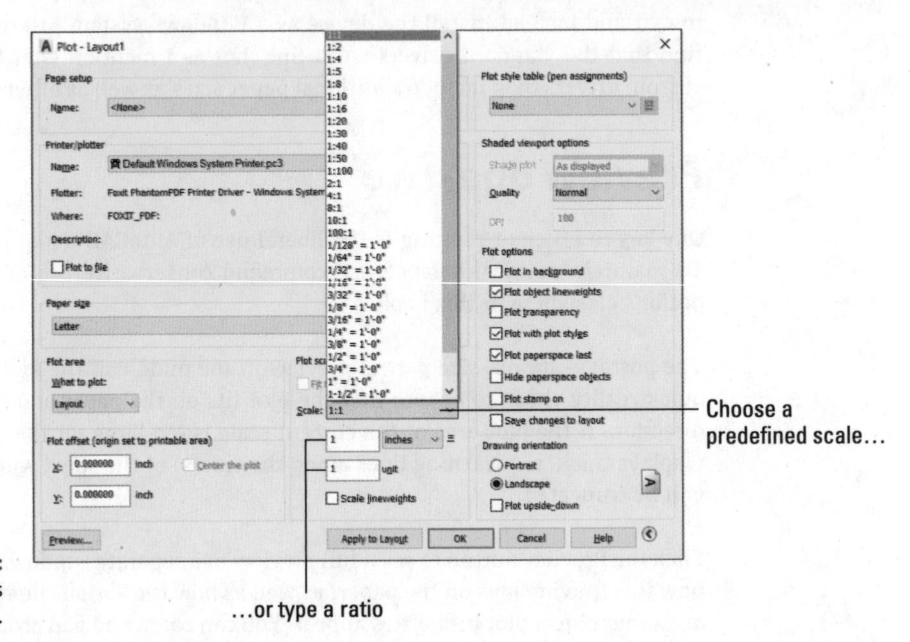

FIGURE 16-5: Lots of ways to scale.

TIP

Your CAD manager may have edited the Scale drop-down list to add uncommon scales or remove scales that your company never uses. If you're designing espresso machines in Milano, for example, you'll probably never need to plot drawings at 1/128" = 1'-0". Yes, that scale is on the AutoCAD list.

Creating half-size plots for certain purposes is common in some industries. To plot a drawing in model space at half-size, you double the drawing scale factor. For example, a 1/8" = 1'-0' drawing has a drawing scale factor of 96, which is equivalent to a plot scale of 1 = 96. To make a half-size model space plot of it, specify a plot scale of 1 = 192 (or choose 1/16" = 1'-0" from the Scale drop-down list).

Even if you work with drawings created to be plotted at a specific scale, plotting with a Fit to Paper scale may be the most efficient way to make reduced-size check plots. For example, drafters in your office might create drawings that are plotted on D-size sheets (24 x 36 inches), whereas you have access to only a laser printer with a B-size (11 x 17 inches) paper tray. By plotting the D-size drawings scaled to fit on B-size paper, you produce check plots that are slightly smaller than half-size. You can't measure distances on the check plots with a scale, but you can give them a visual check for overall correctness, or use them at project progress meetings.

Plotting the Layout of the Land

In the section "Plotting success in 16 steps," earlier in this chapter, I show you how to plot the model space representation of a drawing by making sure that the Model tab is active when you open the Plot dialog box. In most industries, however, model space plotting went the way of the dodo two decades ago. It's paper space you want to plot from today. Paper space gives you many additional options for controlling the look of the output — without having to modify the underlying geometry. So, most of the time, you plot paper space layouts instead.

Paper space is the environment that's specifically designed for outputting hard copy of drawings. For information on paper space or layouts, see Chapter 12.

As I explain in Chapter 12, AutoCAD gives you two ways to switch between full-screen model space and the paper space layouts stored in the drawing. Simply click the appropriate tab to select Model space or a layout.

Plotting a drawing in a paper space layout is much like plotting model space, except that you have to select the tab with the appropriate layout *before* you open the Plot dialog box. Follow these steps:

1. **Move the mouse pointer across the Layout tabs near the bottom left corner of the screen.**

   A preview of each layout appears as you select a tab with your mouse.

2. **Choose a tab to display the layout, and then right-click the tab.**

   A multi-item context menu appears with a number of items related to layouts.

3. **Click the Plot item in the context menu.**

   The Plot dialog box appears with the name of the layout to be plotted displayed on the title bar. In Figure 16-6, I'm plotting a layout named D-Size Layout.

If no layout was previously set up, AutoCAD creates a default layout. If the Show Page Setup Manager for New Layouts setting on the Display tab of the Options dialog box is turned on, you see the Page Setup Manager dialog box first; just click the Close button. The default layout probably isn't useful for real projects, but you can use it to find out about layout plotting. Refer to Chapter 12 for instructions on creating a real layout.

4. **Specify a printer/plotter name and a paper size.**

If you don't have a printer capable of producing sheets larger than letter or tabloid size, you can still experiment by selecting a device that outputs to a file, such as *AutoCAD PDF.pc3*.

5. **In the What to Plot drop-down list, choose Layout (refer to Figure 16-6).**

The Layout option is available only when plotting a layout tab; Limits is available only when plotting the Model tab.

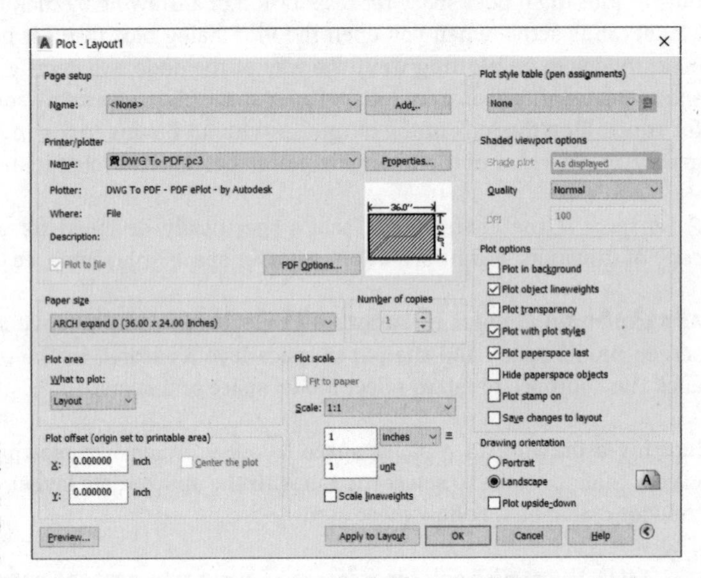

FIGURE 16-6: Settings for plotting a paper space layout.

6. **Specify the plot offset (such as 0 in both the X and Y directions).**

Specifying the plot offset as 0 in both X and Y directions places the lower-left corner of the plotted drawing at the lower-left corner of the printable area.

7. **In the Plot Scale area, select 1:1 from the Scale drop-down list (refer to Figure 16-6).**

One primary advantage of using layouts is that you don't need to know about drawing scale in order to plot the drawing — hence the name *paper* space. Figure 16-6 shows the proper settings for plotting a layout.

TIP

Layouts are almost always printed at full size (1:1). To create a half-size plot of a layout, select 1:2 from the Scale drop-down list. In addition, select the Scale Lineweights check box to reduce lineweights proportionally. I cover plotting lineweights in the next section.

If you find that the layout is too big for the plotter's largest paper size at a plot scale of 1:1, you can choose Extents from the What to Plot drop-down list and then select the Fit to Paper check box in the Plot Scale area. Alternatively, you can close the Plot dialog box and fix the problem to have a paper space layout permanently reflect a different paper size. Use the Page Setup dialog box to modify the layout settings, or copy the layout and modify the new copy.

8. **Click the More Options button and change any additional plot options you want.**

   Refer to Steps 11–13 in the earlier section "Plotting success in 16 steps."

9. **Click the Preview button, ensure that the drawing displays on the paper at the correct orientation and size, right-click, and choose Exit from the menu that appears to return to the Plot dialog box.**

   If you find any problems in the preview, change the plot settings and preview again until it looks right.

10. **Click OK to create the plot.**

Plotting Lineweights and Colors

In previous sections of this chapter, I help you gain some plotting confidence. Those sections show you how to create scaled, monochrome plots with uniform lineweights in model space or paper space. Those skills may be all you need, but if you care about controlling plotted lineweights and colors or adding special effects such as *screening* (plotting shades of gray), read on.

Plotting with style

Plot styles provide a way to override object properties with alternative plot properties. See Chapter 9 for information about object properties. The properties include plotted lineweight, plotted color, and screening. Figure 16-7 shows the full range of options. Plot styles come in two exciting flavors:

- **Color-dependent:** Based on the standard way of plotting in earlier versions of AutoCAD (before AutoCAD 2000)
- **Named:** Provide a newer (but not necessarily better) way

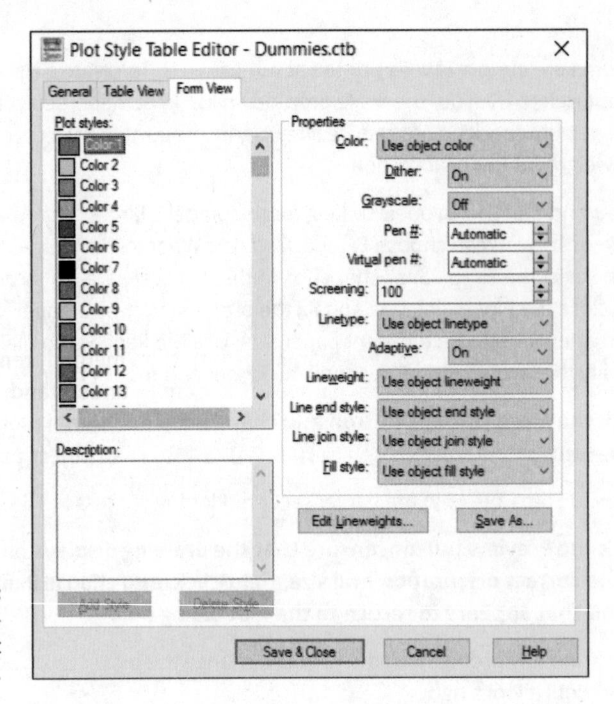

FIGURE 16-7: Editing a color-dependent plot style table.

You may not need to bother with plot styles. If the drawings have layer and object properties (especially lineweights) that specify how you want objects to plot, you can dispense with plot styles. But many people and most drawings use plot styles, so you should at least be familiar with them.

A couple of common reasons for using plot styles are to

- **Map screen colors to plotted lineweights.** If this idea seems completely loony to you, try to suspend judgment until you read the "Plotting through thick and thin" section, later in this chapter.
- **Create screened lines on monochrome plots.** Lines that are screened will be displayed in various shades of gray, not black. Drafters sometimes use screened lines to deemphasize secondary objects that otherwise would overwhelm the main objects in drawings. Screening is expressed as a percentage: 100 percent is completely black, and 0 percent is invisible. How can you tell when your pen has run out of invisible ink? You can see the writing. Anyway, on a subdivision layout, for instance, you might want to deemphasize all property and house outlines except one.

Using plot styles

If you want objects in a drawing to plot with properties that differ from their display properties, you plot with styles. For example, they can plot with different lineweights or colors from the ones you're using for display purposes. Or you may need to map display colors to plotted lineweights. AutoCAD lets you group lists of plot styles into plot style tables, each of which is stored in a separate file.

Color-based plot styles live in *color-dependent plot style table* (CTB) files, and they map the 255 AutoCAD display colors to 255 plot styles. AutoCAD automatically attaches these color-dependent plot styles to every object, based on — you guessed it — the object's color. Are those AutoCAD programmers brilliant or what? Color-dependent plot style tables mimic the old color-mapped-to-lineweight plotting approach of AutoCAD Release 14 and earlier; this method remains the most common one in most companies because it is easy to understand.

The named plot style, introduced in AutoCAD 2000, is an object property, just like color, linetype, and lineweight. Named plot styles live in (wouldn't you know it?) *named plot style table* (STB) files. After you create a new named plot style table, you add one or more plot styles, giving them any names you like. Then you assign the named plot styles to layers or even to individual objects. See Chapter 9 for more information about object and layer properties.

To use the plot styles in a plot style table (whether they're color-dependent or named ones), you must attach the plot style table to model space or a paper space layout. The attached plot style table affects plotting only for that specific model space or layout. This approach lets you plot the same drawing in different ways by attaching different plot styles to multiple layout tabs — one plot *style* per layout tab, one plot style *table* per drawing.

You attach a plot style to model space or a paper space layout by right-clicking a tab, choosing Page Setup Manager, and then clicking the Modify button. Choose the plot style table name from the Plot Style Table (Pen Assignments) area of the expanded Page Setup dialog box. See the section "Controlling plotted lineweights with screen colors," later in this chapter, for an example.

TECHNICAL STUFF

When you start a new drawing by choosing a template drawing in the usual way (see Chapter 4), the template drawing's plot style behavior determines whether you can choose CTB or STB files. If you want to change a drawing from color-dependent plot styles to named plot styles (or vice versa), use the CONVERTPSTYLES command.

The CONVERTPSTYLES command can handle styles that contain spaces in their names.

Creating plot styles

If you're lucky, someone will provide you with the plot files you need. When that's the case, you must put the CTB or STB files in the Plot Styles folder for AutoCAD to recognize them. To find the location of the Plot Styles folder, open the Options dialog box, select the Files tab, and look for the Printer Support File Path ⇨ Plot Style Table Search Path setting.

If you're unlucky, you need to be smart enough to know how to create your own plot style table files. Personally, I'd rather be lucky than smart; hey, it's worked so far in life! Here's how to create plot style table files:

1. **Click the Application button to open the Application menu, click Print, and then choose Manage Plot Styles.**

   Use the tiny down arrow at the bottom of the list to scroll farther down, if necessary. The Plot Styles folder opens in a separate Windows Explorer window.

2. **Double-click the Add-a-Plot Style Table Wizard program shortcut.**
3. **Read the opening screen and then click Next.**
4. **On the Add Plot Style Table – Begin page, choose the Start from Scratch option or one of the other three options if you want to start with settings from another file. Then click Next.**

   The remaining steps assume that you chose Start from Scratch. If you chose another option, simply follow the wizard's prompts.

   If the creator of a drawing provides you with an AutoCAD R14/AutoCAD LT 98 PC2 (version 2) or AutoCAD R12/AutoCAD LT 95 PCP (version 1) file, choose the Use a PCP or PC2 File option. With this option, the wizard imports color-to-plotted-lineweight settings automatically.

5. **On the Add Plot Style Table – Pick Plot Style Table page, choose whether you want to create a color-dependent plot style table (CTB file) or a named plot style table (STB file). Then click Next.**

   Choose Color-Dependent Plot Style Table to map screen colors to plotted lineweights. Choose Named Plot Style Table to create named plot styles that you can apply to layers or objects.

6. **On the Add Plot Style Table – File Name page, type a name for the new CTB or STB file and then click Next.**
7. **Click the Plot Style Table Editor button on the Add Plot Style Table – Finish page.**

   The Plot Style Table Editor dialog box opens to the Form View tab when you're creating a color-dependent plot style table (refer to Figure 16-7), or to the Table View tab when you create a named plot style table.

TECHNICAL STUFF

If you choose a named plot style in Step 5, the Plot Style Table Editor dialog box opens in Table view, with one plot style named Normal in the first data column, a blank column to its right, and Add Style and Delete Style buttons at the bottom. New named plot styles that you create continue to be added in columns to the right of the previous column. For more information, click the Help button in Plot Style Table Editor.

8. **If you created a color-dependent plot style table, assign Lineweight, Screening, or other plot properties to each color that's used in the drawing. If you created a named plot style table, click the Add Style button and then assign plot properties to each of the named styles you create.**

   To determine which colors are used in a drawing, switch to the AutoCAD window and open the Layer Properties Manager palette by clicking the Layer Properties button, located on the Layers panel of the Ribbon's Home tab.

TIP

   To change a setting for all colors or named styles, select them all first by clicking the first color or named style, holding down the Shift key, scrolling to the end of the list, and then clicking the last color or named style. Any subsequent changes you make are applied to all the selected colors or named styles.

9. **Click the Save & Close button to close the Plot Style Table Editor dialog box. Then click Finish to complete the steps in the wizard.**

   The Plot Styles folder now displays the new CTB or STB file.

10. **Close the Plot Styles folder by clicking the X on its title bar.**

Creating a plot style table the first time can be a harrowing experience because you have many options. Just remember that the most likely reason for creating one is to map screen colors to plotted lineweights (as I describe in greater detail in the next section). Also remember that you may be able to minimize your effort by getting a CTB or STB file from the person who created the drawing you want to plot.

REMEMBER

In Chapter 9, I recommend limiting yourself to the first nine standard AutoCAD colors when defining layers, and not the patchwork of the 16.7 million colors that AutoCAD makes available. If you follow my advice, your work to create a color-dependent plot style table is greatly reduced because you have to assign plot properties for only nine colors rather than worry about hundreds of them.

TIP

For systematic testing of CTB files, you can download the file named `plot_screening_and_fill_patterns.dwg` from the AutoCAD 2010 Sample Files group at `www.autodesk.com/autocad-samples`. AutoCAD LT users, help yourselves to this file too; it isn't included with the AutoCAD LT sample files but still works for you. This drawing shows an array of color swatches for all 255 AutoCAD colors.

The layouts (such as Grayscale and Screening 25%) demonstrate how different CTB files attached to the same layout produce radically different results.

WARNING

Named plot styles hold a lot of promise, but in at least a couple of places (such as dimensions and tables, for example), they don't work as well as traditional color-based plotting. Dimension properties allow you to assign different colors to dimension lines, extension lines, and text. The purpose is to allow different parts of a dimension object to print with different lineweights; for example, you can have dimension text print with a medium lineweight, the same as the annotation text, while retaining the fine lineweight of extension and dimension lines. But because named plot styles are based on objects or layers, you don't have that lineweight control over individual dimension components. The same limitation applies to tables, where you can set text to be one color and grid lines to be another.

If you get carried away and decide to take advantage of the 16 million colors (or so) in the AutoCAD True Color or Color Book modes, you can't control lineweights with color-dependent plot styles. CTB plot styles affect the lineweights only of objects that use the traditional 255 colors of the AutoCAD Color Index set. When you want True Color or Color Book colors, use object lineweights or named plot styles to control the plotted lineweight.

Plotting through thick and thin

Long ago, manual drafters developed the practice of drawing lines of different widths *(lineweights)* to distinguish different kinds of objects. Manual drafters did it using technical ink pens with different nib diameters, or with different hardnesses of pencil lead and varying degrees of pressure on the pencil. Because a computer mouse usually doesn't come supplied with mouse balls of different diameters (and I'm going to resist the urge to jump in with a comment about mouse reproduction, especially now that most mice are of the optical variety), AutoCAD developers had to figure out how to let users indicate lineweights onscreen and on plots. They came up with two different ways to indicate lineweight:

- **Mapping onscreen colors to plotted lineweights:** I describe this common approach in Chapter 9.
- **Displaying lineweights onscreen to match what the user can expect to see on the plot:** This approach appeared in AutoCAD 2000.

PLOTTING WITH PLODDERS

Color-as-color and lineweight-as-lineweight seem like great ideas, but Autodesk knew (when it added object lineweights back in 1999) that longtime users of AutoCAD wouldn't abandon overnight the approach of mapping old colors to lineweights. Thus, you can still control plotted lineweight by display color in AutoCAD.

AutoCAD veterans, by and large, have chosen to stick with the old way for now. They've done so for a variety of reasons, including inertia, plotting procedures, and drawings built around the old way, third-party applications that don't fully support the newer methods, and the need to exchange drawings with clients and subcontractors who haven't changed. The ripple effect of those who need to, or want to, continue using colors mapped to lineweights is lasting a long time. Don't be surprised if you go with the flow for a while.

The default setting in AutoCAD is to plot object lineweights, so that's the easiest method if you don't have to consider the historical practices or predilections of other people with whom you exchange drawings. Mapping screen colors to lineweights requires some initial work on your part, but after you've set up the mapping scheme, the additional effort is minimal.

Controlling plotted lineweights with object lineweights

Plotting object lineweights is trivial, assuming that the person who created the drawing first assigned a lineweight property to layers or objects. See Chapter 9 for details. Just make sure that the Plot Object Lineweights check box is selected in the expanded Plot dialog box. You can also deselect the Plot with Plot Styles check box because plot styles can override the object lineweights with different plotted lineweights. You can also make these settings in the Page Setup dialog box for the appropriate Layout or Model tab. To access Page Setup Manager, right-click the Model tab or a Layout tab and then choose Page Setup Manager.

TIP

If you want object lineweights to control plotted lineweights, make sure that the Plot Object Lineweights check box is selected in the Plot Options area of the Plot or Page Setup dialog box. If you *don't* want to plot the lineweights assigned to objects, you must deselect both the Plot Object Lineweights and Plot with Plot Styles check boxes in the Plot or Page Setup dialog box. Selecting Plot with Plot Styles selects Plot Object Lineweights as well.

Controlling plotted lineweights with screen colors

To map screen colors to plotted lineweights, you need a color-dependent plot style table (CTB file). If you're plotting a drawing created by someone else, that person may be able to supply you with the appropriate CTB file, or at least with a PCP or PC2 file from which you can create the CTB file quickly. The creator of the drawing should be able to supply at least a printed chart showing which plotted lineweight to assign to each AutoCAD screen color. Use the steps in the earlier section "Plotting with style" to copy or create the required CTB file.

WARNING

Unfortunately, no industry-wide standards exist for mapping screen colors to plotted lineweights. Different offices do it differently. That's why it's useful to receive a CTB, PCP, or PC2 file with drawings that clients send you. If they use eTransmit, the plot style tables and plotter configuration files are included in the package. For more on eTransmit, see Chapter 20.

WHEN IN DOUBT, SEND IT OUT

Whether you plot to scale or not, with different lineweights or not, or in color or not, consider using a service bureau for plotting. In-house plotting on the output devices in your office is useful for small check plots on faster laser or inkjet printers. Large-format plotting, on the other hand, can be slow and time consuming. If you need to plot lots of drawings, you may spend an afternoon loading paper, replenishing ink cartridges, and trimming sheets. That gets expensive.

Successful plotting service bureaus have large, fast, expensive plotters (that you can only dream about owning, even though *they're* responsible for babysitting, feeding, and fixing these fancy devices). As a bonus, service bureaus can make blueline prints from your plots when you need to distribute hard-copy sets to other people.

The downside is that you must ensure that the service bureau receives all necessary information and that it can deliver all necessary plots on time. Some service bureaus plot directly from DWG files, whereas others ask you to make PLT (plot) files. Some service bureaus specialize in color plotting, and others are more comfortable with monochrome plotting and making blueline copies.

When you're choosing a service bureau, look for one that has traditionally served drafters, architects, and engineers. These service bureaus tend to be more knowledge-able about AutoCAD, and they should have more plotting expertise than the desktop-publishing, printing, and copying shops.

Whomever you choose, send them some test plots well before the day that the important set of drawings is due. Talk to the plotting people and get a copy of their plotting instructions. Have the service bureau create plots of a couple of your typical drawings, and make sure that they look the way you want them to. A large drawing set can be many megabytes, so ask the service how you should send them files, such as by email, a cloud service like Dropbox, or a USB drive.

If you often work with a service bureau, consider charging the fee to your clients as a standard expense (as you would with bluelines and copying).

After you have the appropriate CTB file stored in the Plot Styles folder, follow these steps to use it:

1. **Move the mouse pointer across the Layout tabs at the bottom of the screen.**

   A preview of each layout appears as you hover the cursor on each tab.

2. **Right-click the tab of the layout that you want to plot.**

   A multi-item context menu appears with a number of items related to layouts.

3. **Click the Plot item in the context menu.**

   The Plot dialog box appears with the name of the layout to be plotted displayed on the title bar.

4. **In the Plot Style Table (Pen Assignments) area of the Plot dialog box, select the CTB file from the drop-down list. (See Figure 16-8.)**

   The plot style table (CTB file) is attached to the layout or model space tab that you clicked in Step 2.

   If the drawing uses a named plot style table instead of a color-dependent plot style table, select an STB file instead of a CTB file.

5. **Click the Apply to Layout button.**

   AutoCAD records the plot setting change with the current layout's configuration information. Assuming that you save the drawing, AutoCAD uses the CTB you selected as the default plot style when you (or other people) plot this layout in the future.

6. **Continue with the plotting procedures described earlier in this chapter.**

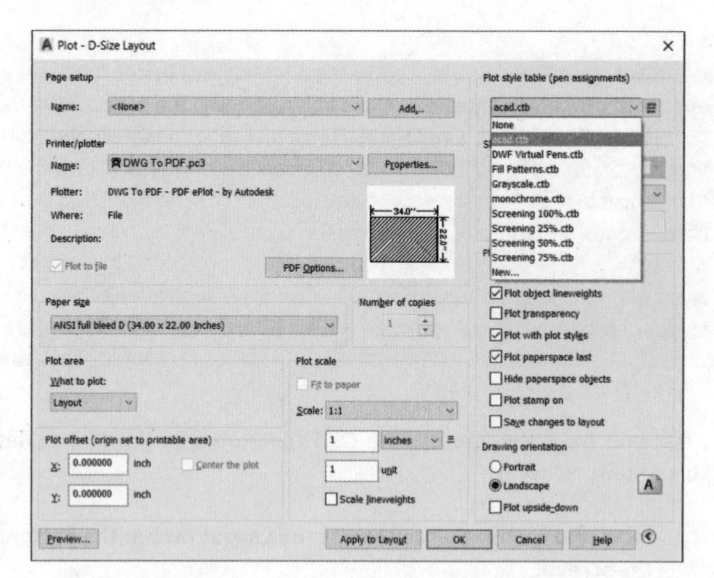

FIGURE 16-8: Selecting a plot style table that maps screen colors to plotted lineweights.

TIP

You can tell whether the current drawing was set up to use color-dependent plot styles or named plot styles by looking at the Properties panel on the Ribbon's Home tab. If the upper-right drop-down list is grayed out, shows ByColor, and doesn't display a tooltip, the drawing uses color-dependent plot styles. If this list isn't grayed out and displays the Plot Style tooltip, the drawing uses named plot styles.

Plotting in color

Plotting the colors you see onscreen requires no special tricks. In the absence of a plot style table (that is, if you selected None from the drop-down list in the Plot Style Table [Pen Assignments] area in the Plot or Page Setup dialog box), AutoCAD sends color information as it appears onscreen to the plotter. As long as the output device can plot in color, what you see should be what you get.

When you attach a plot style table to the layout you're plotting (as described in the preceding section), you can choose to map screen colors to different plotted colors. In most cases, you don't want that kind of confusion. Instead, leave the Color property in the plot style table set to Use Object Color.

WARNING

If your goal is *not* to plot color, verify that the Color property for all plot styles is set to Black. If you try to plot colors on a monochrome device, you may find that objects appear in various shades of gray, with some colors mapped to lighter shades of gray and others to darker shades of gray. This process of mapping colors to shades of gray is *dithering*, and it usually is *not* what you want in a CAD drawing. To override it, use Plot Style Table Editor, as described in the earlier section "Creating plot styles," to set the Color option for all colors to Black. The default setting is Use Object Color. If you don't already have a plot style table to use, select `monochrome.ctb` for color-based plot styles or select `monochrome.stb` for named plot styles (both of which come with AutoCAD) from the drop-down list in the Plot Style Table (Pen Assignments) area of the Plot dialog box.

To see the full range of AutoCAD colors available on a plotter, or to see how a particular plot style table affects plotting, open and then plot the sample file `plot_screening_and_fill_patterns.dwg`, which you can download from `www.autodesk.com/autocad-samples`. The Screening 100% layout in this drawing contains color swatches for all 255 AutoCAD colors. (This file is available to AutoCAD LT users as well.)

It's a (Page) Setup!

Page setups specify the plotter, paper size, and other plot settings that you use to plot a particular layout or the model space of a drawing. AutoCAD maintains separate page setups for model space and for each paper space layout. When you click the Apply to Layout button in the Plot dialog box (or select the Save Changes to Layout check box and then click OK to plot), AutoCAD stores the current plot settings as the page setup for the current layout.

You can also name page setups and save them. Then you can switch quickly between different plot settings and copy plot settings from one drawing layout to another. Named page setups are stored with each drawing, but you can copy them from another drawing into the current one with the Page Setup Manager dialog box (described later in this section).

If you want to get fancy, you can create named page setups to plot the same layout (or model space) in different ways, or to copy plot settings from one layout to another or one drawing to another. Click the Add button in the Page Setup area of the Plot dialog box to create a named page setup from the current plot settings. After you create a named page setup, you can restore its plot settings by choosing it from the Page Setup Name drop-down list.

For even greater control, right-click the current drawing space tab and then choose Page Setup Manager to create, change, and copy page setups. In the Page Setup Manager dialog box, as shown in Figure 16-9, you can create new page setups and modify existing ones. Click the Modify button to open the Page Setup dialog box, which is almost identical to the Plot dialog box. The primary difference is that you're changing plot *settings* rather than actually plotting. The Set Current button copies the page setup that you've selected on the Page Setups list to the current layout tab. With the Import button, you can copy a page setup from another drawing (DWG) or drawing template (DWT) file.

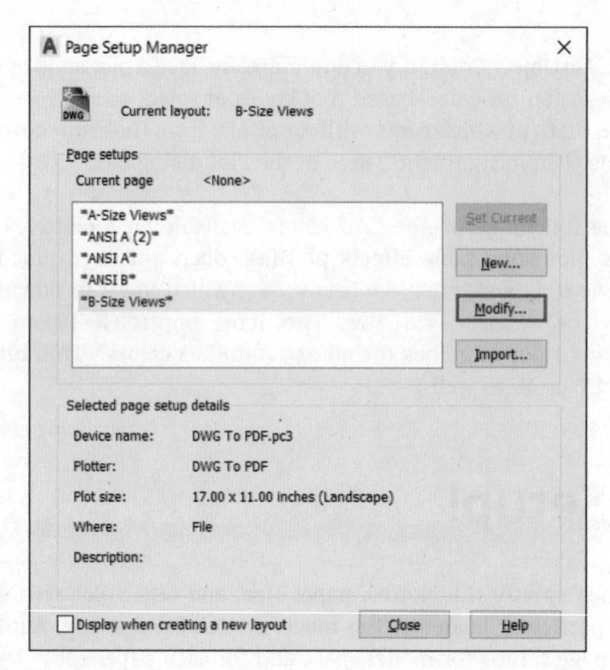

FIGURE 16-9: The Page Setup Manager dialog box.

Continuing the Plot Dialog

Earlier in this chapter, I cover several important options in the Plot dialog box. This section reveals finer points to simplify your plotting life. I don't cover every minute, obscure, useful-only-at-cocktail-party-discussions detail, and if this sounds like your kind of cocktail party, please remind me that I'm busy that night because I could end up monopolizing the conversation. I point out some occasionally useful options to help increase your vocabulary when you're communicating with the Plot dialog box.

TIP

Use the tooltips in the Plot dialog box to find out more about any part of the dialog box:

1. **Hover the mouse pointer over the part of the dialog box that you want to know more about.**
2. **Press F1 or click the Help button at the bottom of the dialog box if the pop-up help isn't helpful.**
3. **For a more conceptual take on plotting, click the Learn about Plotting link to view the Quick Start to Plotting section of the online Help system.**

The following list explains most of the remaining controls, check boxes, and lists in the Plot dialog box:

- **Printer/Plotter:** As I describe in the earlier section "Configuring your printer," you use the Name drop-down list to select the Windows system printer or non-system driver configuration that you want to use for plotting.

  Clicking the Properties button opens the Plotter Configuration Editor dialog box, with which you can change *media* (type of paper) and other properties that are unique to the selected plotter or printer. In particular, you can define custom paper sizes, should you use them.

  WARNING

  As if AutoCAD's Plot dialog box settings weren't overwhelming enough, you may also have to deal with the Plotter Configuration Editor dialog box, depending on the plot device. Some plotter drivers hide important settings in this dialog box. To access it, you typically click the Custom Properties button near the bottom of the Plotter Configuration Editor dialog box. For example, if you're using the enhanced Windows system driver for HP plotters, available at `https://support.hp.com/us-en/products/printers/designjet-printers`, you can click the Custom Properties button and then the More Sizes button to specify which paper sizes are available to you on the Paper Size drop-down list of the main Plot dialog box.

  TECHNICAL STUFF

  If you make changes in the Plotter Configuration Editor dialog box, AutoCAD prompts you to save the changes to a separate PC3 file. You should choose Save Changes to the Following File (that is, create a new AutoCAD-specific configuration that includes the revised settings) and type a configuration name that you'll recognize later. When you want to plot with custom settings, remember to choose the AutoCAD-specific PC3 configuration near the end of the Printer/Plotter Name drop-down list, and not the original Windows system printer configuration near the beginning of the list.

- **Plot to File:** You don't have to plot to a piece of paper. Your plotter choices are PNG, JPG, and PDF, and more. Plotting to a PDF file creates a version of the drawing that can be opened and viewed in the free and widely available Adobe Acrobat Reader DC software. Similarly, PNG and JPG options are useful when you want to include a drawing illustration in a Word or other text document.

  What if you can't justify the cost of a particular plotter for the few times you need it? No problem. Simply install the plotter driver on your computer, select it as the plot destination, and select Plot to File. Now when you click OK to plot, AutoCAD asks you for a plot filename and location. This creates a file to send to a plotting service bureau, and they don't need your drawing file or AutoCAD.

- **Plot Stamp On:** Use this option to turn on and off and configure the contents of a text string that AutoCAD adds automatically to the corner of each plot. The plot stamp can include useful information, such as the drawing's filename and plot date and time.
- **Plot Area:** Specify the area of the drawing to plot. Your choices include Display, Extents, and Window regardless of whether you're plotting a paper space layout or model space. If you defined named views in the drawing, AutoCAD adds a View option. The additional choice is Layout for a paper space layout, or Limits for model space.
  - *Display:* The drawing as it's displayed in the drawing window (including any empty space outside the drawing objects).
  - *Extents:* The rectangular area containing all objects in the drawing.
  - *Limits (full-screen model space only):* If you didn't specifically set the drawing's model space limits (as I demonstrate in Chapter 3), plotting limits produce unpredictable results. Extents is usually a better option.
  - *Layout (Layouts only):* The paper space area you defined when you set up the layout.
  - *Window:* A rectangular area that you specify.
  - *View:* A named view that you select from a drop-down list. Chapter 5 describes named views and how to create them.

Usually, you choose to plot Layout in paper space. For model space, the choice depends on how the drawing was set up and what you want to plot. Use Window or View to plot only a portion of model space.

- **Plot Offset:** A plot offset of X=0 and Y=0 positions the plot at the lower-left corner of the plottable area. If you want to move the plot from this default position on the paper, enter nonzero numbers or select the Center the Plot check box. This check box isn't available if you select Layout from the What to Plot drop-down list because the layout determines the position of the drawing relative to the margins of the paper.
- **Shaded Viewport Options:** If the drawing includes viewports showing shaded or rendered 3D models, use this area to control the plotted appearance.
- **Plot Options:** The Plot Object Lineweights and the Plot with Plot Styles check boxes control whether AutoCAD uses the features described in the "Plotting with style" and "Plotting through thick and thin" sections, earlier in this chapter.

- **Hide Paperspace Objects:** This check box controls whether AutoCAD hides objects that are behind other objects when a 3D model is displayed in a paper space viewport.
- **Plot Upside-Down:** Select this check box to rotate the plot 180 degrees on the paper. This option is handy for plotting in the southern hemisphere and for preventing having to cock your head at an uncomfortable angle as you watch plots emerge from the plotter.

AutoCAD normally takes over the program for the duration of the plot. If you have a reasonably fast computer with adequate memory, turn on Background Processing in the Options dialog box: Type **OP**, press Enter, and click the Plot and Publish tab; in the Options area, select Plotting.

TIP

If you want to automate plotting for a batch of drawings, check out the sheet sets feature in AutoCAD and AutoCAD LT. One of the tasks that sheet sets are designed to accomplish is the publishing of a set of drawing sheets at one fell swoop. If this sounds useful, go to the online Help system and search for *Work with Sheets in a Sheet Set* in the Search Help Resources box.

The Plot Sickens

No matter how many times you read this chapter or how carefully you study the AutoCAD documentation, you occasionally run into plotting problems. You're especially likely to encounter problems when trying to plot other people's drawings, because you don't always know what plotting conventions they had in mind. A *plotting convention* isn't a place where spies meet; it's a standardized approach to plotting issues.

WARNING

If a drawing contains object property overrides instead of ByLayer, you'll have plotting problems. For example, you want to plot an architectural drawing to show the client the general room layout. You don't want to confuse them with all the dimensions, so you freeze the Dimension layer. Oops, only half of the dimensions go away, but so do several walls, a couple of windows, and the toilet. Objects aren't on their correct layers, so use the LAYMCH (layer match) command to fix them.

Here are some common plotting problems (in italic) and solutions.

Nothing comes out of the plotter (system printer driver). Check the following:

- Verify that the printer is plugged in and turned on, that it has paper (and the paper isn't jammed) and plenty of toner or ink, and that it's connected to the computer.

- When the drawing is a large one, it may take a long time to rasterize. Look for the Printing icon on the AutoCAD status bar or a flashing busy light on the plotter.
- Check whether you can print to the device from other Windows applications. If not, the problem is not an AutoCAD problem.

TIP

HP provides its Print and Scan Doctor utility for finding problems troubling its printers and scanners. The software is available free from `http://support.hp.com/us-en/document/c03286146`.

Nothing comes out of the plotter (non-system printer driver). Choose Plotter Manager on the Plot panel of the Ribbon's Output tab, double-click the plotter configuration, and check the settings.

Objects don't plot the way they appear onscreen. Check for a plot style table with weird settings, or try plotting without a plot style table.

Objects appear ghosted or with washed-out colors. In the plot style table, set Color to Black for all colors.

Scaled to Fit doesn't work correctly in paper space. Change the What to Plot drop-down list from Layout to Extents.

The Windows system driver that you downloaded doesn't have the correct paper sizes. In the Plot dialog box's Printer/Plotter area, click the Properties button to display the Plotter Configuration Editor dialog box, click the Custom Properties button (near the bottom), and then click the More Sizes button to specify the standard and custom paper sizes.

Something else is wrong. Check the plot log: Click the Plot/Publish Details Report Available icon near the right end of the status bar and look for error messages.

4

Advancing with AutoCAD

IN THIS PART . . .

Use blocks, arrays, and xrefs in drawings effectively.

Apply geometric and dimensional constraints and formulas — parametrics — to control the behavior of objects.

Share AutoCAD files through email, online storage, and Autodesk cloud services.

- » **Creating block definitions with Block**
- » **Inserting blocks with Insert**
- » **Adding attributes to blocks with ATTdef**
- » **Removing unwanted block definitions with PUrge**

Chapter 17
The ABCs of Blocks

In Chapter 11, you can see how to copy objects in a drawing or even *to* another drawing. That's one way to use AutoCAD to improve drafting efficiency. Another way is to make a copy of a DWG file and then modify it to create a different but similar drawing. This can be an even better productivity booster, as long as you're in the habit of making similar drawings. But those are all baby steps compared with the techniques I cover in this chapter and in Chapter 18: By treating parts of drawings as modules or symbols that can be reused and updated. AutoCAD calls these modules *blocks.*

This block treatment applies also to entire drawings, drawings saved in web format (*Design Web Format,* or *DWF*), PDFs, Autodesk Inventor 3D models, MicroStation DGN files, raster images, and others.

In this chapter, I present the ABCs of blocks, including the basic creation of blocks and their insertion into drawings, adding attributes (data) to them, and getting rid of block definitions you no longer need or want. In Chapter 18, I show you how to make even more of existing drawing data, including dynamic blocks, associative array objects, and several flavors of external reference files (including PDFs and MicroStation DGN files) and raster images.

Rocking with Blocks

A *block* is a collection of objects grouped to form a single object, like the lines and arcs that make up the drawing of a chair. You can insert this collection more than once in the same drawing, and when you do, all instances of the block are identical — that is, the drawing has lots of identical chairs.

By redefining the chair block definition, you automatically change all instances of the block's insertions at one time. You might do this, for instance, if your boss decides to save money by changing all armchairs to armless chairs. Although a block lives in a specific drawing, you can transfer copies of it to other drawings. Draw the chair once in one drawing, and you can use it in any other drawing.

And you can add fill-in-the-blank text fields, or *attributes,* to blocks. These attributes might be information about the chair, such as its price, where it was purchased, and its current owner. The data can be exported to Excel files and placed in drawings as tables. You can create single-line or multiline attributes; when multiline, they have more than one line of text and many of the formatting options of multiline text, including fields. And blocks both with and without attributes can be defined as annotatively scaled objects to boot. See Chapter 13 for a rundown on annotative objects.

To use blocks in drawings, you need two elements: a block *definition* and one or more block *insertions*. AutoCAD doesn't always make clear the distinction between these two terms, but you need to understand the difference to avoid terminal confusion about blocks. Maybe this syndrome should be called *blockheadedness*?

A block definition lives in an invisible area of the drawing file: the block table. It's one of those sets of named symbols that I describe in Chapter 9. The block table is a book of graphical recipes for making different kinds of blocks. Each block *definition* is a recipe for making one kind of block. Each time you insert the block, as described in the later section "Inserting Blocks," AutoCAD creates special objects called block *insertions.* The insertion refers to the recipe that tells AutoCAD, "Hey, draw me according to the instructions in this recipe," and this is called a block *reference.*

Although a block may look like a collection of stored objects that is given a name, it's a graphical recipe (the block definition) — plus, one or more pointers to that recipe (one or more block references). Every time you insert a particular block, you create another pointer to the same recipe.

You benefit from using blocks because you can

» **Group objects when they belong together logically.** You can draw a bolt, using lines and arcs, and then make a block definition from all these objects. When you insert the bolt block, AutoCAD treats it as a single object for copying and moving, for example.

» **Save time and reduce errors.** Inserting a block is, of course, much quicker than drawing the same geometry again. The less geometry you draw from scratch, the less opportunity you have to make a mistake and the faster you complete your designs.

» **Gain efficiency of storage when you reuse the same block repeatedly.** If you insert the same bolt block 15 times in a drawing, AutoCAD stores the detailed block definition only once. The 15 block references that point to the block definition occupy much less space than 15 copies of all the lines, polylines, and arcs. No matter how large the block definition, each insertion adds only about 100 bytes to the drawing file size.

» **Edit all instances of a specific symbol in a drawing by simply modifying its block definition.** If you decide that the design requires a different kind of bolt, you simply redefine the bolt's block definition. With this new recipe, AutoCAD then replaces all 15 bolts automatically. That's a heck of a lot faster than erasing and recopying 15 bolts!

» **Vary the specification of block references by using dynamic blocks.** If the design requires a different kind of bolt, different lengths for example, you simply change the view of the bolt to the other kind, assuming, of course, that you've defined the bolt as a dynamic block. Every instance of the bolt block in the drawing can show a different length of bolt. And *that's* a heck of a lot more efficient than creating 15 different block definitions. For the lowdown on creating, inserting, and manipulating dynamic blocks, see Chapter 18.

Blocks may not be best for common drawing elements used in multiple drawings, however, especially when several people are working on and sharing parts of drawings. If you copy and paste a block from one drawing to another, AutoCAD creates a new block definition in the target drawing and there is no connection back to the source drawing. A later modification to a block definition in one drawing does *not* automatically modify all other drawings using that block. If you use a block with the company logo in a number of drawings and there is a change in corporate ownership, you must make the change in each drawing that uses the block. To overcome this problem, use external reference (xref) files. I cover XREFs in Chapter 18.

If you need to group objects only so that you can more easily select them for copying or moving, for example, use the AutoCAD Group feature (described more fully in Chapter 10). Type **Group** and press Enter, or simply click the Group button in the Groups panel on the Home tab. Then select some objects and you're done. When you're editing drawings that contain groups, press Ctrl+Shift+A to toggle "group-ness" on or off. If you've toggled group-ness on, picking any object in a group selects all objects in the group. If you've toggled it off, picking an object selects only that object, even if it happens to be a member of a group. For more information, refer to Chapter 10 or visit the online Help index.

Blocks — along with external references, DWF and PDF underlays, raster images, and other files — enable you to reuse your work and the work of others, giving you the potential to save tremendous amounts of time — or to cause tremendous problems when you change a file that other people's drawings depend on. To save time, use these features when you can, but do so in an organized and careful way so as to avoid problems.

TIP

How you use blocks and xrefs depends a lot on your profession and office in which you work. Some disciplines and companies use these drawing organization features heavily and in a highly organized way, but others don't. Ask your colleagues about the local customs and follow them.

Creating Block Definitions

To create a block definition from objects in the current drawing, use the Block Definition dialog box. The other way to create a block definition is to insert another drawing file into the current drawing as a block, which I explain in the later section "Inserting Blocks." The following steps show you how to create a block definition by using the Block Definition dialog box:

1. **On the Insert tab of the Ribbon, click the Create Block button on the Block Definition panel, or type** Block **and press Enter.**

   The Block Definition dialog box appears, as shown in Figure 17-1.

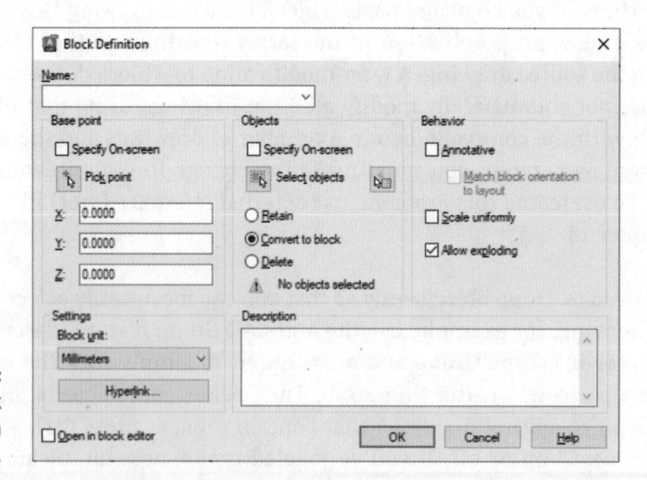

FIGURE 17-1: The Block Definition dialog box.

TIP

Pay attention to layers when you create the objects that make up a block. As a rule, block geometry created on nearly all layers retains the color, linetype, lineweight, transparency, and plot style properties of those layers. There is an exception to the rule: Object geometry created on Layer 0 (zero) takes on the properties of the layer into which you insert the block.

2. **Type the block definition's name in the Name text box.**

   If you type the name of an existing block definition, AutoCAD warns you when you click OK at the end of the process. This isn't always a bad thing, because AutoCAD asks whether you want to replace that existing block definition with the new objects you select. This is a legitimate process known as *block redefinition.* For example, if you have a drawing with a large number of block insertions, such as the bathtubs in a 1,000-room hotel, you can use a simplified block representation of tubs during the design process and then redefine it to a more complex tub definition just before final plot time.

TIP

To see a list of the names of all current blocks in the drawing, open the Name drop-down list.

3. **Specify the *base point* (also known as the insertion point) of the block, using either of the following methods:**

   - Enter the coordinates of the base point in the X, Y, and Z text boxes.
   - Click the Pick Point button, and then specify a point on the screen. In this case, use an object snap or another precision technique, as described in Chapter 8, to grab a specific point on an object.

   The *base point* is the point on the block by which AutoCAD knows to insert it later, as I describe in the next section.

TIP

Use an obvious and consistent point for the base point on the group of objects, such as the lower-left corner or the center, so that you know what to expect when you insert the block. If you do not define a base point, AutoCAD uses 0,0,0; when you later insert the block, it may appear in an unexpected location, such as near Tuktoyaktuk.

4. **Click the Select Objects button and then select objects that are to become part of the block.**

   AutoCAD makes the block definition from the selected objects and creates a thumbnail image next to the block's name. Figure 17-2 shows the base point and a group of selected objects that were specified during the process of creating a new block definition.

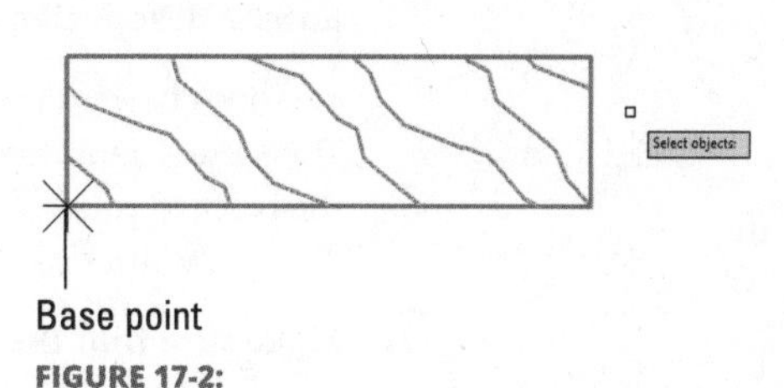

FIGURE 17-2: Building a block.

TIP

The components of a block definition can include other block insertions and so on and so on, for as many levels deep as you need. If you try to create a circular reference, however, AutoCAD will object. For example, block A can contain block B, which can contain block C as long as block C doesn't contain block A.

5. **In the Objects area, select a radio button to tell AutoCAD what to do with the objects that are used to define the block: Retain them in place, convert them into a block instance, or delete them.**

   The default choice, Convert to Block, is usually the best. See Step 9 for a description of what happens with each choice.

6. **Specify the insertion units by which the block will be scaled in the Block Unit drop-down list.**

   As you learn later in this chapter, you can copy an existing block from another drawing and then insert it into the current one. When you do so, the units you specify here and the units of the drawing you're inserting into control the default block's scale factor. The list contains 17 different units, from angstroms to parsecs, but doesn't include fathoms or furlongs, unfortunately.

TIP

Three additional options in the Block Definition dialog box give you even more control over what happens to blocks as they're inserted:

- *If the Annotative check box is selected:* The block insertion is scaled to suit the current drawing scale, and additional scales can be applied so that the insertion is scaled differently to suit other drawing scales. This feature would typically be applied only to symbols (such as north arrows) and not to insertions of blocks depicting objects, such as toilets or motors. I explain the nuts and bolts of annotative objects in Chapter 13.
- *If the Scale Uniformly check box is selected:* A block is inserted with the same X, Y, and Z scale factors. Scale Uniformly is selected automatically if Annotative is selected.
- *If the Allow Exploding check box is selected:* A block can be exploded (as explained in Chapter 11) during or after its insertion in a drawing, turning it back into its constituent objects.

7. **Enter a description of the block in the Description text area.**

   You don't have to enter a description to create a block, but it isn't a bad idea. Think like a database manager by entering a useful description that identifies the block to yourself and others, such as "Sink, Kitchen, 2D." Note that this book now includes just about everything, including the kitchen sink.

8. **Make sure that the Open in Block Editor check box is deselected.**

   You don't need to use the Edit Block Definition dialog box unless you add dynamic features to the block. See Chapter 18 to find out more about dynamic blocks.

9. **Click OK to complete the block definition process.**

AutoCAD stores the block definition in the current drawing's block table. This list describes the behavior of the radio buttons from in Step 5:

- If you select the *Convert to Block* radio button (the default) in Step 5, AutoCAD creates a block reference pointing to the new block definition. The objects look the same onscreen, but they are now an instance of the block rather than the original, separate objects. Most of the time, this is the best choice.
- If you select the *Retain* radio button, the objects remain in place but aren't converted into a block reference. They remain individual objects with no connection to the new block definition. The Retain option is useful when you want to define another block using these objects.
- If you select the *Delete* radio button, the objects disappear, but the block definition still gets created. This option is useful when you don't want to insert the block quite yet.

When you define a block, you can include a special kind of variable text object: an *attribute definition.* When you insert a block that contains one or more attribute definitions, AutoCAD prompts you to fill in values for the text fields. Attributes are useful for elements like variable title block information (sheet number and sheet title, for example), electric motor nameplate information, and symbols that contain different codes or callouts. I describe how to create and use attribute definitions later in this chapter.

Store commonly used symbol drawings in one or more specific folders that you set aside for this purpose. You can use one of these techniques to develop a library of frequently used symbols:

» **Create a separate DWG file for each block.** Insert these symbols by using the Insert command or by dragging them from Windows Explorer into the drawing area.

» **Store several symbols as block definitions in one master drawing.** Then use DesignCenter to import block definitions from this drawing whenever you need them.

» **Use the WBLOCK (WriteBlock) command to extract a block definition from a drawing and then write a copy of it to disk as a new DWG file.** Then you can insert the block definition into other drawings as required.

You can clean up a drawing by using the Wblock command to write an entire drawing to disk. It writes out only what it needs, but be careful because it also cleans out any unused custom layers, styles, and block definitions.

Inserting Blocks

AutoCAD provides a number of ways to insert a block or a whole drawing file, but the most commonly used and most flexible method is using the Insert button on the Ribbon menu. To insert a block, follow these steps:

1. **Set an appropriate layer current, as described in Chapter 9.**

Insert each block on a layer that has something to do with the block's geometry or purpose:

- *If all objects in the block definition reside on one layer,* you usually should insert the block on that layer.
- *If the block geometry spans several layers,* choose a suitable one on which to insert the block. For example, a block that shows a gearbox made of geometry on layers named Casing, Armature, Shaft, and Fasteners might be inserted on a Drives layer.

If any of the block definition's geometry was created on Layer 0 (zero), the geometry inherits the color, linetype, and other object properties of the layer on which you insert the block. It's like chameleons changing color to match their surroundings or politicians changing their position to match the day's opinion polls. Any block components on layers other than layer 0 retain their respective layer's properties.

2. **On the Home tab of the Ribbon, click the Insert button on the Block panel.**

A slideout panel drops down, displaying a thumbnail view of all block definitions in the current drawing.

3. **Click the block definition you want, and then click an insertion point for it.**

If you want to do something fancier, in older AutoCAD releases you click the More Options button at the bottom of the thumbnail preview slideout panel to produce the Insert dialog box, as shown in Figure 17-3, left. In AutoCAD 2020 and later, click the Recent Blocks button of the preview panel to launch the Blocks tool palette, as shown in 17-3, right.

I won't bore you with the minute details of every option, but here are a few to give you an idea.

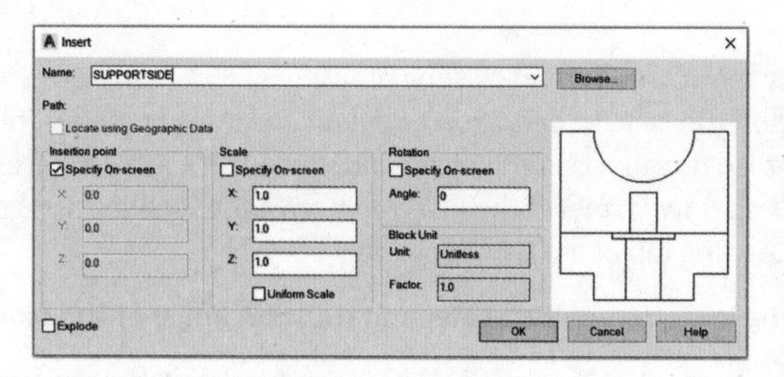

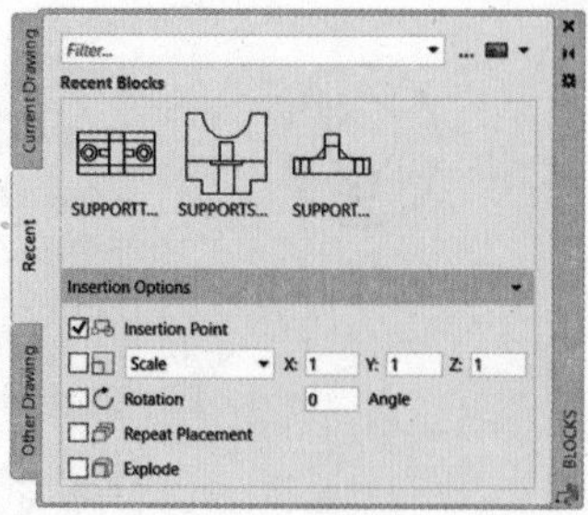

FIGURE 17-3: The Insert dialog box, with previous releases on the left and AutoCAD 2020 on the right.

1. **Enter the block definition name (or external filename) by using one of these methods:**

   - *Use the Name drop-down list* to select from a list of block definitions in the current drawing or click the appropriate preview image.

     The Name drop-down list supports autocomplete. As you begin to type the name of a block, AutoCAD searches through the drawing's list of block names and finishes filling in the name for you.

   - *Click the Browse button* to select an external DWG file and have AutoCAD create a block definition from it.

TIP

Browse to any existing drawing and then insert it into the current drawing to create a new block definition in the host drawing. Note that nothing is magical about the incoming drawing. *Any* existing drawing can be inserted into any other drawing. The new definition retains no connection to the source drawing.

Ah, but here's the real magic of inserting an existing drawing to create a block definition. A great many manufacturers and suppliers provide standard AutoCAD drawings of their products, often online. Building a conveyor system and need a particular motor-gearbox unit? No need to draw it from scratch; just download it from the manufacturer's website and plug it into your drawing.

You can use an external drawing to replace a block definition in the current drawing. If you click Browse and choose a file whose name matches the name of a block definition that's already in the drawing, AutoCAD asks you to confirm the update and, if you agree, updates the block definition in the drawing with the current contents of the external file. In this process, *block redefinition,* AutoCAD automatically updates all block insertions that point to the block definition.

Earlier in the chapter, I mention the example of using a simplified representation of a bathtub during the design process and then replacing it with a detailed version later. Block redefinitions by inserting an existing drawing is the fastest and easiest way to do this.

TIP

Not only can you import an existing drawing, but you can also browse in it and just import the specific definition or definitions that you need. This feature can be accessed through Design Center and, since AutoCAD 2020, from the bottom of the drop-down panel of the Insert button of the Ribbon menu or from the Other Drawing tab of the Blocks palette.

2. **Enter the insertion point, scale, and rotation angle of the block.**

   You can either select the Specify On-Screen check box in each area to specify the parameters onscreen at the command prompt or type the values you want in the text boxes in the Insertion Point, Scale, and Rotation areas.

TIP

   Select the Uniform Scale check box to constrain the X, Y, and Z scaling parameters to the same value (which is what you want in almost all cases).

3. **If you want AutoCAD to create a copy of the individual objects in the block instead of a block reference pointing to the block definition, select the Explode check box and click OK.**
4. **If you select the Specify On-Screen check box for the insertion point, scale, or rotation angle, answer prompts that appear on the command line to provide these parameters.**

After you insert a block, all the objects displayed in the block reference behave as a single object. When you select any object in the block reference, AutoCAD highlights all the objects in it.

AutoCAD gives you three more ways to insert a block:

» **Drag and drop:** Drag a DWG file from Windows Explorer and drop it anywhere in the current drawing window. AutoCAD then prompts you to choose an insertion point and (optionally) change the default scale factor and rotation angle. If you drag the drawing file to AutoCAD's title bar, the file is instead opened as a new drawing.

  You can import Inventor part (IPT) and assembly (IAM) models directly into AutoCAD. Inventor is AutoCAD's 3D parametric and associative sibling, intended primarily for mechanical design. Imported Inventor files come across as dumb 3D solids that lose their parametrics and associativity, but they can be modified and edited like any other AutoCAD 3D solid. I cover the 2D parametrics of AutoCAD in Chapter 19, and introduce you to 3D modeling in Chapters 21–23.

» **DesignCenter palette:** Drag a block definition from the Blocks section of the DesignCenter palette and drop it into the current drawing window. Turn to Chapter 9 to find out how to use DesignCenter.

» **Tool Palettes palette:** As is true of using the tool palette for hatching (described in Chapter 15), you first must create and configure the appropriate tools. The easiest method is right-clicking a drawing in DesignCenter and choosing Create Tool Palette. A new tabbed page is added to the Tool Palettes window containing all block definitions from the drawing. Simply click and drag a tool to insert its corresponding block into a drawing.

Dragging blocks from a tool palette doesn't give you the chance to specify a different insertion scale, nor can you use all the precision tools in AutoCAD to specify the insertion point precisely, so you may need to move the block into place after inserting it. First master the other block insertion methods described in this chapter, especially using the Insert dialog box and DesignCenter palette. Then, when you find yourself inserting the same blocks frequently, consider creating a tool palette containing them. Check out the Add Content with DesignCenter topic in the AutoCAD online Help system for more information.

TECHNICAL STUFF

Although the preceding paragraph refers to the Tool Palettes palette, palettes in AutoCAD aren't like regular dialog boxes. They're *modeless*, so they can stay open while you carry out other tasks outside them. The official programmer-ese term for the palette is *enhanced secondary window*, or ESW. I stick with *palette*.

WARNING

Be careful when you insert one drawing into another. If the host (or *parent*) drawing and the inserted (or *child*) drawing have different definitions for layers of the same name, the objects in the inserted drawing take on the layer characteristics of the host drawing. For example, if you insert a drawing with lines on a layer named Walls that's blue and dashed into a drawing with a layer named Walls that's red and continuous, the inserted lines on the Wall layer turn red and continuous after they're inserted. The same rules apply to linetypes, text styles, dimension styles, table styles, multileader styles, and block definitions that are nested inside the drawing you're inserting. As you may recall, parents rule! Yeh, I wish.

LOCATE USING GEOGRAPHIC DATA OPTION

Geographic data refers to locating drawing geometry with reference to specific locations on planet Earth (so far) defined in one of a number of recognized global coordinate systems. This feature can be useful in civil engineering and large architectural projects. If you have a GPS-enabled laptop or tablet, it can display your current position on a site plan as you wander about. If geographic coordinates sounds like an option you need, look in the online Help system.

TIP

To modify a block definition after you've inserted at least one instance of it, use the BEdit (Block Editor) command: Choose Block Editor in the Block panel on the Home tab, or simply double-click a block insertion. Look up *BEDIT* in the online Help system in AutoCAD.

Attributes: Fill-in-the-Blank Blocks

You may think of attributes as the good (or bad) qualities of your significant other, but *attributes* in AutoCAD are fill-in-the-blank text fields that you can add to blocks. When you create a block definition and then insert it several times in a drawing, all the ordinary geometry elements (lines, circles, and regular text strings, for example) in all instances are exactly identical. Attributes provide a little more flexibility in the form of text strings that can be different with each block reference.

Suppose that you frequently designate parts in drawings by labeling them with distinct numbers or letters enclosed in a circle. If you want to create a block for this symbol, you can't simply draw the number or letter as regular text by using the mText (T) or TEXT (DT) commands. If you were to create a block definition with a regular text object (for example, the letter *A*), the text is the same in every instance of the block (always the letter *A*). That's not much help in distinguishing between parts.

Instead, you create an *attribute definition*, which acts as a placeholder for a text string that can vary every time you insert the block. You set up the attribute definition when you create the block definition. Refer to the "Creating Block Definitions" section, earlier in this chapter. Then every time you insert the block, AutoCAD prompts you to fill in an *attribute value* for each attribute definition.

When the block attribute value was first introduced, and for a long time afterward, it was limited to a single line of text with a maximum of 255 characters. AutoCAD 2008 and later supports multiline attributes; in addition to offering more than one line, multiline attributes have many of the formatting options of multiline text. For more information on creating and inserting blocks with multiline attributes, look up *Define Block Attributes* in the online Help system.

REMEMBER

The AutoCAD documentation and dialog boxes often use the term *attribute* to refer indiscriminately to an *attribute definition* or an *attribute value*. I attribute a lot of the confusion about attributes to this sloppiness. Just remember that an attribute *definition* is the placeholder in the block definition, and an attribute *value* is the specific text you provide when you insert the block.

Creating attribute definitions

You use the Attribute Definition dialog box to create attribute definitions. Clever, eh? The steps are similar to creating a text string, except that you must supply a little more information. Create attribute definitions by following these steps:

1. **Change to the layer on which you want to create the attribute definition.**
2. **To run the ATTdef command, click Define Attributes in the Block Definition panel on the Insert tab.**

   The Attribute Definition dialog box appears, as shown in Figure 17-4.

   You rarely need to use any of the first four Mode settings (Invisible, Constant, Verify, or Preset). Leave them deselected. If you're curious about what the modes do, hover the mouse pointer over an item; if that doesn't give you enough information, click the Help button in the dialog box to find out more.

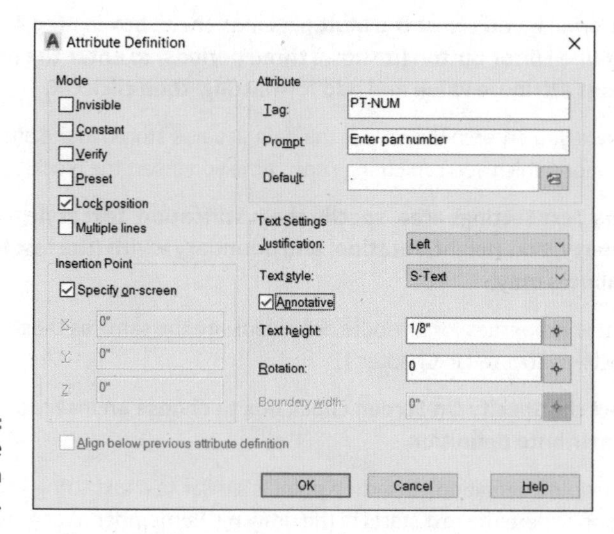

FIGURE 17-4: The Attribute Definition dialog box.

3. **Select or deselect the Lock Position check box.**

   If Lock Position is selected, the attributes are fixed in the block reference and the whole thing is treated as a single object. Deselecting Lock Position allows attributes to be moved around the drawing by dragging their grips, without moving the block reference itself.

4. **Select or deselect the Multiple Lines check box.**

   Selecting Multiple Lines in the Mode area disables the Default text box and displays a button to open the Multiline Editor. Initially, you don't see the whole panoply of formatting options of the mText command's In-Place Text Editor.

You can overscore or underscore text; and from a right-click menu, you can import text, assign a background mask, or choose from a number of other options. Setting the value of the system variable ATTIPE (Attribute In-Place Editor) to 1, however, enables *all* formatting options in the In-Place Text Editor. See the online Help system for more information.

5. **In the Attribute area, type values for the *tag* (a unique identifier for the attribute), a prompt for users, and a default value.**

REMEMBER

The name you type in the Tag text box can't contain spaces. The Prompt and Default text boxes may contain spaces, though. The Prompt tells users what kind of text they should enter, such as a name or a price. The Default is a typical value.

6. **(Optional) Click the Insert Field button to the right of the Default text box to insert a field.**

Attribute values can include fields that update automatically, such as date, filename, and system variable settings. See Chapter 13 for more information on fields.

7. **(Optional) If you select the Multiple Lines check box in Step 4, click the Multiline Editor button (it shows three periods) to enter the multiline default attribute value and add formatting; then click OK.**

The text you enter in this step is the default value stored in the attribute definition, which you can change each time you insert the block.

8. **In the Text Settings area, specify the justification, text style, annotative property, text height, rotation, and boundary width (the last for multiline attributes only).**

The text properties for attribute definitions are the same as those for text objects, as covered in Chapter 13.

9. **Select the Specify On-Screen check box to choose an insertion point for the attribute definition.**

An attribute definition's insertion point is similar to a text string's base point: It's the spot where the text starts in the drawing. Remember to use snap, object snap, or another precision tool when you want the eventual attribute values to be located at a precise point.

10. **Click OK to create the attribute definition.**

11. **Repeat Steps 1–10 for all additional attribute definitions.**

TIP

To create a series of attribute definitions in neat rows for a single block, create the first one by following Steps 1–9, and then select the Align Below Previous Attribute Definition check box for the subsequent definitions. To make a series of nonadjacent attributes, create the first one by using Steps 1–10, and then copy the first attribute definition and edit the copy by using the Properties palette. You can prevent attributes from being dragged around the block by selecting the Lock Position check box in the Attribute Definition dialog box.

Defining blocks that contain attribute definitions

After you create one or more attribute definitions and any other geometry that you want to include in the block, you're ready to create a block definition that contains them. Follow the steps in the earlier section "Creating Block Definitions." At Step 4 in that section, you can select any attribute definitions before or after you select the other geometry. You should, however, select each attribute definition one by one (clicking each attribute definition rather than selecting multiple attributes with a selection window) in the order you want the attribute value prompts to appear in the Edit Attributes dialog box. They appear in the order you pick them, which might be important to you. (See Figure 17-5.) If you don't select the attributes one by one, the block and attributes still work, but the order of the attribute prompts in the Edit Attributes dialog box may not be what you want.

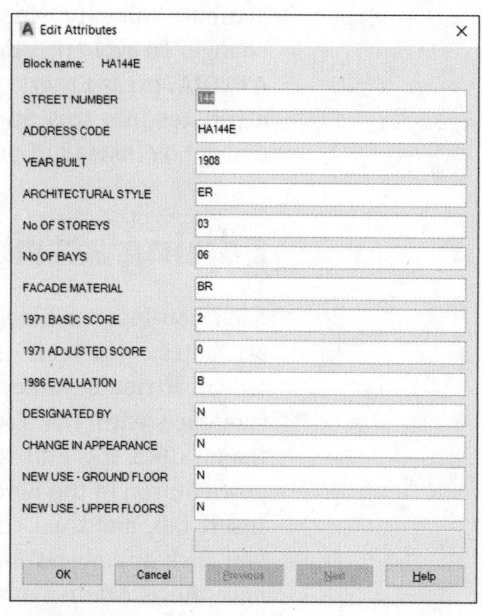

FIGURE 17-5: The Edit Attributes dialog box.

TECHNICAL STUFF

You can use Block Attribute Manager to reorder the attribute definitions in a block definition. Choose Attribute ➪ Block Attribute Manager in the Block panel slide-out on the Home tab to start the BATTMAN command. Yes, Robin, there actually is such a command name. You can also use this dialog box to edit other attribute definition settings such as the prompt, text style, or layer.

Inserting blocks that contain attribute definitions

After you create a block definition that contains attribute definitions, you insert it the same way you insert any other block. Follow the steps in the earlier section "Inserting Blocks." At the end of the steps, AutoCAD displays the Edit Attributes dialog box (refer to Figure 17-5). The dialog box contains one row for each of the attribute definitions and has the default values filled in for you. You simply edit the values as necessary and then click OK.

The ATTDIA (ATTribute DIAlog box) system variable controls whether AutoCAD prompts for attribute values in a dialog box (ATTDIA=1) or at the command line (ATTDIA=0). If you insert a block and see command line prompts for each attribute value, type a value and press Enter for every attribute value you want to change. To avoid this in the future, when you return to the command prompt, type **ATTDIA**, press Enter, type **1**, and press Enter again. When you insert blocks with attributes into this drawing in the future, AutoCAD displays the Edit Attributes dialog box instead of prompting you at the command line.

Editing attribute values

After you insert a block that contains attributes, you can edit the individual attribute values in that block reference with the EATTEDIT command. Click the Edit Attribute drop-down button in the Block panel on the Insert tab, and then choose Single. If you click any object in a block reference, AutoCAD displays the Enhanced Attribute Editor dialog box with the current attribute values, as shown in Figure 17-6. The most common attribute editing operation is to edit the *text value,* which is the text string displayed by inserted blocks. You also can change properties of the attributes, such as their layer and text style.

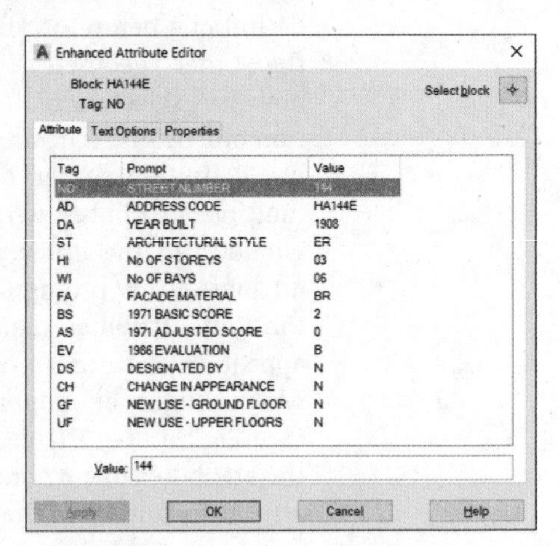

FIGURE 17-6:
The Enhanced Attribute Editor dialog box.

Many people use attributes only in the way I describe in this chapter, as fill-in-the-blank text fields in blocks. But attributes also can work with data extraction tools. For example, you can export attribute values, such as part numbers and quantities, to a table object in AutoCAD or to a text, spreadsheet, or database file for further analysis and reporting.

Extracting data

Selecting Extract Data from the Linking & Extraction panel on the Insert tab of the Ribbon starts the Data Extraction Wizard. You can find out much more about this specialized function in the online Help system. Check out User's Guide ⇨ Share Data between Files ⇨ Extract Data from Drawings and Spreadsheets. And if you use AutoCAD LT and don't have this wizard, you can still extract attribute information to space- or comma-delimited text files with the Attribute Extraction dialog box (ATTEXT). These data files can be imported into spreadsheets and database programs for further processing. For more information, visit the online Help system.

Exploding Blocks

In a regular block definition (not a dynamic block), the objects in each block reference act like a well-honed marching squadron: If you move or otherwise edit one object in the block reference, all objects move or change in the same way. Usually this cohesion is a benefit, but occasionally you need to break up the squadron to modify a single object and not affect the others.

To *explode* a block reference back into the individual objects, click Explode in the Modify panel on the Home tab; or type **X**, press Enter, and select the block reference. When you explode a block reference, AutoCAD replaces it with all objects (the lines, polylines, and arcs, for example) specified in the block's definition. You then can edit the objects individually or perhaps use them to make more block definitions. Only the block you select is exploded.

When you explode a block that contains attributes, the attribute values are lost, and become attribute definitions (usually not a desirable change). When you truly need to explode the block reference, you may want to erase the attribute definitions and draw regular text strings in their place. You can perform this task automatically with the BURST command: Type **BURST** and press Enter (not available in AutoCAD LT).

Both AutoCAD and AutoCAD LT have an NCOPY command that you can use to copy objects contained in blocks, without having to explode the block. AutoCAD users who are familiar with the Express Tools may already be familiar with NCOPY; the command is now in the core of the program and therefore available to LT users. To access the Copy Nested Objects tool, go to the Modify panel slideout on the Home tab or simply type **NCOPY**.

Purging Unused Block Definitions

Each block definition increases the size of the DWG file, as do other named objects, such as layers, text styles, and dimension styles. In the case of large, complex block definitions, the size increase can be significant. If you delete (or explode) all block references that point to a particular block definition, the block definition no longer serves any purpose.

You should run the PUrge command periodically in each drawing to purge unused block definitions. To display the Purge dialog box, click the Application button to open the Application menu, choose Drawing Utilities on the left side, and then choose Purge — or enter **PU** at the command line. Click the Purge All button to purge all unused named objects in the current drawing.

PUrge isn't only for blocks: It also removes empty layers, blank text strings, zero-length lines, empty groups, and unused style definitions. In Figure 17-7, the Purge dialog box shows the unused dimension, multileader, and text styles that would be removed from this drawing with a click of the Purge All button.

Figure 17-7 shows the rearranged Purge dialog box introduced in AutoCAD 2020. The dialog box in earlier releases has the same basic functions but is arranged differently.

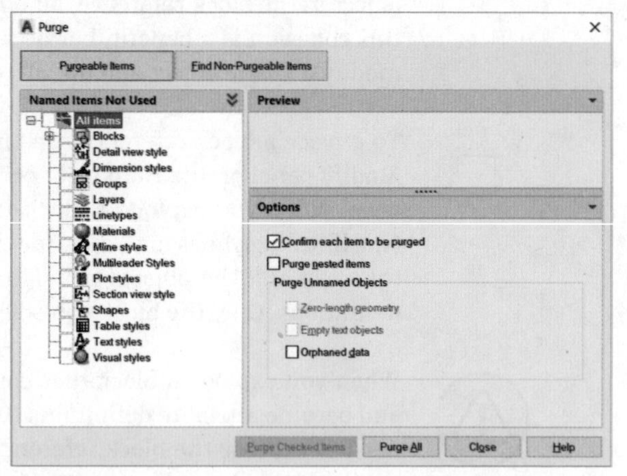

FIGURE 17-7: Purging the drawing of unneeded named objects.

Note the Purge Nested Items check box. Some items may not become unused until other items are purged. For example, a block definition can contain one or more other block definitions as components, and they in turn can contain other components. If the parent contains the only use of a child or grandchild, purging the parent will orphan the child and so on. When Purge Nested Items is selected, AutoCAD tunnels down through the nestings, looking for more orphans. These may include block definitions or items such as layer, text style, and dimension style specifications.

As I mention earlier in this chapter, you can insert any existing drawing into the current one where it becomes a block definition. In a common scenario, you know that another drawing of a 1,000-room hotel contains a block definition of a bathtub you want to use in the master bathroom of a house drawing. No problem: Simply insert the hotel drawing into the house drawing, and then erase or delete the hotel drawing insertion. The drawing now contains a block definition for the bathtub. Oops! Here's a problem. The house drawing file size has suddenly increased to government-spending numbers because it now contains the hotel drawing and all its block definitions, along with elements such as layer, text style, and dimension style specifications. It's PUrge to the rescue!

images, and dynamic blocks

- Creating associative arrays with ARray
- Attaching and managing xrefs, DWFs, and PDFs
- Controlling xref paths
- Importing other file types
- Attaching and managing raster image files
- Authoring dynamic blocks

Chapter 18
Everything from Arrays to Xrefs

Simple arrays have served AutoCAD users well for many years. They're basically just a quick way of producing multiple copies of drawing objects in regular patterns. If you're not familiar with simple arrays, then flip to Chapter 11.

Associative arrays are *dynamic:* That is, you can change the number of items, change the objects being arrayed, or substitute a different object set for one or more object sets in an array.

An *externally reference drawing* — xref — is like an industrial-strength block. An xref is a pointer to an independent drawing file outside the drawing you're working on. When you attach a reference drawing, it appears onscreen and on plots as part of your drawing, but continues to live as a separate document on your hard drive or network. When you edit the externally referenced drawing, its appearance changes in all drawings that reference it. Like a block insertion (more on that shortly), an xref attachment increases the drawing file only slightly, regardless of the size of the xref file.

A *raster image* file stores a graphical image as a series of dots. PNG and JPG images seen on the internet or from digital cameras are typical examples. Raster files are good for storing photographs, logos, and other images. Read more about raster images in "Mastering the Raster" later in this chapter. Comparatively, AutoCAD *vector* files are good for storing geometrical objects, such as lines and arcs, along with text and other annotations for describing the geometry. Sometimes combining raster images with AutoCAD vector files by attaching them to your drawing files is handy, and the External References palette makes the process straightforward.

TECHNICAL STUFF

Before AutoCAD 2007, *external references* meant AutoCAD drawing files (and only AutoCAD drawing files) that you attached to your current drawing. Images were (and still are) raster graphics files that you attached in a similar way but with a different command. In AutoCAD 2007 and later, you use the External References palette to attach and manage not only externally referenced AutoCAD drawings (xrefs) and 2D MicroStation DGN drawing files but also image files and DWF or PDF underlays. *DWF* stands for Design Web Format; a DWF is a lightweight version of a DWG file intended for design review or posting on a website. I discuss attaching DWFs and PDFs as external references at the end of this chapter and describe the web functions of DWFs in Chapter 20.

Both AutoCAD and AutoCAD LT allow you to import DGN files into and export DGN files out of the current drawing. As well, you can attach DGN files via the External References palette. *DGN* files are design files created by one of AutoCAD's competitors, MicroStation from Bentley Systems. If you're new to AutoCAD, you're not likely to encounter DGN files, unless you're working for a company that exchanges these files with partners and consultants, such as in transportation projects. For more information on DGN files in AutoCAD, enter *DGN Files* in the Search box at the top of the online Help home page.

WARNING

As I mention in Chapter 17, blocks, external references, DWF underlays, PDF underlays, raster images, and DGN files enable you to reuse your work and the work of others, giving you the potential to save tremendous amounts of time — or cause tremendous problems if you change a file on which other people's drawings depend. Use these features to save hours of time, but do so in an organized and careful manner to avoid problems.

In Chapter 17, I introduce *blocks* as collections of object geometry that are treated as single entities in AutoCAD. In that chapter, I also describe giving blocks a little more oomph by adding *attributes*, which are editable text tags that can differ with each insertion. Other than that kind of slight variation, though, the blocks that I describe in Chapter 17 tend to be pretty static.

Blocks, however, don't have to be static creations. Instead of having a half dozen blocks for a half dozen door sizes, you can create a single *dynamic block* that includes all the sizes. Unlike a regular block, in which every instance of a particular block is geometrically identical, each instance of a dynamic block can contain geometric variations. For example, you can insert one furniture block three times and change one instance to be displayed as a sofa, one as a loveseat, and one as an armchair. I talk about *block authoring* — creating and editing dynamic blocks — later in this chapter.

Arraying Associatively

Things used to be so simple back around . . . oh, AutoCAD 2011. ARray was unequivocally a modifying command, just like COpy or ROtate. You filled in values in a dialog box, and after a bit of tweaking, you ended up with multiple copies of the source object(s), neatly arranged in geometric patterns: either in evenly spaced rows and columns, or evenly distributed radially about a center point. Then Autodesk programmers decided that old-fashioned arrays weren't enough, so Autodesk took the old ARray command into the lab, wired electrodes to its brain, and threw the switch — and up arose the new super array!

Follow these steps to create a rectangular array that's associative with the ARRAYRECT command:

1. **Start a new drawing and then create a couple of simple objects — say, a line passing through a circle — in the lower left of the screen.**
2. **Type** ARRAYRECT **and press Enter.**
3. **Select the objects (a line and the circle, if you followed my lead) that you drew in Step 1 and then press Enter.**
4. **AutoCAD displays a three-row-by-four-column preview with six blue grips.**
5. **Click and drag the grips to change the number of rows and columns in the array, change the spacing between the rows and columns, and relocate the array.**
6. **Press Enter to create the associative array.**

TIP

To see what associative means, click any of the objects in the array.

Magic! The blue grips come back, and you can change the number of rows and columns and their spacing by dragging the grips.

Comparing the old and new ARray commands

The results of using the ARray command in AutoCAD 2012 and later are so different from those in previous releases that I really wish that Autodesk had given it a different name. I'm going to do my best to clarify the muddy waters, so here's the terminology I use in this book:

- **Associative array:** This is an object type created as a single object from separate source objects, using the current (AutoCAD 2012 and later) ARray command. Use this command for creating associative rectangular, polar, path, and 3D arrays of existing objects. ARRAYRECT, ARRAYPATH, and ARRAYPOLAR are subsets of the ARray command.
- **Simple or non-associative array:** Individual copies of source objects are created in regular rectangular or polar patterns or by using the -ARray command (note the hyphen in front of the command name). You can create the exact equivalent of a simple array by repeatedly using the COpy command.
- **ARRAYCLASSIC:** If you type this command, the Classic Array dialog box appears. It produces simple, non-associative arrays exactly like those produced by -ARray.

Some of the differences between simple arrays and associative arrays are

- **A simple array is just multiple copies.** When you array objects with -ARray or ARRAYCLASSIC, the source objects are simply copied in rectangular or radial patterns. They remain as separate lines or circles or whatever the original entity type.

TIP

An array with a large number of objects in the array and a large number of objects in each copy can rapidly reach government spending–sized file sizes. In fact, AutoCAD limits arrays to a maximum of 100,000 individual elements. Best practice is to create a block of the objects first. I discuss blocks in Chapter 17. A further advantage of this practice is that if you change the block definition, all insertions of it in the array also update.

- **An associative array is a single object.** When you create an associative array, the source objects are deleted from the drawing and replaced by a new associative array object. An array object is effectively a block. I discuss blocks in Chapter 17.

WARNING

- **The two array types interact differently with layers.** When you create a simple array, the copies of the selected drawing objects are placed on the same layers as the source. On the other hand, an associative array follows the same rules as any block insertion: The new object is placed on whatever layer happens to be current. Source objects located on layer 0 (zero) take on the properties of the current layer, while source objects on other layers retain the properties of their layers.

- **There are two types of simple arrays but three associative types.** Simple arrays can be rectangular (that is, in evenly spaced rows and columns) or polar (evenly arranged around a center point). Associative arrays have a Path option as well as Rectangular and Polar patterns.
- **Associative arrays can be edited.** An associative array is a single object, and so it can be edited as such. Associative arrays (think of them as *intelligent* arrays) have far more editing options than non-associative arrays do. Okay, anything above *pretty well none* counts as *more*. With an associative array, you can change the quantity values (rows, columns, spacing, and so on), you can move and suppress individual items in the array, and you can replace individual items with different ones. You can redefine the source objects, and every repetition updates. The really good news is that you can always put the array back like it was if you mess up too badly.

WARNING

Associative array objects are recognized as such in AutoCAD 2012 and later and can be edited only by using the ARRAYEDIT command. If you save a drawing to an AutoCAD 2007 or earlier drawing file format and then open your drawing in an earlier AutoCAD release, the associative arrays are treated as anonymous blocks, and you can edit the content only by exploding the array. This turns them into simple arrays. If you then open a drawing again in AutoCAD 2013 or later, the array isn't associative anymore. I discuss exploding in Chapter 11.

Hip, hip, array!

When associative arrays were introduced in AutoCAD 2012, many reviewers were startled to see that the dialog box used for simple arrays had gone away and seemed to be replaced by command line input, just like the good old days. The disadvantage to command line input is that you must follow a step-by-step sequence, and if you make a mistake or change your mind, you have to cancel and start over again. The good news is that the ARray command displays a live preview so you can see exactly how your array will turn out. The bad news is that AutoCAD 2012's interface was a little clunky, much like the old command line version. The good news is that it was improved in 2013.

The following steps show you how to create multiple copies of a set of drawing objects, neatly arranged in a grid pattern of several rows and columns.

1. **Open a drawing containing some objects you want to array, or draw some simple geometry for your source objects.**

2. **Click Rectangular Array from the Array drop-down button on the Modify panel of the Home tab.**

   Choosing the specific array type from the drop-down list saves you from answering one prompt you'd get if you used the ARray command.

   REMEMBER

   When drop-down lists are available, the Ribbon remembers the last selection chosen. The Array button may show any one of three different icons. If you like the one that's showing, you can simply select it instead of initiating the drop-down list.

3. **At the** `Select Objects` **prompt, select one or more objects that you want to array.**

   When you finish selecting, AutoCAD displays a three-row-by-four-column preview with six blue grips (noted in Figure 18-1) and prompts:

   ```
   Select grip to edit array or [ASsociative Base point COUnt
      Spacing COLumns Rows Levels eXit]<eXit>:
   ```

   If Dynamic Input is turned on, the cursor prompt displays:

   ```
   Select grip to edit or <eXit>:
   ```

   The Ribbon changes to display the Array Creation contextual tab, as shown in Figure 18-1.

4. **Enter the number of rows and columns of your source objects that you want to be arrayed and also their spacing.**

   As I mention earlier, when AutoCAD 2012 introduced associative arrays, many people commented on what appeared to be a regressive step away from a dialog box and back to command line input. The good news is that AutoCAD 2013 gave you most of the advantages of both methods but few of the disadvantages and added yet another variant: the Ribbon.

5. **Grip-edit your array.**

   As you play with the following options, observe how the dynamic preview constantly updates.

   Each grip (see Figure 18-1) controls a different aspect of your array, as follows:

   - *Row 1, column 1:* Base. You can move your array to a different location as you create it. After you select this grip, you can drag and drop it to a new location (object snaps work), or you can enter X and Y values at the dynamic cursor prompt.

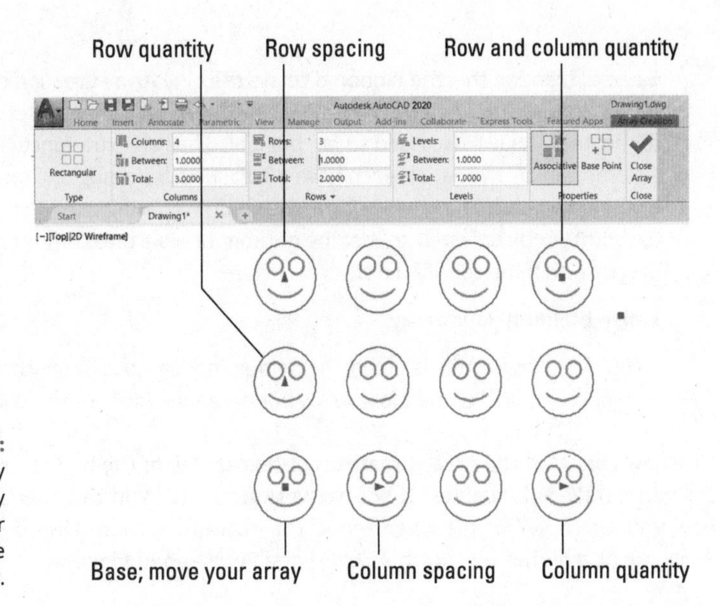

FIGURE 18-1: You're ready to dynamically define your associative array.

- *Row 1, column 2:* Column spacing. Drag this grip (or enter a value at the dynamic prompt) to change the spacing of all columns.

TIP

  You can enter a negative value, or you can drag to the left of row 1, column 1 to make your array grow in the opposite direction.

- *Row 1, column 4:* Column quantity. Drag this grip (or enter a value at the dynamic prompt) to change the quantity of columns.
- *Column 1, row 2:* Row spacing. Drag this grip (or enter a value at the dynamic prompt) to change the spacing of all rows.

TIP

  You can enter a negative value, or you can drag below row 1, column 1 to make your array grow in the opposite direction.

- *Column 1, row 3:* Row quantity. Drag this grip (or enter a value at the dynamic prompt) to change the quantity of rows.
- *Row 3, column 4:* Row and column quantity. Drag this grip (or enter values at the dynamic prompt) to change the quantity of rows and columns at the same time.

6. **Check out the other methods of specifying your array.**

The AutoCAD 2013 change to the Array feature gave you three basic methods for defining your array, one of which is grip editing. The interesting option is that the command line continues to show almost all the prompts. Typically, command line input forces you to follow a specific sequence, but ARray now allows you to enter any command line option at any time.

Earlier, I mention that the Ribbon displays the blue Array Creation contextual tab while you're defining your array. At any time, you can enter specific values for anything I have discussed so far, but it also has an extra functionality: The Columns and Rows panels each have a Total field, referring to a total distance. If you enter values in either window, the column spacing or row spacing (or both) will be adjusted so that the number of rows or columns (or both) fit exactly within the specified distance.

7. **Finish building your array.**

   If you don't have a grip selected, the default mode is eXit. Simply press Enter or the spacebar, or click the Close Array button on the Ribbon, and you're done!

TIP

The Rows panel of the Array Creation contextual tab of the Ribbon includes options for vertical (Z-axis) values. If you're working in 3D, you can specify elevations to easily create a set of bleacher seats, for example, where the higher-numbered rows are at a higher elevation so everyone gets a good view.

Creating polar arrays

AutoCAD also offers the ARRAYPOLAR command. This has nothing to do with the arctic but instead creates polar (circular) arrays, such as bolt holes in a pipe flange, horses on a carousel, or chambers in a revolver. As I indicate earlier, it may be hidden in the drop-down list under the Array button on the Modify tab of the Home panel of the Ribbon.

The basic principles of a polar array are the same as a rectangular one:

- Each polar array is a single, associative object.
- Objects in an array can be edited separately. I cover this a little later in this chapter.
- As you create them, you see a dynamic preview of what the final array will look like.
- You can specify your input to the command through grip edits, the command window, the dynamic cursor, and the Polar version of the Array Creation tab of the Ribbon.

But polar arrays also have a few differences:

- When you finish selecting the objects to be arrayed, you are asked for a center point for the array. Note that normal object snaps can be most helpful here.
- After you select the center point, a six-item dynamic preview with editing grips appears, along with the polar variant of the Array Creation tab of the Ribbon.

» Input specifications include

- *The radius of the array.*
- *The number of items in the array.*
- *The angle to fill.*

  A gear sector, for example, might need only ten teeth in a 60-degree segment.
- *The number of rows.*

  Although this seems like strange terminology for a circular pattern, it's like the growth rings of a tree: You can specify the number of concentric repeats of the circular pattern and the radial distance between them.
- *Whether items are rotated (for example, gear teeth) or not (Ferris wheel seats) as the array is produced.*
- *The number of levels and the Z direction spacing between them.*

Down the garden path

Associative arrays have a third type that isn't available when you use ARRAYCLASSIC or -ARray to produce a simple array: the path array.

Unlike rectangular and polar arrays, which require only source objects and some input parameters, path arrays require one additional piece of drawing geometry — a path that guides the placement of arrayed copies, like laying down stepping stones. A path can be as simple as a line or a circle, or it can be a spline or a 2D or 3D polyline. Other than that, path arrays behave very much like rectangular and polar arrays.

In the following steps, I show you how to use the ARRAYPATH command to lay stepping stones along a path.

1. **Open a drawing containing some objects you want to array along a path (or draw some simple geometry for your source object, such as a circle). Draw a line or an open or a closed spline or polyline for your path.**

   Draw the path starting from the circle.

2. **Click Path Array from the Array drop-down button on the Modify panel of the Home tab.**

   AutoCAD prompts:

```
Select objects:
```

3. **At the `Select Objects` prompt, select one or more objects that you want to array along a path.**

   You can select any and all AutoCAD object types, including block insertions and text.

WARNING

   Object arraying is done based on the position of the objects relative to the starting end of the path object. If you want the arrayed objects to land on the path, they must be on the path before you select them. If they aren't, the spacing between them will probably become uneven.

   When you finish selecting objects, AutoCAD prompts:

```
Select path curve:
```

4. **Select the object you want to serve as the path.**

   Valid object types include straight lines, open or closed polylines and splines, arcs, circles, and ellipses, as well as helixes and 3D polylines.

   You don't have to press Enter after you select the path. AutoCAD immediately responds with a lengthy, multi-option prompt, displays a dynamic preview, and displays the path version of the Array Creation contextual Ribbon tab.

5. **Enter the specifications for your array.**

   As with rectangular and polar arrays, you can specify the number of items, the distance between them, and the length of a portion of the path to fill. A bit of experimentation will show that these options are often interrelated; changing one forces another one to change. You can also specify whether the arrayed objects stay horizontal or rotate to stay parallel to the path.

   Again, just like the other array types, you can jump back and forth between grip editing, Ribbon entries, the command line options, and Dynamic Input tooltip.

6. **Finish building your array.**

   If you don't have a grip selected, the default mode is eXit. Simply press Enter or the spacebar, or click the Close Array button on the Ribbon, and you're done!

TIP

Put the path object on a separate layer. After you complete the array, freeze that layer, and the path object becomes invisible. I cover layers in Chapter 9.

Figure 18-2 shows the finished path array.

Associatively editing

If you click any arrayed element (not the path of a path array), AutoCAD displays the Array contextual tab on the Ribbon. From here, you can revise pretty much every aspect of the array definition. In addition, the Options panel appears near the end of the Array tab.

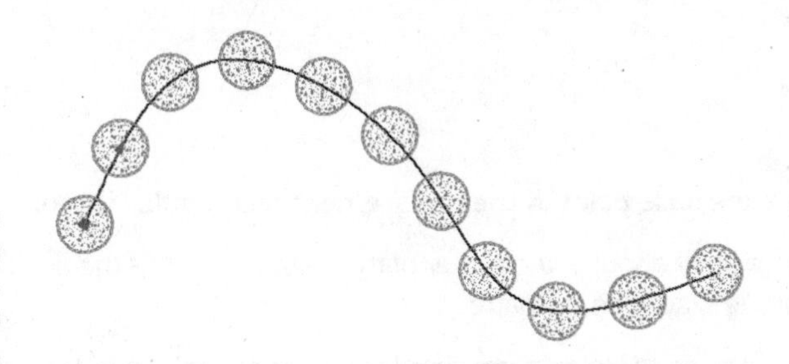

FIGURE 18-2: Down the garden path, using a path array.

You need a special tool for tinkering with the entrails of associative arrays, and AutoCAD provides it with the ARRAYEDIT command. The three items on the Options panel link to the three options of this command. Start with a typical scenario: You just finished the drawing of the garden path when the client comes in and announces that the steppingstones need to be bigger.

No problem. Select any one of the steppingstones (but not the spline path curve itself) in the array. Now click the Edit Source button in the Options panel of the Array tab, and then select any one of the arrayed items. Click OK when AutoCAD asks if you want to edit the source objects. Edit it to your heart's content, including adding and moving objects, changing layers, and deleting objects. When you're finished, select Save Changes from the Edit Array panel at the end of the current tab of the Ribbon, and all the arrayed items update.

So far, so good, but now the client wants a square steppingstone to replace one of the circular ones halfway down the path (see Figure 18-3). No problem. Step through the following steps (pun intended):

1. **Draw the new steppingstone.**

   The easiest way to do this is to use the RECtang command, as I discuss in Chapter 6.

2. **Select the array you want to edit.**

   Select any object in the path array.

   WARNING

   Don't select the path itself, or you'll be editing the path, not the array.

3. **Click Replace Item on the Options panel of the Array contextual tab.**
4. **Select the replacement stone. Press Enter to continue.**

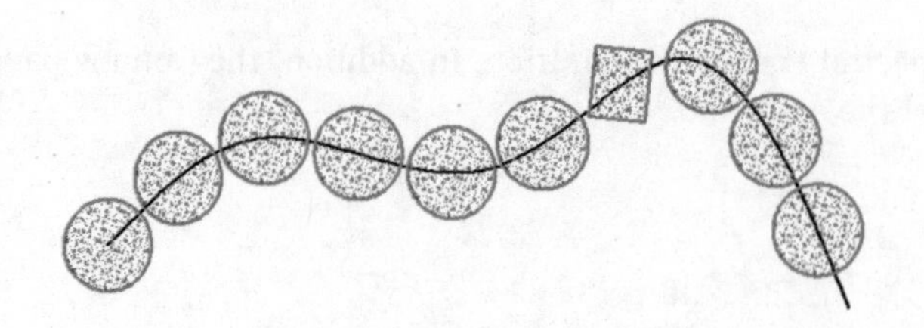

FIGURE 18-3: Going down a different path: The results of editing an associative array.

5. **Select the base point of the replacement rectangular stone.**

   Select a point about in the center of the rectangle. This is the point that will end up sitting exactly on the path.

6. **Select one or more existing circular stones to replace in the array. Press Enter to continue.**

   The objects you select in the array are replaced with the new object.

7. **Press Enter or select eXit from the context menu.**

Oops! You replaced the wrong steppingstone. No problem. Select the array, and then select Reset Array from the Options panel of the Array tab. *Voilà!* Your changes are undone. Now you can replace it with the correct one.

REMEMBER

Earlier, I emphasize the importance of selecting an arrayed item in a path array and not the path. The reason for this is simple: The path can be edited separately. Simply select it and then edit it like the object that it is. When you're finished, the path array updates to match the revised path.

And no matter how much tinkering you do, you can always revert to the original state by clicking Reset Array on the Options panel of the Array tab.

Going External

In AutoCAD, an *xref* (external reference) is a reference to an *external* file — one outside the current drawing — that acts as though it's part of your drawing. It's a way to see two or more files at the same time, all superimposed on one another. Technically, a reference is a simple pointer from one file to another. The xref is the actual pointer, but many people call the combination of the pointer and the external file "the xref." By the way, xref is pronounced exactly as it's spelled. Just fooling; it's pronounced "ex ref."

In both AutoCAD and AutoCAD LT, external drawing (DWG) files are just one of several file types you can attach to your current drawing. You can attach externally referenced AutoCAD drawings (DWG), DWF and DWFx files, PDF files, MicroStation DGN drawing files, as well as most raster image files, such as BMP, JPG, TIF, and GIF. External files do not need any special preparation. Just choose one or more, and attach them as-is.

When you attach an external file to your current drawing, you become the host of the external file. No need to break out the cocktails and canapés, though, because it's your current drawing that's doing the hosting. In AutoCAD, this drawing is called (what else?) the *host drawing.* You can think of the attached xrefs as guests, but most of the time they're pretty well behaved ones — and like the best of guests, they go away as soon as you want them to.

Xrefs have a big advantage over blocks: If a drawing is *inserted as a block* into another drawing, its geometry doesn't change if the original drawing is changed in any way. It always looks the way it looked when it was inserted. If that drawing is *attached as an xref,* however, AutoCAD automatically updates the host drawing to which it's attached. The update occurs each time the host drawing is reopened. You can manually reload an xref if you receive an alert that it has changed.

When you open a drawing that contains xrefs, AutoCAD displays a little symbol (looks like papers with a binder clip) on the right end of the status bar. This symbol alerts you that some of the things you see in the drawing are actually parts of other, xref'ed drawings. If someone changes an xref while you have it open in the host drawing, the status bar's xref symbol displays an `External Reference Files Have Changed` balloon notification. Simply click the Reload link in the balloon notification to show the updated xrefs. If you want to change how AutoCAD checks for changes, look up *XREFNOTIFY* in the online Help.

Another advantage that xrefs have over blocks is that their contents aren't stored in your drawing even once. The disk storage space taken up by the original drawing (the xref) isn't duplicated, no matter how many host drawings reference it. This characteristic makes xrefs much more efficient than blocks for larger drawings that are reused several times. Sure, large hard drives are cheap, so the storage issue isn't crucial.

The key benefit of xrefs is that they enable you to organize your drawings. Changes you make to a single drawing file ripple through all the host drawings into which it's xref'ed automatically. This benefit is even greater on larger projects involving multiple drafters, each of whom is working on a part of a large project, and whose work is incorporated in part or in whole in the work of others.

You can use xrefs to your advantage in almost any type of drawing. For example, several drafters might be working on the components for a large machine, while one master drawing brings them all together in the final assembly. Similarly, a set of separate architectural drawings could show electrical, plumbing, HVAC, and so on with everyone referencing the same floor plan. The host file would contain nothing but a title block and xrefs to the other files.

As a bare minimum, the border and title block of almost every drawing should be an xref. This saves a tremendous amount of disk space in the long run, plus should corporate ownership (and hence the corporate logo) change yet again, you just need to revise the xref master files instead of every drawing individually.

The automatic update feature of xrefs is a big advantage only if you're organized about how you use xrefs. Suppose that an architect creates a plan drawing showing a building's walls and other major features that are common to the architectural, structural, plumbing, and electrical plan drawings. The architect then tells the structural, plumbing, and electrical drafters to xref this background plan into their drawings so that everyone is working from a consistent and reusable set of common plan elements. If the architect decides to revise the wall locations and updates the xref-ed drawing, everyone will see the new wall configuration and be able to change their drawings. But if the architect absentmindedly adds architecture-specific objects (toilets and furniture, for example) to the xref'ed drawing or shifts all the objects with respect to 0,0, everyone else will unnecessarily see the furniture or all the walls shifted over. If people in your office share xrefs, create a protocol for who is allowed to modify which file when and what communication needs to take place after a shared xref is modified.

Becoming attached to your xrefs

Attaching an external reference drawing is similar to inserting a block, and almost as easy. Just use the following steps:

1. **Set an appropriate layer current, as described in Chapter 9.**

   Insert xrefs on a separate layer from all other objects.

   Note that if you freeze the layer on which an xref is inserted, the entire xref disappears. This behavior can be a handy tip, a nasty surprise, or a fun trick.

2. **Click the Attach button of the Insert tab's Reference panel to open the Select Reference File dialog box.**

   Use the drop-down list at the bottom of the dialog box to select the type of file that you want to attach. Choices include several types of image files and other CAD drawing files.

3. **Browse to find the file you want to attach, select it, and then click Open.**

   The Attach External Reference dialog box appears, as shown in Figure 18-4.

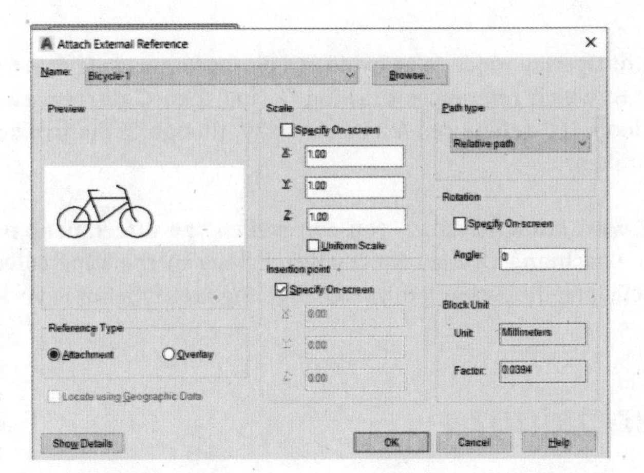

FIGURE 18-4: Use the Attach External Reference dialog box to attach an xref.

4. **Specify the parameters for the xref in the dialog box.**

   Parameters include the insertion point, scaling factors, location based on geographic data, and rotation angle. I describe geographic location briefly in Chapter 17. You can set these parameters in the dialog box or specify them onscreen, just like you can do when inserting a block, as described in Chapter 17.

   The Path Type drop-down list provides more flexibility in how the xref path is stored. See the "Forging an xref path" section, later in this chapter, for more information.

   I recommend that you choose Relative Path instead of Full Path. This makes it easier to move a full set of drawings to another location if all the xrefs are located below the host. Autodesk must have read previous editions of this book because, as of AutoCAD 2018, the default switched from Full to Relative.

5. **Click OK.**

   The externally referenced file appears in your drawing.

You can select either the Attachment or Overlay option to tell AutoCAD how to handle the xref. The choice matters only if you create a drawing that uses xrefs, and then your drawing is, in turn, used as an xref. *Attachment* is the default choice, and it dictates that the xref'ed file will always be included with your drawing when someone else uses your drawing as an xref. *Overlay*, the other choice, means that you see the xref'ed drawing, but someone who xrefs *your* drawing won't see the overlaid file. By choosing Overlay, you can xref a map, for example, to your drawing of a house, but the map won't show up when someone else xrefs your house drawing. That person can xref the map, if needed. I recommend that you use the default Attachment reference type unless you have a specific reason not to.

TIP

One use of Overlay mode is to avoid circular references. If drawing A references drawing B, which references drawing C, and then C references A, you have an infinite loop. If C references A as an overlay, though, C doesn't see B, and so the loop is broken.

Starting with AutoCAD 2014, you can switch the attachment type for an xref between Attach and Overlay by double-clicking in the Type column. Better yet, a right-click menu option lets you change the xref type for several selected xrefs all at once.

Layer-palooza

When you attach or overlay an xref, AutoCAD adds new layers to your current drawing drawn from the layers in the xref'ed DWG file. The new layers are assigned prefixes based on the xref's file name. For example, if you xref the drawing `MYBOLT.DWG` (which has the layer names GEOMETRY, TEXT, and so on), the xref'ed layers will be named MYBOLT|GEOMETRY, MYBOLT|TEXT, and so on. By separating layers from the xref'ed file, AutoCAD eliminates the problem of objects from the xref merging with the current drawing; in this way, they are kept separate. It solves the problem with blocks that I warn you about in Chapter 17, when layers have the same name but different color or linetype in the two drawings.

A host drawing with many xrefs can end up with some pretty cluttered dialog boxes and palettes. Starting with AutoCAD 2014, xref linetypes are not displayed in the linetype list of the Properties palette or of the Ribbon. The Ribbon still displays xref layers, but you can control their visibility.

Editing an external reference file

You can do three kinds of editing to an xref:

- **You can alter its appearance in the host drawing,** including moving its location, changing its size and rotation angle, and fading its look. Fading can be useful when you want to see the xref but don't want it to overwhelm you. A typical situation is an xref'ed site plan used as the background for a foundation drawing.
- **You can alter the content of the xref file.**

WARNING

Any changes made to the content of an xref file will reflect through to all other drawings that point to the same xref!

- **You can alter the xref's relationship to the host.** You can unload it so it isn't visible but the link is still lurking in the background, you can reload it so it becomes visible again, you can update it to match any changes made to the xref, you can bind it so that it is inhaled into the host as a normal block definition and the link is lost, or you can detach it completely as though it never existed.

The xref'ed file appears in the host drawing as a single object, very much like a block insertion. In other words, if you click any object in the xref, AutoCAD selects the entire xref. If the selected insertion is a CAD file, you can measure or object snap to the xref'ed geometry, but you can't modify or delete individual objects in the xref. You need to open the xref drawing itself to edit its geometry.

To open an xref for editing, you can either click any element of it, which then opens a new External Reference tab on the Ribbon, or you can click the small diagonal arrow at the right-hand end of the Reference panel of the Insert tab of the Ribbon. If in doubt, try right-clicking an entry in the External References palette to see the alternatives.

Forging an xref path

When you attach an xref, one option is to have the host drawing store the xref's *full path* — that is, the drive letter and sequence of folders and subfolders all the way down to the folder in which the DWG file resides — along with the filename. This behavior corresponds to the Full Path setting in the Path Type drop-down list. Figure 18-5 shows the three xref path options. Full Path works fine as long as you never move files on your hard drive or network and never send your DWG files to anyone else — which is to say, it almost never works fine!

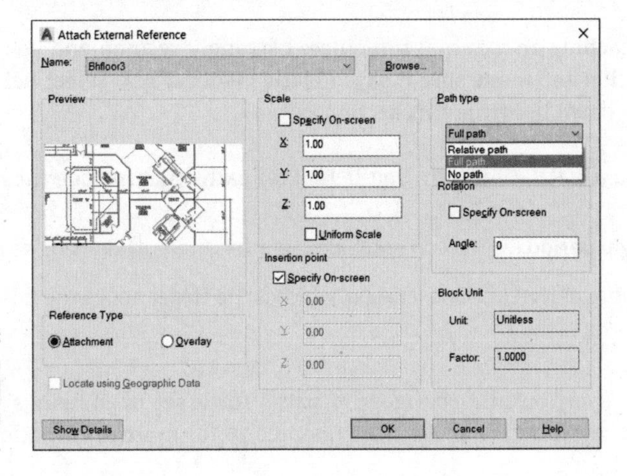

FIGURE 18-5: Follow the path less traveled when you attach an xref.

At the other end of the path spectrum, the No Path option causes AutoCAD to not store any path with the xref attachment: Only the filename is stored. This is the easiest and best option if the host drawings and the xrefs reside in the same folder. However, if the host and the xref are in different folders, you have to browse to find the reference file every time you open the host drawing.

If you prefer to organize the DWG files for a particular project in more than one folder, you'll appreciate AutoCAD's Relative Path option, as shown in Figure 18-5. This option permits xref-ing across more complex folder structures but avoids many of the problems that the Full Path option can cause. For example, you may have a host drawing

```
H:\Project-X\Plans\First floor.dwg
```

that xrefs

```
H:\Project-X\Common\Column grid.dwg
```

If you choose Relative Path, AutoCAD will store the xref path as

```
..\Common\Column grid.dwg
```

instead of

```
H:\Project-X\Common\Column grid.dwg
```

This way, if you decide to move the \Project-X folder and its subfolders to a different drive (or send them to someone else who doesn't have an H: drive), AutoCAD can still find the xrefs.

When attaching an external reference, I strongly recommend you check the Path Type section and make sure it says Relative Path. If not, select Relative Path from the drop-down list from this section.

When you use Relative Path, you'll see xref paths that include these special codes:

- **Single period** (.): The current folder, which is the host drawing's folder
- **Double period** (..): One level up, which is the folder above the host drawing's folder

You can report on and change xref paths for a set of drawings with Reference Manager (not in AutoCAD LT). See Chapter 20 for more information.

TIP

If all these path options and periods leave you feeling punchy, you can keep your life simple by always keeping host and xref drawings in the same folder and using the No Path option when you attach xrefs. On the other hand, it can get a little messy when a file, such as a standard title block and border, is xref'ed into many different drawings for several different projects and you want to keep each project's files in separate folders. The best solution is usually a bit of each; the title block and border can have a specific path in each host, but the other project files don't.

Managing xrefs

After you attach xrefs, the External References palette includes many more options for managing them. Many of these options are hiding in right-click menus. Important options include

- **List of external references:** You can change between the List view and Tree view of your drawing's external references by clicking the appropriate button at the top of the palette You can resize the columns by dragging the column dividers or re-sort the list by clicking the column header names, just like in Windows or File Explorer.
- **Unload:** Right-click an xref and then choose Unload to hide the selected xref from the onscreen display of your drawing and from any plots you do of it but still retain the pointer and attachment information. Right-click again and choose Reload to redisplay an unloaded xref.
- **Reload:** Right-click an xref and then choose Reload to force AutoCAD to reread the selected xref'ed DWG file from the disk and update your drawing with its latest contents. This feature is handy when you share xrefs on a network, and someone has just made changes to a drawing that you've xref'ed.
- **Detach:** Right-click an xref in the External References palette and then choose Detach to completely remove the selected reference to the external file from your drawing.
- **Bind:** Right-click an xref and then choose Bind to bring the selected xref into your drawing and turn it into a block definition. The link to the xref will be broken. You might use this function, for example, to roll up a complex set of xrefs into a single archive drawing, or to send the entire set to a client without having to worry about the path structure.

  REMEMBER

  If you bind an xref, any changes made to the xref drawing will no longer reflect through to the host.

WARNING

In many offices, binding xrefs without an acceptable reason for doing so is a crime as heinous as exploding blocks indiscriminately. In both cases, you're eliminating an important data management link. Find out what the policies are at your company. When in doubt, keep yourself out of a bind. And even when you do have a good reason to bind, you generally should do it on a *copy* of the host drawing.

- **Path:** Right-click an xref in the External References palette and then choose Path to bring up a submenu to switch between no path (Remove), Relative, and Absolute.
- **Open:** Right-click an xref and then choose Open to open one or more xref drawings in separate drawing windows. After you edit and save an xref drawing, return to the host drawing and use the Reload option in the External References palette to show the changes.

None of these options (other than opening and editing the xref) has any effect on the xref'ed drawing itself; it continues to exist as an independent DWG file. If you need to delete or move the DWG file that the xref refers to, do it in Windows or File Explorer.

REMEMBER

The fact that the xref'ed drawing is an independent file is a potential source of problems when you send your drawing to someone else; that someone else needs *all* the files that your drawing depends on, or it will be useless. Make sure to include xref'ed files in the package with your drawing. See Chapter 20 to find out how the ETRANSMIT command helps.

Both AutoCAD and AutoCAD LT include an additional xref feature called *xref clipping.* You can clip any kind of externally referenced file or block insertion so that only part of it appears in the host drawing. Use the CLIP command to trim away unwanted parts of xref drawings, DWF and DWFx files, raster images, MicroStation DGN files, or PDFs. You can even use CLIP on block insertions. All clips can be reversed to return to the full xref or block. For more information, look up CLIP in the online Help's Command Reference section.

Blocks, Xrefs, and Drawing Organization

Blocks and xrefs are useful for organizing sets of drawings to use and update repeated elements. It isn't always clear, however, when to use blocks and when to use xrefs. Some applications for xrefs follow:

- Ensuring that the parts of a title block are the same on all sheets in projects

- Referencing elements that must appear in multiple drawings (for example, wall outlines, site topography, and column grids)
- Including individual component parts going into a final assembly drawing for a machine or mechanism
- Inserting assemblies that are repeated in one or more drawings, especially if the assemblies are likely to change together (for example, a motor, coupling, speed reducer, and base plate assembly in large machines)
- Pasting several drawings (for example, details or a couple of plans) onto one plot sheet
- Temporarily attaching a background drawing as a reference or for tracing

On the other hand, blocks remain useful in simpler circumstances. A few situations in which you might stick with a block follow:

- Inserting components that aren't likely to change
- Working with simple components
- Using an assembly that's displayed repeatedly but in only one drawing
- Including *attributes* (variable text fields) that you can fill in each time you insert a block; xrefs cannot contain attributes

Blocks let you include attribute definitions; xrefs don't. Okay, technically you can have attributes in xrefs, but they won't display in the host drawing. Refer to Chapter 17 for the lowdown on attributes.

Everyone in a company or workgroup should aim for consistency as to when and how they use blocks and xrefs. Check whether guidelines exist for using blocks and xrefs in your office. If so, follow them; if not, it would be a good idea to develop some guidelines.

Mastering the Raster

AutoCAD includes four more xref-like features: the capability to attach raster images, DWF files, PDF files, DGN files, point cloud files, and Navisworks files to drawings. I cover DWF and PDF files in the next section and refer you to the online Help should you find yourself working with DGN, Navisworks, or point cloud files. The image feature is useful for adding a raster logo to a drawing title block or placing a photographed map or scene behind a drawing. A *raster*, or *bitmapped*, image is one that's stored as a field of tiny dots.

Raster images often come from digital cameras or from other programs, such as Photoshop. Raster images can also come into the computer from scanners that read a blueline print, photograph, or other image.

AutoCAD drawings are vector files. A *vector* drawing is a graphic file defined by storing geometrical definitions of a bunch of objects. Typical objects include a line (defined by its two endpoints) and a circle (defined by its center point and radius). Vector-based images are typically smaller (in terms of the disk space they occupy) and more flexible than raster images but also are less capable of displaying visually rich images, such as photographs.

Whether you're doing your own scanning or having a service bureau do it for you, you need to know that AutoCAD handles most popular image file formats, including the Windows BMP format; the popular web graphics formats GIF, JPEG, and PNG; common print formats, such as PCX and TIFF; the less popular DIB, FLC, FLI, GP4, MIL, RLE, RST, TGA formats, and several even more obscure ones.

Three scenarios in which you could incorporate raster images in your drawing include

- **Small stuff:** You can add logos, special symbols, and other small images that you have in raster files.
- **Photographs and maps:** You can add photographs (such as the building site) and maps (for example, showing the project location).
- **Vectorization:** To convert a raster image into a vector drawing by tracing lines in the raster image, you can attach the raster image in your drawing, trace the needed lines by using AutoCAD commands, and then detach the raster image. This procedure is okay for a simple raster image; add-on software is available, from Autodesk and others, to support automatic or semiautomatic vectorization of more complex images. In the long run, however, it is often easier, faster, and more accurate to simply redraw the image in AutoCAD.

Your raster image may be distorted a bit; for example, the aspect ratio (ratio of width to height) may not be correct, but AutoCAD won't let you change it. No problem; simply attach the raster image, make a block that contains it, and then insert the block with different X and Y scale factors. I love cheating.

Using raster images is much like using external references. The raster image isn't stored with your drawing file, though. Instead, a reference (link) to the raster image file is established from within your drawing, like an xref. You can clip the image and control its size, brightness, contrast, fade, and transparency. These controls fine-tune the appearance of the raster image onscreen and on a plot.

REMEMBER

When you attach raster images, you have to make sure that you send the raster files along when you send your drawing to someone else. Raster images are simply referenced from the drawing and aren't part of it. If you don't send the raster image file along with the drawing, the drawing displays a rectangle containing the name of the missing file in its place. The best way to make sure that you get all the required files when sending a drawing file to someone else is to use the ETRANSMIT command, which I cover in Chapter 20.

WARNING

Some raster image files that you find on the internet may not be appropriate as AutoCAD attachments.

Attaching a raster image

Follow these steps to bring a raster image into AutoCAD:

1. **Set an appropriate layer current, as described in Chapter 9.**

   TIP

   Insert raster images on a separate layer from all other objects.

   Note that if you freeze the layer on which an image is inserted, the entire image disappears.

2. **Click the Attach button of the Insert tab's Reference panel to open the Select Reference File dialog box.**

   Use the drop list at the bottom of the dialog box to select All Image Files.

3. **Browse to find the file you want to attach, select it, and then click Open.**

   The Attach Image dialog box appears, as shown in Figure 18-6.

   TIP

   Click the Show Details button in the Attach Image dialog box to see more information about the resolution and image size of the image you're attaching.

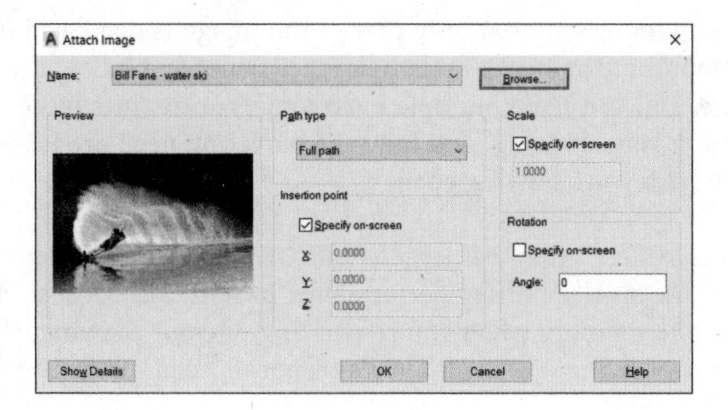

FIGURE 18-6: The Attach Image dialog box.

4. **Specify the parameters for the attached image in the dialog box.**

   Parameters include the insertion point, scale factor, and rotation angle. You can set these parameters in the dialog box or specify them onscreen, similar to what you can do with blocks and external references, as described earlier in this chapter.

   The Attach Image dialog box includes the same Full Path, Relative Path, and No Path options as those for attaching xrefs. See the "Forging an xref path" section, earlier in this chapter.

5. **Click OK.**

   The image appears in your drawing.

6. **If you need to ensure that the raster image floats behind other objects in the drawing, select the raster image's border, right-click, choose Draw Order, and then choose Send to Back.**

The DRaworder command provides additional options for determining which objects appear visually on top of other objects. Drawing order applies even to block insertions, text, xrefs, attachments, and so on. You can use DRaworder, for instance, to make sure attachments are not underneath other parts of the drawing. Note that raster images in particular will blank out anything behind them. If you need this kind of flexibility, look up the DRaworder command in the AutoCAD online Help system.

Maintaining your image

You manage the images in your drawing via the External References palette. You can view a list of image files that appear in the current drawing, detach (remove) image references, and unload and reload images when needed. You can't bind an image to your drawing, though; it always remains an external file.

You can clip images so that only part of the image is displayed in your drawing. Choose Clip from the Insert tab's Reference panel and follow the prompts to clip the image. You can have multiple overlapping or distinct pieces of any number of images in your drawing, and only the parts you need are loaded into memory when you have your drawing open.

Raster image files often are larger than DWG files of corresponding complexity; raster file size can affect performance in AutoCAD because the entire raster file loads into memory when you're working on your drawing. If you find raster attachments slowing down AutoCAD's response, then follow these tips:

- Attach raster images late in the production process.
- Create a lower-resolution version of the raster file, just large enough to create the desired effect in your drawing.
- Right-click an image in the External References palette and choose Unload from the context menu to temporarily hide an image without losing the attachment information.

In addition, raster files can also dramatically increase the time that AutoCAD takes to generate plots (and the plot file sizes). Before you settle on using large raster files in your AutoCAD drawing, do some testing on zooming, editing, and plotting.

Starting in AutoCAD 2014, you can insert a link to a mapping service, such as Google Earth, to place maps as underlays in your drawing.

You Say PDF; I Say DWF

The Adobe PDF (Portable Document Format) format has been around for a very long time, and for a while it seemed like Autodesk was trying to supplant it with its own universal file format: DWF (Design Web Format). That didn't happen, but if you can't beat 'em. . . . AutoCAD includes a very acceptable PDF printer driver, and both PDF and DWF are suitable candidates for external reference attachments.

You could think of a DWF as DWG Lite because it looks just like a drawing file and contains some of the actual drawing file data. I talk more about the web side of DWFs in Chapter 20; in this section, I explain how you can use DWFs as well as PDFs as reference files in your own drawings. Some people call DWF files "dwiffs," but I'm going to hold off on that one until I start hearing DWG files called "dwiggs." And why is *phonetic* spelled that way?

You create both DWFs and PDFs from AutoCAD in one of two ways. Either choose Plot and then select a DWF, DWFx, or PDF option in the Printer/Plotter name list, or choose Print ➪ Batch Plot from the Application menu and then choose DWF, DWFx, or PDF from the Publish To area of the Publish dialog box.

All three file types are compact and secure. Because you can't normally edit PDF, DWF, or DWFx files in AutoCAD, both formats are ideal for two purposes:

- You can post DWFs or PDFs on the web.
- You can send your drawings to consultants and clients in a form that they can't mess up.

WARNING

Oops, AutoCAD 2017 changed part of the foregoing a bit. It is now possible to edit a PDF file in AutoCAD, but there are quite a few limitations, as I describe in Chapter 24.

You can attach PDFs and DWFs to your drawing files in pretty much the same way you attach drawings as external references. DWFs and PDFs attached to drawing files are referred to as DWF underlays or PDF underlays.

As with DWF and DGN attachments, you can object-snap to entities in the PDF underlay by enabling the Snap to Underlays function in the Reference panel on the Insert tab.

The previous sections show how to attach a DWG and a raster image. Follow these steps to attach a DWF or PDF file as an underlay:

1. **If the External References palette isn't already open, click its icon on the Palettes panel of the View tab to open it.**

   Use the toolbar at the top of the palette to attach an external file as an xref, a raster image file, or a DGN or DWF underlay. I cover attaching xrefs and images earlier in this chapter. See the online Help for information on attaching DGN files.

2. **Click Attach DWF or Attach PDF and then locate the file you want to attach.**

   The Select Reference File dialog box appears.

3. **Browse to find the file you want to attach, select it, and then click Open.**

   The Attach DWF Underlay or the Attach PDF Underlay dialog box appears.

4. **Specify the parameters for the DWF or the PDF in the dialog box.**

   Parameters include specifying a sheet, the insertion point, scaling factors, rotation angle, and path type, as shown in Figure 18-7. You can set these parameters in the dialog box or specify them onscreen, just like you can do when inserting a block, attaching an xref, or attaching an image, as described earlier in this chapter.

5. **Click OK.**

   The externally referenced DWF or PDF file appears in your drawing.

TIP

When using object snaps to locate points in DWFs, you may see the word `approximate` on the Object Snap tooltip. Neither PDF nor DWF files are as precise or contain as much data as DWGs, so that's why their file sizes are a lot smaller. If this is a problem, you should increase the precision of the DWF file when you next create it.

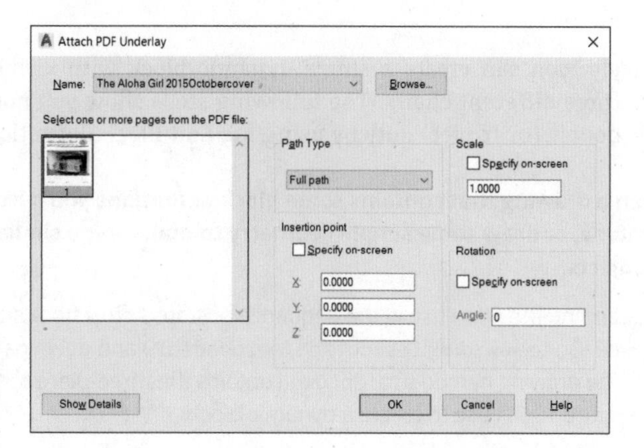

FIGURE 18-7: The Attach PDF Underlay dialog box.

In addition to the formats discussed in this chapter, AutoCAD allows you to import another 20 file formats into AutoCAD drawings, most of which are 3D modeling formats. They are imported into model space (and cannot be attached or xref'ed), and are used as the basis for additional 3D modeling and editing, and for generating 2D working drawings from them. I discuss 3D to 2D in Chapter 23.

Theme and Variations: Dynamic Blocks

You can add variety to your blocks by making them dynamic. The two most useful applications for dynamic blocks are multiple presentations of similar objects and manipulation of components in individual block inserts.

AutoCAD's dynamic blocks feature offers you a great deal of flexibility in block creation and insertion, but using them is also a very complicated system with its own set of commands and system variables. I recommend that you become *very* familiar with regular block creation and insertion techniques for creating and inserting blocks (which I describe in Chapter 17) before you tackle dynamic blocks.

TIP

Spend some time planning your dynamic blocks. Sketch out the geometry for each variation in appearance (known as a *visibility state*) and decide where the common base point should be. Unless you're a lot smarter than I am, you'll probably find that creating dynamic blocks is complex enough without trying to design your blocks as you go.

Now you see it

If you have drawings that show, for example, three different kinds of chairs, one approach is to create three different standard blocks to represent them all.

Alternatively, you can create a single dynamic block with visibility states that cover the three different chairs. The following steps show you how to make your blocks do double (or triple?) duty by using the Edit Block Definition dialog box:

1. **Open a drawing that contains some block definitions you'd like to combine, or draw some simple geometry to make some similar types of objects.**

   You can find the files I use in this sequence of steps at this book's companion website. Go to `www.dummies.com/go/autocadfd19` and download `afd18.zip`. The drawing named `afd18b.dwg` contains the three-piece furniture suite (see Figure 18-8) I use to create a dynamic block.

   You can create dynamic blocks from scratch, or you can work with existing standard (nondynamic) block definitions. Figure 18-8 shows a drawing with three nondynamic blocks.

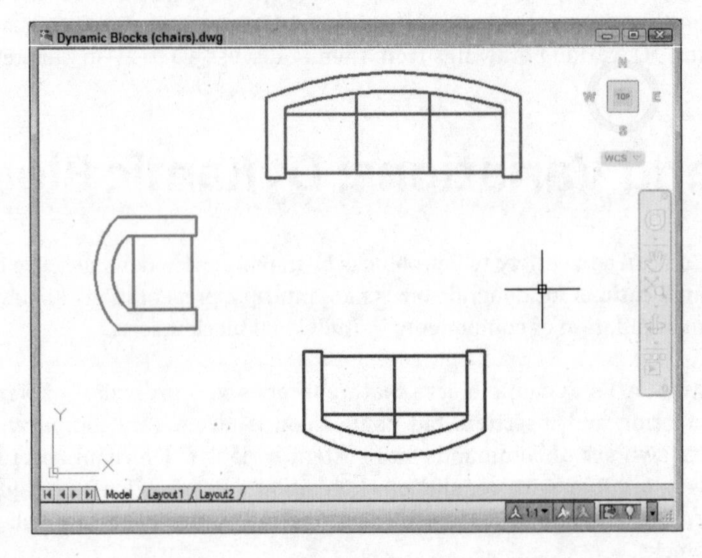

FIGURE 18-8: Three blocks to make three seats.

2. **On the Block panel of the Home tab, choose Block Editor to open the Edit Block Definition dialog box.**
3. **In the Block to Create or Edit box, specify a new block name or select Current Drawing; then click OK to display the Block Editor window.**

   Block Editor is a special authoring environment with its own Ribbon tab, plus a passel of command-line commands. You also have access to the rest of the Ribbon tabs, so you can draw and edit just like you would in the regular drawing window. The background color is different from the drawing editor's background color to help you remember where you are.

The Block Editor tab's Geometric, Dimensional, and Manage panels (see Figure 18-9) are elements of the AutoCAD parametric drawing feature. AutoCAD LT doesn't fully support parametric drawing, so LT Block Editor lacks the Geometric and Dimensional panels and gets a miniversion of the Manage panel. In this book, I don't have room to cover parametric constraints in dynamic blocks, but I do cover parametric drafting in Chapter 19. The concepts are pretty similar to adding parametric features to dynamic blocks.

If you enter a new block name, AutoCAD displays an empty block-authoring environment where you draw geometry or insert existing blocks. If instead you select Current Drawing, AutoCAD places all drawing objects inside the block-authoring environment.

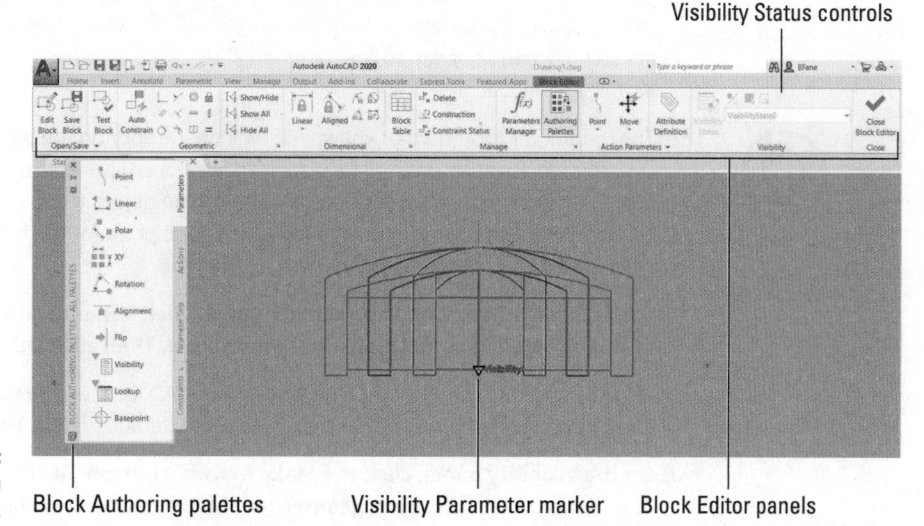

FIGURE 18-9: Three seats in Block Editor.

4. **Draw some geometry for the first visibility state. Alternatively, click the Home tab, choose Insert on the Block panel, and select an existing block definition to serve as the first visibility state.**

   When drawing geometry from scratch, pay attention to where the common base point should be. Although you use different blocks to assemble a multiple-view block, they should all have the same base point. Either 0,0 or exactly the same center are good ones for blocks. You don't want your chairs jumping around between different insertion points!

5. **If you inserted an existing block in Step 4, deselect all three Specify On-Screen check boxes, make sure that the Explode check box is *not* selected, and then click OK.**

6. **Repeat Steps 4 and 5, drawing or inserting all the necessary geometry.**

At this point, your drawing screen may look pretty strange (see Figure 18-9). Don't worry; you're going to fix it in the next steps.

7. **Click the Parameters tab of the Block Authoring Palettes and then click Visibility, as shown in Figure 18-9.**

   If the palettes aren't open, click Authoring Palettes on the Manage panel on the highlighted Block Editor contextual tab of the Ribbon.

   AutoCAD prompts you to specify the parameter location.

8. **Click to place the parameter marker somewhere other than the base point location you chose in Step 4.**

   AutoCAD places a parameter marker at the selected point and returns to the command line. As shown in Figure 18-9, the label `Visibility1` appears next to the Visibility Parameter marker, and a yellow Alert symbol indicates that no action has been assigned to the parameter yet. The controls on the Ribbon's Visibility panel become active.

TIP

   The parameter location that you specify will be the spot on the block where the dynamic block option grip will be displayed. It's not crucial where you locate this point, but try to pick a sensible location on the object. If you specify the same point for the parameter location as the base point for the block, you may have a hard time selecting the dynamic option grip.

9. **Click Visibility States on the Visibility panel. Click Rename and change VisibilityState0 to something more descriptive. Then click OK.**

   As is the case with other named objects in AutoCAD, best practice is to assign useful, descriptive names rather than accept the default generic labels.

10. **On the Visibility panel, click the Make Invisible button. At the** `Select Objects` **prompt, select the geometry or block inserts that should be invisible in the current visibility state — that is, those that are *not* associated with the current visibility state — and then press Enter.**

    Enter **c** at AutoCAD's prompt:

```
Hide for current state or all visibility states [Current
  All] <Current>: C
```

11. **Click Visibility States again, and then click New to create a new visibility state.**

12. **In the New Visibility State dialog box, enter a descriptive name. Also select the Show All Existing Objects in New State option and then click OK.**

    All your geometry should reappear.

13. **Repeat Steps 9 and 10 to create additional visibility states associated with the remaining geometry or blocks.**

    The geometry or block insert associated with the last created visibility state should be visible onscreen.

14. **Click OK to close the Visibility States dialog box, and then click Close Block Editor on the Ribbon. Save the changes to your new block or to Current Drawing.**

    AutoCAD displays an alert box asking whether you want to save changes to your block. Your choices are

    - *Cancel:* Click Cancel to return to Block Editor without saving your changes.
    - *Discard changes:* Clicking Discard the Changes closes Block Editor without saving your changes.
    - *Save:* Click Save the Changes to update the block and exit.

    AutoCAD closes the block-authoring environment and returns to the standard drawing editor window.

TIP

The Block Editor tab includes a Test Block tool you can use to see what your finished product will look like without the hassle of closing the editor and inserting or manipulating the block inside the drawing editor. Test Block displays the geometry and lets you change the parameters to see the result; then you can easily return to Block Editor to tweak your masterpiece. You'll find the Test Block button on the Open/Save panel of the Block Editor contextual tab.

Lights! Parameters! Actions!

You can modify the appearance of individual instances of the same block by defining parameters and actions to move, rotate, flip, or align parts of them. You can adjust the block's appearance while you insert it or at any time afterward. The following steps show you how to use Block Editor to add some action to a block definition:

1. **Open a drawing that contains some block definitions whose appearance you'd like to spice up a little, or else draw some simple geometry that might make a suitably dynamic block.**

   Action parameters are most effective in block definitions that contain groups of related objects: for example, an office desk and chair or a furniture arrangement.

2. **On the Block panel of the Ribbon's Home tab, choose Block Editor to open the Edit Block Definition dialog box.**

3. **In the Block to Create or Edit text box, type a new block name or select Current Drawing. Then click OK.**

4. **Create some geometry or insert some blocks.**

REMEMBER

When you insert blocks, make sure that the Explode check box at the lower-left corner of the Insert dialog box is *not* selected. Then click OK.

Draw the geometry or insert the blocks in a group so that you can insert the finished arrangement into your drawings. For example, Figure 18-10 shows the creation of a dynamic block for one cell of a telephone boiler-room call center.

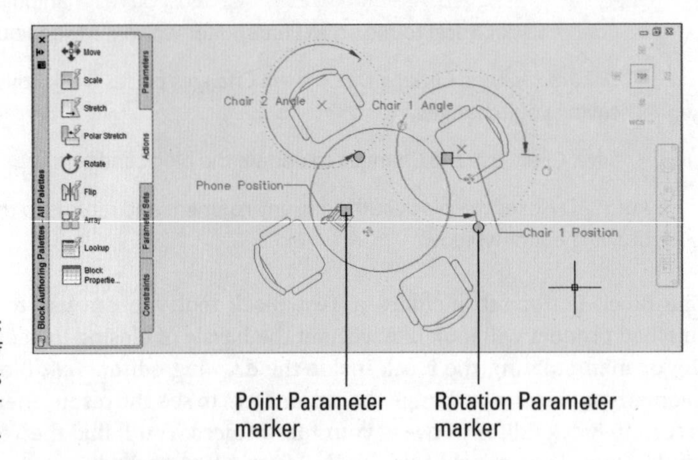

FIGURE 18-10: A passel of parameters ready for action.

5. **Repeat Step 4 until you've drawn all the needed geometry or inserted all the necessary blocks.**

6. **Click the Parameters tab of the Block Authoring Palettes and then click Rotation Parameter.**

   AutoCAD prompts you to specify the parameter location.

7. **Click to place the parameter marker somewhere on the object geometry other than the base-point location.**

   If you specify the same point for the parameter location as the base point for the block, you may have a hard time selecting the dynamic option grip.

8. **AutoCAD places a parameter marker at the selected point and returns to the command line, prompting you to specify the radius of the rotation parameter and the default rotation angle.**

   The parameter marker's label appears next to the rotation parameter marker.

9. **Click the Actions tab of the Block Authoring Palettes, and then click Rotate. Select Rotate Parameter, select the objects that should be modified when the grip is used, and specify a point for the action's label.**

   AutoCAD returns to the command line. At this point, it's fine to go with default values and onscreen pick points.

10. **Repeat Steps 6–9, trying different parameters and actions.**

   For example, choose a Point Parameter and a Move Action.

   Figure 18-10 shows a set of block components, several of which have action parameters assigned. After the block is inserted, you can manipulate the components to which you've added parameters to vary the appearance of the blocks. I explain how in the next section.

11. **Click Close Block Editor on the Ribbon and save or discard your changes.**

   AutoCAD closes the block-authoring environment and returns to the standard drawing editor window.

Manipulating dynamic blocks

Once a dynamic block is inserted in a drawing, you can select it and modify its display through a special set of *custom* grips. That's what AutoCAD calls them, so I'm following suit.

When you select a nondynamic block, you see a single grip at the insertion point. When you select a dynamic block, however, you see at least two (and maybe more) custom grips, as well as the insertion point grip. The custom grips usually look different from the regular object grips, but not always, so be careful while clicking grips.

The following steps show you how to make your dynamic blocks do the things you just spent all this time teaching them to do:

1. **Insert a few blocks that contain some dynamic parameters, such as visibility or action parameters.**

   The Tool Palettes palette contains many dynamic blocks; they are indicated by the yellow lighting symbol.

   If your block inserts don't have any action parameters, go to Step 4.

2. **Select a block that includes some action parameters.**

   The block insert displays a number of grips (see Figure 18-11). If the insert displays only a single square grip, it isn't a dynamic block.

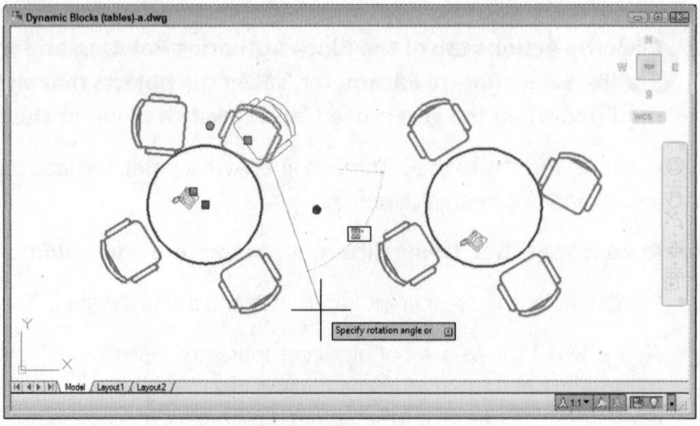

FIGURE 18-11: Variations on dynamic blocks (original on the right, and a handy grip tip on a chair at left center).

3. **Click one of the custom grips.**

   For example, clicking a round grip opens the rotation parameter of the object. Then you rotate the component as required.

4. **Select a block that includes a visibility parameter and then click the visibility grip. Choose the desired visibility state from the menu that appears.**

   For additional information on manipulating actions and visibility states, look up Dynamic Blocks in the online Help system.

Dynamic blocks, as I suggest more than once, are a powerful — and complicated — feature. Prior to AutoCAD 2010, the Block Authoring Palettes contained 10 selectable parameters, 8 actions, and 20 parameter sets. AutoCAD 2010 added 12 more geometric constraints and 6 dimensional constraints. I could write a book on the topic. Well, I did: See "Tailoring Dynamic Blocks" at `www.worldcadaccess.com/ebooksonline/2015/07/tdb.html`. AutoCAD's online Help system is another resource for more information on the possibilities of dynamic blocks. The quickest and most direct way to AutoCAD's own help on the subject is to type **Dynamic Blocks** into the Search box on the program title bar.

» Understanding constraints

» Applying geometric and dimensional constraints

» Editing parametric drawings

Chapter 19
Call the Parametrics!

A decade ago, Autodesk added parametric tools to AutoCAD, and I can't think of a more significant new feature in recent years. In fact, if you work primarily in 2D, and *especially* if you're in the manufacturing business, parametrics in AutoCAD (and, to a very limited degree, in AutoCAD LT) might just be the most significant new feature since the beginning of orthographic engineering drawings, generally attributed to the Roman architect Marcus Vitruvius in the 1st century BC. Contrary to popular opinion, I am not so old that I knew him personally.

Parametric tools allow you to create relationships between geometry and control the size of geometry in drawings. These relationships constrain changes to designs, so parametric tools are also known as *constraints*. The power of parametrics comes in their capability to quickly generate variations of a design. A single design of, say, a bicycle can be output with bigger (or smaller) wheels and longer (or shorter) handle bars, once all parametric parameters are set up correctly.

If you know what I'm talking about, you're probably going to be pretty interested in finding out more about what you can do with parametrics in AutoCAD. If, on the other hand, you think parametrics are the folks who show up when you dial 911, the following paragraphs should straighten you out.

Imagine this scenario: Your boss has been in a meeting all morning with an important client. At two minutes before noon, they emerge. Your boss hands you a marked-up printout and commands you rather imperiously: "We need these twelve dimensions changed and the drawing updated by the time we return from lunch." The drawing involves multiple views, several cross sections, and a couple of details at different scales.

By the time the pair put on their coats, you've met them at the door and handed them a printout of the revised drawing. The client invites you to join them for lunch.

This chapter shows you how to make this scene happen in real life.

Maintaining Design Intent

Parametric (rule-based) drawing is by far the best way to enforce design intent in 2D drafting. *Design intent* in AutoCAD (and in any other engineering software) means that when drawings are edited to make "this part" wider and "that hole" larger, all the attached or related objects behave in a predictable way that honors the designer's intent in creating the drawing.

Before AutoCAD 2010, there was simply no practical way to maintain the design concepts behind an AutoCAD drawing. You could use AutoCAD drawing and editing commands to draw accurate, precise plans, sections, and details, but they were, to AutoCAD, simply a bunch of lines and circles.

Suppose that an engineer has taken a second look at a baseplate (see Chapter 3) and determined that its 1½" bolts should be changed to 1¾". To revise the drawing in AutoCAD's traditional way, you would draw a new, larger circle for the bolt and erase the old one — or maybe use the OFFSET command. But now the nut is too small, and so is the hole through the plate. Maybe you can't see it, but you *know* — and I know — that the problems are there. And you have four of these changes to make. That's a lot of editing to fix a simple drawing. Even so, the good news is that CAD editing is still a lot easier than using paper and pencil and eraser.

By applying parametric constraints, you can add *intelligence* to those lines and circles in AutoCAD. Rather than try to explain what the *intelligence* means, I show you.

AutoCAD does not support 3D constraints, such as those found in Inventor. The constraints in AutoCAD are 2D only.

Unfortunately, the following exercise doesn't work in AutoCAD LT, although it can display and edit constraints added in full AutoCAD.

Follow these steps to constrain a line to a circle with parametrics:

1. **Start a new drawing.**
2. **Draw a circle and a line anywhere in the drawing.**

   Specifying exact sizes and alignments isn't critical. You'll see why in a moment, and I explain even more a little later in this chapter.
3. **Apply a tangent constraint between the circle and the line.**

   Parametric functions are found on the Ribbon's Parametric tab. Nothing in AutoCAD should surprise you by now. Click the Tangent button, and then select the circle and the line. The sequence doesn't matter. Observe how the second object you selected jumps to be tangent to the first one you selected. Note the little gray icon (called a *constraint bar*) that indicates the type of constraint.
4. **Grip-edit the line and the circle.**

   You can move the circle or line, change the diameter of the circle, or move either end of the line — but no matter what you do, the other object obediently — stubbornly — remains tangent. I cover grip editing in Chapter 11. Note that *tangent* doesn't necessarily mean that the objects need to touch.
5. **Apply a diameter constraint dimension to the circle.**

   Don't use the normal DimDIameter or DIM command, described in Chapter 14. Instead, click the Diameter button in the upper-right corner of the Dimensional panel of the Parametric tab on the Ribbon. When you select the circle as prompted, AutoCAD places a gray-colored dimension that reads something like this: `dia1=5.3716`. The gray look of the dimension tells you that it's a constraint; the *dia1* text tells you that it's the first (1) diameter (dia) constraint placed in the drawing.
6. **Change the diameter of the circle.**

   It has the same background color as the mText edit box (see Chapter 13); in fact, that's exactly what it's inviting you to do. Enter a different value, say, 7.5. The circle resizes to match the diameter value you entered, and the line adjusts to remain tangent. Magic!

   It's the opposite of associative dimensions (see Chapter 14). Here, the value of constraints determines the size and positions of objects; with associative dimensions, objects determine the values and positions of regular dimensions.
7. **Apply an aligned constraint dimension to the line.**

   Again, be sure to click the Aligned button from the Dimension panel on the Parametric tab, and not use the DIM command. Select each end of the line in turn or press Enter and then select the line. This time, when AutoCAD places the dimension, enter **=dia1*2** including the equal sign and omitting spaces. The line automatically resizes to become twice as long as the diameter of the circle.

8. **Double-click the diameter dimensional constraint and type a new value, such as 5, and press Enter.**

   Surprise! The size of the circle and the length of the line change, so that the line is still twice as long as the new diameter of the circle. That is one smart drawing. Imagine a similar effect on your productivity in creating and editing drawings.

   You can still grip-edit the position of the line and circle, but you can no longer change the diameter of the circle or the length of the line using grips editing, nor can you stretch the line length any more. Constraints are in control!

Defining terms

The word *parameter* is derived from these two Greek roots:

» *para:* To work with, to assist, to work alongside. Think of *paramedic* ambulance attendants. You may have thought "ambulance" when I mentioned *parametrics* earlier in this chapter.

» *metros:* To measure (hence *meter* and *metric*). For the purposes of making drawings in AutoCAD, a *parameter* is a rule that works alongside AutoCAD objects. You can think of *parametric* drawing as rule-based drawing.

AutoCAD's parametric rules (officially called *constraints*) fall into two categories:

» **Geometric:** Constrains the sizes of object and the relationships between them; different object types have varying constraints you can apply to them. In the example in the preceding step list, I show you how to apply a Tangent constraint between the line and the circle — as you move or change one, the other one follows along.

» **Dimensional:** Constrains the size of objects and the distances between objects. In the earlier example, you see how to apply a dimensional constraint to the circle. Changing the dimension value changes the circle size. Better yet, you can establish a relationship between the circle and the line so that changing the circle diameter also changes the length of the line.

Both AutoCAD and AutoCAD LT support constraints, but (as is usually the case) the feature is limited in LT. If you use the full version of AutoCAD, you can create and modify geometric and dimensional constraints, as I describe in the steps in this chapter. If you're using LT, you can't *create* constraints, but you can work with existing constraints in drawings that were created in AutoCAD. Figure 19-1 shows the difference between the AutoCAD and AutoCAD LT Parametric tabs.

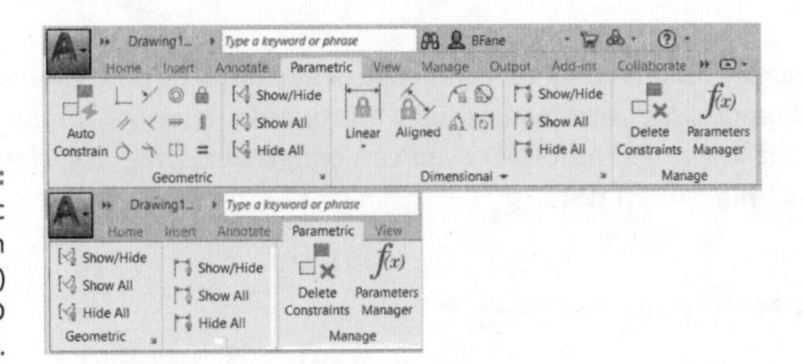

FIGURE 19-1: Parametric tabs in AutoCAD (top) and AutoCAD LT (bottom).

If you're using AutoCAD LT, or if you just want to check out some ready-made parametric possibilities, you can download the sample drawing datasets from either `www.autodesk.com/autocad-samples` or `www.autodesk.com/autocadlt-samples`. The parametric samples are `architectural_example-imperial.dwg`, `civil_example-imperial.dwg`, and `mechanical_example-imperial.dwg`.

Forget about drawing with precision!

"Are you serious, dude? Throughout this book, you have continually nagged me about the need for drawing with precision, and now you want me to forget about it?"

I'm sure that you're familiar with, or at least have come across, computer acronyms and abbreviations, such as RAM, DVD, USB, and, of course, CAD. Okay, here's a new one that I coined for a magazine article a few years back: NAD, or Napkin-Aided Design. It has been my observation that the vast majority of design projects start life as sketches on coffee-shop napkins.

A major strength of CAD is that drawings can (and should) be so precise that you can confidently make use of the information in them, such as volumes, areas, and lengths. On the other hand, a major weakness of CAD is that you're forced to be precise, even early in the design process, when design ideas are not yet fixed and circumstances often change. Yes, commands such as Move and Stretch make editing relatively easy to use, but *relatively* in a relative way, and *easy to use* is easy to say.

With parametrics, you can start out with a NAD sketch that shows the general idea of your design. You don't have to decides on the exact size and placement of objects from the get-go. Remember my parametric motto: Close enough is good enough. By the way, the motto also applies to horseshoes, hand grenades, and dancing. Then apply geometric and dimensional constraints and formulas to those vaguely placed objects to progressively fine-tune their values until you arrive at your final design.

You can apply dimensional or geometric constraints to new geometry as you create it, or you can assign constraints afterwards, such as to existing geometry, even in old drawings. AutoCAD even can convert existing associative dimensions to dimensional constraints.

Constrain yourself

The following sections describe the types of constraints available in each of the two categories, geometric and dimensional. I start with geometric constraints because I've found that it's usually best to pin down objects (pun intended) before applying dimensional constraints so that the drawing objects behave in a predictable manner when dimension values are changed.

Understanding Geometric Constraints

Adding geometric constraints to object geometry is quite simple because it's much like using regular object snaps. In fact, in several cases, AutoCAD's geometric constraints have equivalent object snaps. AutoCAD has tangent, parallel, concentric (center), and perpendicular object snaps, and there are tangent, parallel, concentric, and perpendicular geometric constraints. The difference between them is that constraints are persistent, while object snaps are in effect only at the moment you select a geometric feature. Think of constraints as "sticky" object snaps as opposed to "momentary" ones (as described in Chapter 8).

For example, in the example earlier in this chapter, you can see how the line always remains tangent to the circle, no matter what. The tangent is sticky, and the objects "remember" how to get along with their siblings. If I could only figure out a way to make this principle work with my kids.

Twelve geometric constraints are at your disposal. The easiest way to apply them to drawing objects is to click the buttons on the Geometric panel of the Ribbon's Parametric tab.

Table 19-1 presents the list of the geometric constraints and explains what each one does. In most cases, the second object (or geometric feature) you pick jumps to match up with the first one you pick. In some cases, the constraint that results depends on where on an object you make your pick, such as one endpoint or the opposite endpoint.

TABLE 19-1 Geometric Constraints

Button	Constraint Name	What It Does
	Coincident	Forces two or more points, such as the endpoints of two arcs, to coincide; can also constrain a point to lie anywhere on an object.
	Collinear	Forces two or more lines to lie along an infinitely long projection of the first line that's selected; they do not necessarily end up touching. Works with line and polyline segments, text and mText, and the major and minor axes of ellipses and elliptical arcs.
	Concentric	Forces the centers of two or more arcs, circles, ellipses, or arc segments of polylines to be concentric by sharing the same center point. The second curve selected is made concentric to the first one.
	Fix	Locks the location of an object or a point on an object to a specific location in the drawing; object can rotate around the fixed point.
	Parallel	Forces two lines to be parallel; the second line selected becomes parallel to the first line. Works with line and polyline segments, text and mText, and the major and minor axes of ellipses and elliptical arcs.
	Perpendicular	Forces two lines to be perpendicular to one another. Note that *perpendicular* doesn't mean only "one line horizontal and the other vertical." *Perpendicular* means "at right angles to" no matter the angular orientation, and the lines do not need to touch or cross. The second line selected is made perpendicular to the first. Works with line and polyline segments, text and mText, and the major and minor axes of ellipses and elliptical arcs.
	Horizontal	Forces a line or two points to be horizontal relative to the current coordinate system. Works with line and polyline segments, text and mText, and the major and minor axes of ellipses and elliptical arcs.
	Vertical	Forces a line or two points to be vertical relative to the current coordinate system. Works with line and polyline segments, text and mText, and the major and minor axes of ellipses and elliptical arcs.
	Tangent	Forces an object to be tangent to a selected arc or circle. Lines can be tangent to arcs and circles, and arcs and circles can be tangent to each other. The second object selected is made tangent to the first one.
	Smooth	Joins a selected spline object with another spline, line, arc, or polyline while maintaining curvature (also known as G2) continuity. The second object selected is joined smoothly to the first one.
	Symmetric	Forces two objects or two points on objects to be symmetrical about an imaginary line and is similar to a persistent Mirror command. The second object selected is mirrored to the first one.
	Equal	Forces two lines or linear polyline segments to be the same length; forces two arcs or circles, or an arc and a circle, to have the same radius. The second object selected is sized equal to the first one.

TIP

Text objects can also be geometrically constrained. You can apply horizontal or vertical constraints to text, and you can make text parallel, perpendicular, or collinear with lines. Anyone who has had to label utility lines on civil engineering drawings in earlier releases so that the text aligns with the linework will enjoy this feature.

The three buttons at the right side of AutoCAD's Geometric panel (and the *only* buttons on AutoCAD LT's Geometric panel) let you control the visibility of geometric constraint markers:

- **Show All** shows all constraint markers.
- **Hide All** hides all constraint markers.
- **Show/Hide** displays the constraints on selected objects.

You can remove geometric constraints from objects by hovering your cursor over a constraint marker and clicking the tiny x that appears.

Applying a little more constraint

In the following steps, I show you how to get started with geometric constraints in AutoCAD (not AutoCAD LT). These steps may seem tedious, but I want you to become familiar with the basic properties of each type of constraint. Later, I show you a couple of shortcuts to speed things up.

You can find the files I use in this sequence of steps at this book's companion website: Go to `www.dummies.com/go/autocadfd19` and download `afd19.zip`. The drawing named `afd19c.dwg` contains the unconstrained geometry, and drawing `afd19d.dwg` contains the end product.

Follow these steps to use geometric constraints:

1. **Start a new drawing. Make the Ribbon's Parametric tab current, and turn off Snap, Ortho, Polar, Osnap, and Otrack on the status bar.**

   For production drawings, you will want to make sure that you continue to use precision drafting aids, but in this example, you gain a better sense of parametrics with an *extremely* imprecise drawing.

2. **Draw some linework by using the PLine command, and then add a couple of circles.**

   Draw it to look similar to Figure 19-2 if you want to follow these steps closely.

   Every drawing is tackled differently, so look ahead and figure out the most efficient way to apply the constraints you need in order to maintain your design intent. In this example, you end up with two concentric circles in the middle of a square.

   FIGURE 19-2: Random shapes in need of constraining.

   TIP

   As I mention earlier, applying constraints is easier if at least one point on the geometry is fixed in space. In many cases, consider applying the Fix constraint first — it constrains a point to a single location in the drawing area, locking it into place. This topic has nothing to do with tomcats.

   In this example, I show you how to lock one endpoint of the polyline into position.

3. **Click Fix in the Geometric panel, and then click a point that you want to fix in space.**

   As you move the cursor, a padlock icon is displayed beside the pickbox to remind you that Fix constraint is active. As you move the cursor over an object, a red, circled X marker shows you where you can place the constraint (see Figure 19-3). In this example, I clicked an endpoint on the polyline. Once the constraint is placed, the constraint bar shows the padlock icon, indicating that the endpoint of the polyline segment is fixed in place.

   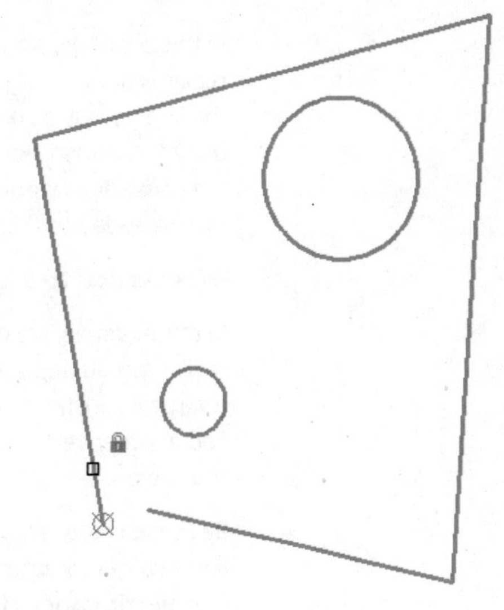

   FIGURE 19-3: Locking down an object in drawing space.

With at least one point on the geometry fixed in place, you can start constraining the geometry by closing that big gap in the linework.

4. **Click Coincident in the Parametric tab's Geometric panel.**

   You use a coincident constraint to make two points coincide, such as the end points of two disconnected lines. A coincident icon appears near the pickbox. As you move the cursor over an object, the red X marker appears at points on which you can place the constraint. In the case of lines or polyline segments, they're at the endpoints and the midpoint.

   TIP

   Use drawing commands appropriate for your design intent. I used a polyline in this example because the ends of the segments are already coincident-constrained by being part of a single polyline object. If you had drawn this shape with lines, you'd have to apply individual coincident constraints to each corner among the lines.

5. **Click an endpoint on the first polyline segment that you want to connect, and then click an endpoint on the second segment.**

   The endpoint of the second polyline segment jumps to the endpoint of the first line, and a small blue square (the marker for coincident constraints) appears at the intersection where this particular constraint was applied. If you don't see the little blue square, click Show All in the Geometric panel.

6. **Apply some orthographic parameters so that the linework starts to look a little more like a rectangle. Click Horizontal in the Parametric tab's Geometric panel, and then click a line or polyline segment that you want to align with the drawing's X-axis.**

   In this example, we picked the bottom segment of the polyline. The segment realigns horizontally from the endpoint nearest to the spot where you picked the line (unless a Fix constraint is added first), and a horizontal constraint marker (a constraint bar) appears near the object. A single item is still called the constraint bar because, if you add additional constraints to the object, they're added to the existing one, like a toolbar.

7. **Click Vertical to align a line or polyline segment with the drawing's Y-axis.**

   In this example, we pick the left vertical(-ish) polyline segment. The segment realigns at 90 degrees to the horizontal segment, and a constraint bar showing a vertical constraint appears. With one line segment horizontally constrained and another vertically constrained, you're halfway to a geometrically precise rectangle.

8. **Because rectangles have parallel and perpendicular sides, apply those constraints to your linework. Click Parallel in the Parametric tab's Geometric panel, click the vertically constrained line segment, and then click the line segment opposite to it.**

Because a vertical constraint is already applied to one segment, it doesn't matter which line you pick first. If neither line had an existing constraint, the second line you picked would become parallel to the first line.

Always prioritize your design intent as you consider which constraints to apply and in which order. As a general rule, start with the most important ones, and drill down to the least important. As I remind you earlier in this chapter, applying a fix constraint early in the game can prevent your geometry from rearranging itself in ways you don't expect.

To make the final side of the almost-rectangle orthogonal with the other three, you could use another parallel constraint and click the bottom line. However, you don't want to wear out the Parallel button, so use Perpendicular on the final segment.

9. **Click Perpendicular on the Geometric panel, click either vertical side, and then click the non-orthogonal side.**

   Again, because three of the four sides are already constrained to horizontal and vertical, it doesn't matter which segment you pick first.

Perpendicular means "at right angles to." It isn't restricted to "a vertical line extending upward from one end of a horizontal line." Lines drawn at an angle can still be perpendicular to one another, and the lines don't even have to touch.

Horizontal and vertical constraints work on one object, but parallel, tangent, and perpendicular constraints need two objects two work. When you roll the cursor over a constraint applied to two objects, its soul mate also lights up.

To delete a constraint, move the mouse pointer over the constraint marker to display the constraints bar (see Figure 19-4), right-click, and choose Delete from the menu that appears.

From four non-orthogonal, not-even-closed line segments, applying a handful of constraints yields a perfect rectangle. However, you really want a square, which is where the Equal constraint comes in.

10. **Click Equal on the Geometric panel, click the bottom side, and then click either of the vertical sides.**

    It's a perfect square! Now you have to constrain those circles.

    The design intent in this example is to make the circles concentric and to locate their centers in the exact center of the rectangle. First, the easy part: Make the circles concentric.

11. **Click Concentric in the Geometric panel, click one circle, and then click the other.**

    The two circles are concentrically constrained. Try saying that ten times fast. A new constraint bar appears in their vicinity. Move one circle by clicking it, and watch the other tag along.

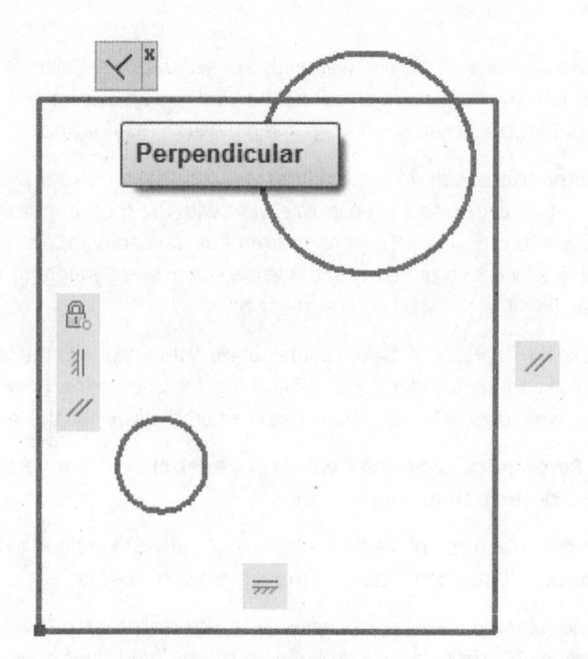

FIGURE 19-4: Constraining to orthogonal.

As I suggest, concentric is the easy constraint. I'd love to be able to use the Mid Between 2 Points object snap and simply click two diagonally opposite corners to locate the circles dead center in the rectangle. However, because you can constrain only objects or points on objects, you have to add some construction geometry to maintain the design intent.

12. **Draw a line between diagonally opposite corners using Endpoint object snaps, and then apply coincident constraints between the endpoints of this line and the corners of the rectangle (see Figure 19-5).**

TIP

Draw construction geometry on a separate layer, and set that layer to NoPlot in Layer Properties Manager. If you don't want to even see your construction geometry (or plot it), turn off the construction geometry layer, or freeze it. The geometric constraints will still work.

13. **Click Coincident in the Geometric panel. Click either circle so that the parametric marker appears in the center, move the pickbox over the construction line, and click when the parametric marker is over the midpoint of the line.**

You're done! You can test your design intent by using the Stretch command on the corner of the rectangle that doesn't display a square, blue coincident icon. As you drag the corner, you should see the two circles moving, too, always maintaining their position in the middle of the rectangle (see Figure 19-6).

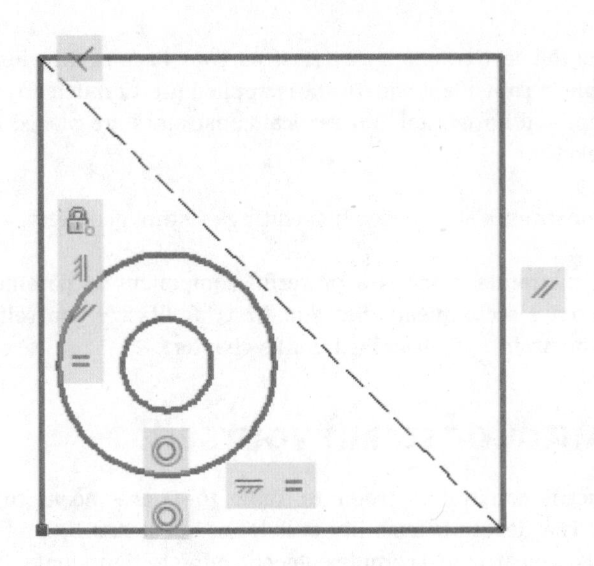

FIGURE 19-5: Adding construction geometry.

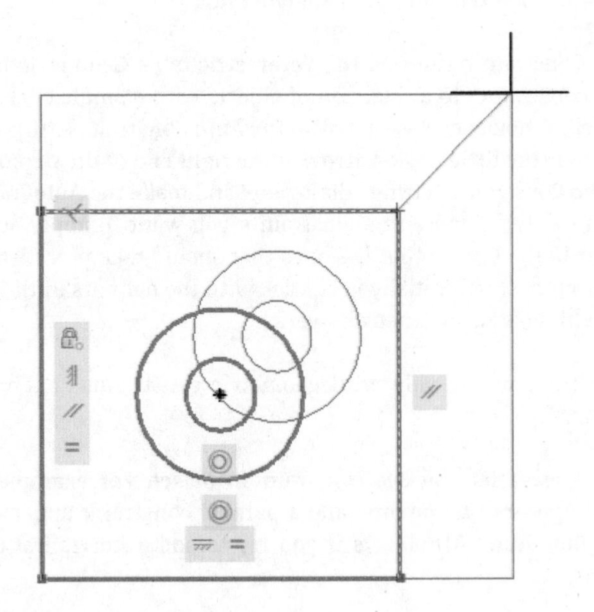

FIGURE 19-6: Full-on geometric constraints.

Using inferred constraints

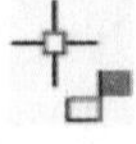

Parametrics were a new feature in AutoCAD 2010 and were considerably enhanced in AutoCAD 2011 by the addition of *inferred constraints*. A status bar button (to the left of the Snap button) toggles Infer Constraints mode off and on. When it's on, you don't have to specifically apply many of the geometric constraints.

To produce the best results, go at least partway back to drawing with precision. For example, a coincident constraint is applied pretty much any time you use an object snap, and horizontal and vertical constraints are placed if polar or ortho modes are on.

TIP

Inferred constraints also work when editing existing geometry.

Inferred Constraints mode is a powerful component of parametric drafting in AutoCAD, but I recommend that you first familiarize yourself with manually applying constraints, as described in this chapter.

You AutoConstrain yourself!

The geometric constraints (refer to Table 19-1) are meant to be applied one at a time. That level of flexibility is helpful, but it can be pretty darned time-consuming to go around a complex object, trying to figure out what the best constraint option would be and then applying it.

The Auto Constrain button on the Parametric tab's Geometric panel can apply a whack of constraints to a selection of objects with a single click. Before you execute that click, however, take a look at the Auto Constrain settings. Click the dialog box launcher (the little angled arrow at the right end of the Geometric panel label) to open the Constraint Settings dialog box and make the AutoConstrain tab active (see Figure 19-7). Choose the constraints you want to apply automatically, and their order in the list, so that the ones that should take precedence have a higher priority. Your best bet initially is to stick with the defaults until you become more familiar with how Auto Constrain works.

REMEMBER

Auto Constrain works only with geometric constraints, not with dimensional constraints.

AutoCAD doesn't let you overconstrain an object. For example, if a horizontal constraint appears on one line, and a parallel constraint appears between it and a second line, AutoCAD objects if you try to add a horizontal constraint to the second line.

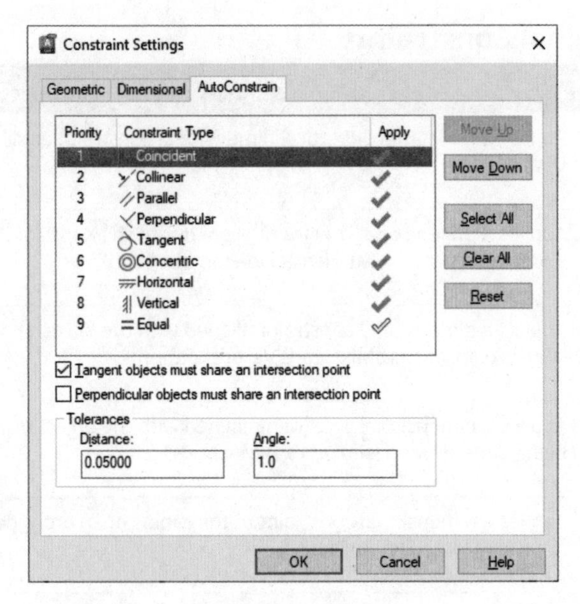

FIGURE 19-7: Auto-constrainable geometric relations in the Constraint Settings dialog box.

Understanding Dimensional Constraints

The normal practice in AutoCAD is first to create some geometry using, of course, the precision techniques I discuss in Chapter 8, and then apply dimensions as I describe in Chapter 14. Assuming that you're using fully associative dimensions, you can edit the geometry and watch the dimensions update automatically. The length of the line or the radius of the circle is in control of the dimensions. These dimensions are known as *driven dimensions* because their values are updated by AutoCAD when you change the size or position of the objects.

Dimensional constraints, unlike regular AutoCAD dimensions, are *driving dimensions* — when you change the value of a dynamic dimension on a line, the line changes to match. In other words, the dimension is driving the length of the line, not the other way around. For instance, when you apply a dimensional constraints on the length of a line, you can move and rotate the line, but you can no longer change its length by dragging grips. Instead, you change its length by changing the value of the dimensional constraint.

There are only eight dimensional-constraint options, but they cover all the bases, as described in Table 19-2.

TABLE 19-2 Dimensional Constraints

Button	Constraint Name	What It Does
	Linear	Applies a horizontal or vertical dimensional constraint (similar to DIMLINear)
	Horizontal	Applies a dimensional constraint aligned with the X-axis of the coordinate system (similar to DIMLINear's Horizontal option)
	Vertical	Applies a dimensional constraint aligned with the Y-axis of the coordinate system (similar to DIMLINear's Vertical option)
	Aligned	Applies a dimensional constraint aligned with the object or with two points being dimensioned (similar to DIMALIgned)
	Radial	Applies a dimensional constraint to the radius of an arc, a polyarc, or a circle (similar to DIMRADius)
	Diameter	Applies a dimensional constraint to the diameter of an arc or a circle (similar to DIMDIAMeter)
	Angular	Applies a dimensional constraint to the angle between two lines or three points (similar to DIMANGular)
	Convert	Converts an existing associative dimension object to a dimensional constraint

Practice a little constraint

The dimension-looking objects you add to a drawing from the Dimensional panel aren't the same as the dimension objects you add from the Annotate tab. Dimensional constraints are *driving dimensions* — when you change the value of one of these dimensions, the linked geometry changes.

A lot is happening behind the scenes as you apply parametric constraints. You can get a sense of how these constraints work at keeping drawing objects in order by using the Stretch command on objects after you apply a constraint to them.

You can find the files I use in these steps at this book's companion website: Go to `www.dummies.com/go/autocadfd19` and download `afd19.zip`. The drawing named `afd19a.dwg` contains the unconstrained geometry, and the drawing named `afd19b.dwg` contains the end product.

These steps present a simple example of dimensional constraints:

1. **Start a new drawing and make the Ribbon's Parametric tab current.**
2. **Turn on the appropriate precision drawing aids on the status bar, such as Snap, Ortho, and Osnap.**
3. **Draw some reasonably precise geometry by applying the precision techniques I describe in earlier chapters.**

   In the following example, I use the RECtang and Circle commands to draw the geometry you see in Figure 19-8. The rectangle is ten units square, and the 2.5-unit-radius circle is deliberately drawn away from the middle of the square.

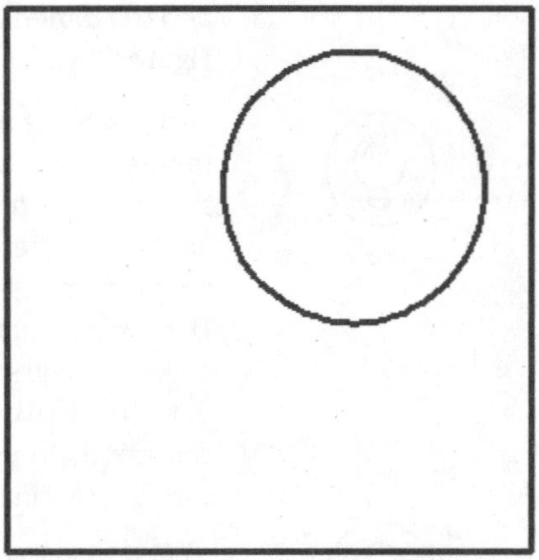

   FIGURE 19-8: Simple geometry, badly in need of constraining.

4. **On the Dimensional panel of the Parametric tab, click the top part of the Linear split button.**

   A linear dimension icon appears beside the pickbox, and AutoCAD prompts you to either specify the first constraint point or pick an object.

   Like the DimLInear command, the Linear dimensional constraint tool is *inferential* — you create a horizontal or vertical dimension depending on which way you drag the cursor. Also, as with DimLInear, you can press Enter at the command line and select an object to dimension.

TIP

   Hover the mouse over the Linear button to see (unlike its neighbor, Aligned) that the button is split into two parts. To force a linear dimensional constraint to be either horizontal or vertical (rather than dependent on the direction you drag the cursor), click the bottom part of the Linear button and choose from the drop-down menu.

5. **Press Enter at the command line to confirm that you want to select an object, and then select the bottom horizontal line segment.**

REMEMBER

   If you see red markers at the midpoint and endpoints of the bottom line, you didn't press Enter — and AutoCAD is in Point Selection mode rather than Object Selection mode.

   AutoCAD generates a preview of a dimensional constraint and prompts you for a location.

6. **Click to locate the dimension position.**

   AutoCAD draws a dimensional constraint with a highlighted text field displaying the dimension name (*d1,* in this example) and the value returned by AutoCAD.

7. **Press Enter to confirm the value and the dimension location (see Figure 19-9).**

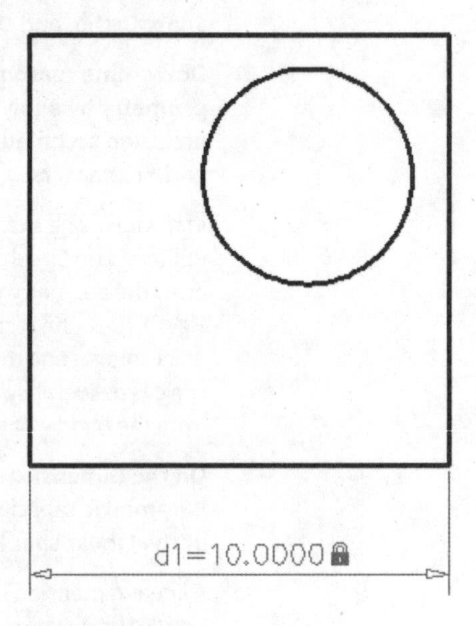

FIGURE 19-9: Placing a dimensional constraint.

TIP

Best practice is usually to accept the default value for all dimensional constraints as they are applied and then edit them later to the exact values that are desired. This avoids problems like trying to make the width of a slot greater than the width of the piece that contains the slot, which can turn your drawing inside out.

TIP

If dimensional constraints disappear as soon as you place them, click the Show All button on the Parametric tab's Dimensional panel (refer to Figure 19-1).

REMEMBER

Because dimensional constraints aren't regular dimension objects, you can't apply dimensions styles and you can't plot them, so it doesn't matter where you put them or what they look like. In fact, AutoCAD automatically resizes them appropriately as you zoom in and out of the drawing so that you can always read the text. I show you in the next section how to turn dimensional constraints into properly styled, plotable dimensions.

8. **Repeat Steps 4–6 and add a dimensional constraint to the rightmost vertical edge of the rectangle.**

   AutoCAD draws a second dimensional constraint, this one named *d2*.

Making your drawing even smarter

If your drawing were a traditional mechanical drawing that followed the rules of drafting, the placement of a horizontal and vertical dimension in the example of the previous section would be enough. Whoever reads your drawing understands that if sides look parallel and perpendicular, a dimension on one side applies to the opposite side as well. But in this chapter, I talk about *intelligent* drawings that respect design intent — not dumb collections of lines and circles, even if they *do* follow the rules of drafting.

If you stretch a rectangle in various ways (from the upper-left corner, for example), you can see that only the bottom and right sides are constrained to 10 units in length. You can add two more linear dimensional constraints to the unconstrained sides, but then you'd have to remember to edit both dimensions. Rather than constrain both sides to be 10 units long, you maintain design intent by making both sides equal in length.

TIP

You can make the two sides equal by using either dimensional or geometric constraints. You should usually apply most, or all, of the geometric constraints to the drawing before adding dimensional constraints. This way, the objects behave in a more predictable manner when you change the dimension values.

Follow these steps to apply three geometric constraints:

1. **On the Geometric panel of the Parametric tab, click Fix (the padlock icon) and then click the lower-left corner of the rectangle.**
2. **On the Geometric panel, click Horizontal and click the bottom side of the rectangle. Then click Vertical and click the left side.**

   Icons appear near the drawing geometry, showing that those three geometric constraints are active.
3. **On the Dimensional panel of the Parametric tab, click Linear and add a dimensional constraint to the top horizontal line.**
4. **Click to locate the dimension, type d1, and then press Enter when AutoCAD prompts you for the dimension text.**

   Rather than display numeric values, like the first two linear constraints, this new dimensional constraint displays `fx: d3=d1`.

   TECHNICAL STUFF

   The main part of this expression sets the d3 dimension to equal the value of the d1 dimension. The `fx:` is short for "formula" and reminds you that other variables in other dimensions are being referenced.
5. **Repeat Steps 3 and 4, and this time add a constraint to the vertical line at the left. Then enter d2 as the dimension text.**

   All four dynamic dimensional constraints display their names and a value or expression for each one.

Dimensional constraints have names as well as values; the values can be expressions or formulas. You can set the default appearance of dynamic dimensional constraints by clicking the dialog box launcher (the little arrow at the right end of the Dimensional panel label) to open the Constraint Settings dialog box, with the Dimensional tab active (see Figure 19-10).

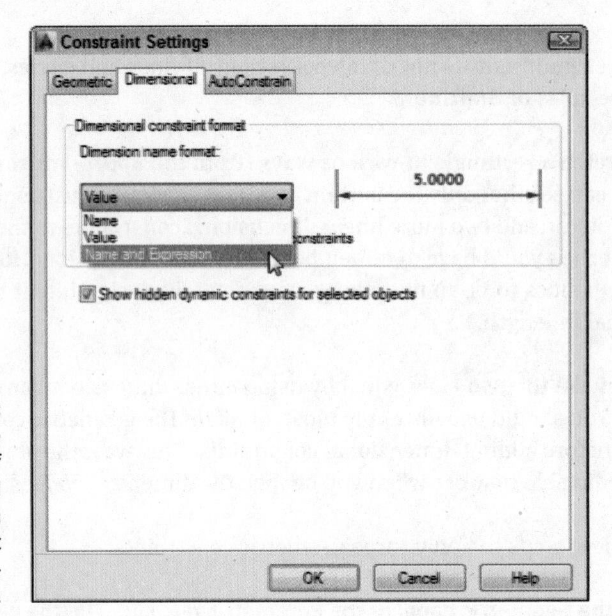

FIGURE 19-10: Format the appearance of dimensional constraints in the Constraint Settings dialog box.

The options are as follows:

- **Name:** The first linear dimension is named *d1*, the second *d2*, and so on. You use the dimension names in expressions.
- **Value:** The numeric value that you enter into the dimensional constraint (or that AutoCAD enters if you don't override it).
- **Name and Expression:** The dimension name shown as equal to an expression. The expression can be a value, as in this example, or it can be a formula that includes other dimension names.

Using Parameters Manager

Both AutoCAD and AutoCAD LT include the Parameters Manager palette, accessible from the Manage panels of the Ribbon's Parametric tab. You can use Parameters Manager to give all those dimensional constraints more sensible names than `d1` and `d2`, but you can (even more usefully) enter expressions instead of plain numeric values, as I explain in these steps:

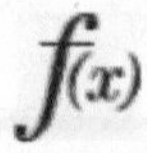

1. **Click Parameters Manager in the Manage panel.**

   The Parameters Manager palette appears, showing a list of the dimensional constraints applied in the drawing (see Figure 19-11).

In Figure 19-11, the Expression column shows the numeric values that I specified for d1 and d2 and the expressions I entered for d3 and d4. The read-only Value column shows the calculated value. You can't change a value in the Value column (it's grayed out to remind you). You can only edit the cells in the Expression column.

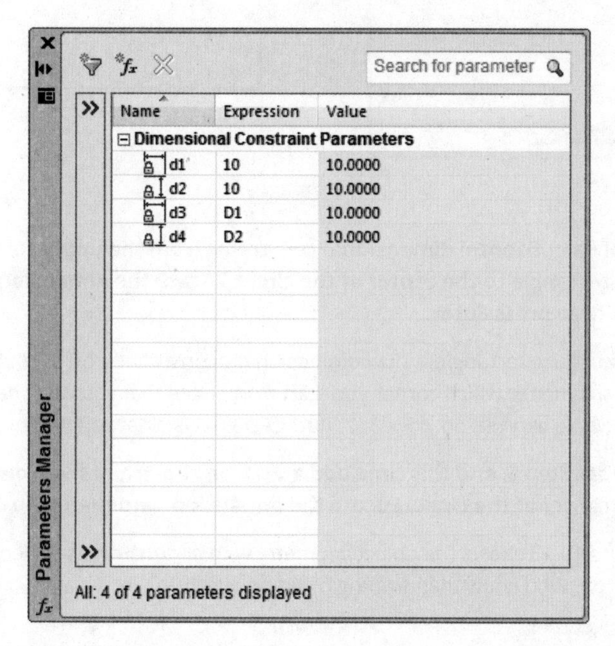

FIGURE 19-11: The Parameters Manager palette.

2. **In the d1 row, click in the Expression field to highlight the current value (10, in this example). Then click again and type a new value. For example, type 13 and press Enter.**

   The rectangle resizes in the drawing editor, and because the d3 constraint on the top side was made equal to the d1 constraint on the bottom side, both sides change equally.

   Next, you use an equation as an expression.

3. **In the d2 row, click in the Expression field to highlight the current value, and then type an expression. For example, type d1*0.75 and then press Enter.**

   The read-only Value column and the drawing geometry show that the new d1 distance of 13 has been multiplied by 0.75 and is now 9.75 (see Figure 19-12).

4. **Close Parameters Manager.**

   Constrain the circle so that its center is always at dead center in the rectangle, no matter how the rectangle's size changes.

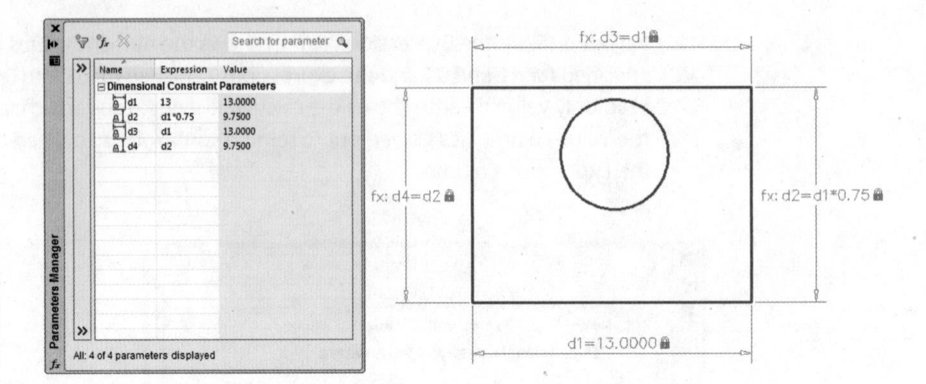

FIGURE 19-12: Editing constraints in Parameters Manager.

5. **Apply a horizontal dimensional constraint from the upper-left corner of the rectangle to the center of the circle. Locate the dimension, type d3/2, and then press Enter.**

   Because the rectangle is now dimensionally constrained on all four sides, it doesn't matter which corner you start from. Note that you don't have to type the whole expression: d5=d3/2. AutoCAD knows what you mean!

6. **Repeat Step 5, and this time add a vertical constraint from one corner to the center of the circle. Locate the dimension, and then type d4/2.**

   Figure 19-13 shows the object geometry with all constraints added in this section. Who knew that drafting could be such fun?

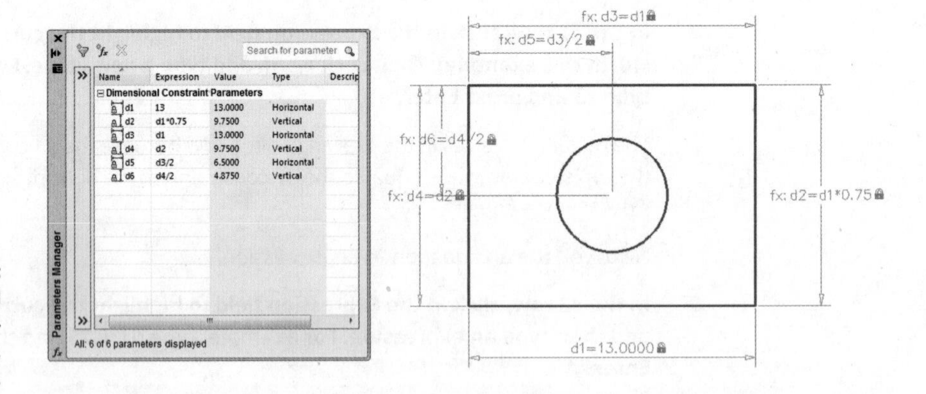

FIGURE 19-13: All locked down — dimensionally, at least.

If your drawing becomes overrun with parameters, you can add parameter filters in the Parameters Manager palette. Right-click any parameter and choose Show Filter Tree, or click the double right arrow at the left of the palette to open the Filters pane. Click the funnel icon on the toolbar to create a new filter group, and then simply drag and drop parameters into the group. Figure 19-14 shows the two constraints (refer to Steps 5 and 6 of the preceding step list) dragged into a new group filter.

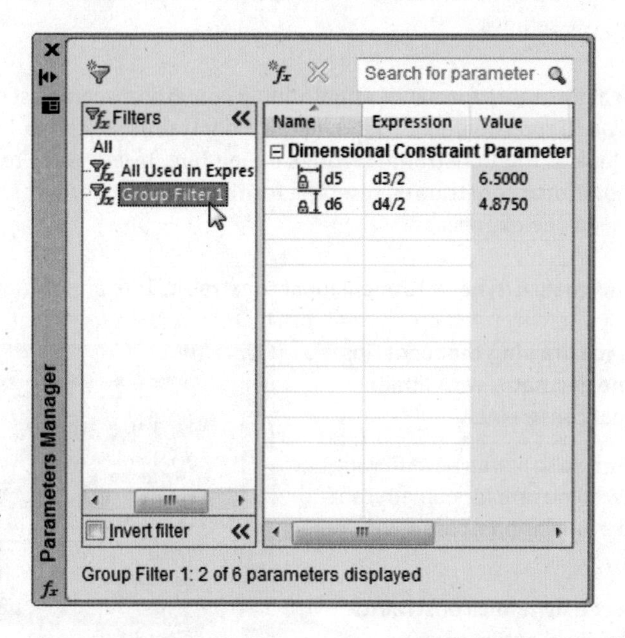

FIGURE 19-14: Filter parameters to keep them organized.

Dimensions or constraints? Have it both ways!

After all the hard work of adding dimensional constraints to a drawing, it would be a downright shame to have to go back and apply regular dimensions, wouldn't it? Well, you don't have to. You can make dimensional constraints look and behave like regular dimensions. You can go the other way, too, and make your regular dimensions act like dimensional constraints.

Dimensional constraints are available in two flavors:

- **Dynamic:** In this default form, a dynamic constraint is gray, with a padlock icon next to it in the drawing area. You can make the constraint appear and disappear by clicking Show All in the Dimensional panel. Dynamic constraints don't plot, and they're always legible because they resize as you zoom in and out of the drawing.

» **Annotational:** This form is controlled as an object property, so you have to set it in the Properties palette. Though annotational constraints plot, they don't resize as you zoom in and out, and they don't disappear as you toggle the Show All button on and off. Annotational constraints conform to dimension style settings.

The dimension name format of annotational constraints can be set to *Name*, *Value*, or *Name and Expression*, just like dynamic constraints. Refer to Figure 19-10 for another look at the Constraint Settings dialog box. If you want to plot a drawing with annotational constraints, reset the format so that it doesn't show the dimension name or the expression.

Here's how to turn dynamic dimensional constraints into annotational constraints:

1. **Open a drawing that contains some geometry with dimensional constraints.**

   You can also start a new drawing, draw some simple geometry, and add a dimensional constraint or two.

2. **Select a dynamic constraint, right-click, and choose Properties.**

   The Properties palette opens with the object properties of the selected dimensional constraint listed in table form.

3. **Click in the Constraint Form field, and in the drop-down list, change Dynamic to Annotational (see Figure 19-15).**

   The dynamic constraint becomes annotational and takes on the appearance of the current dimension style. If you change the dimension style in the Properties palette, the annotational constraint updates to the new dimension style format.

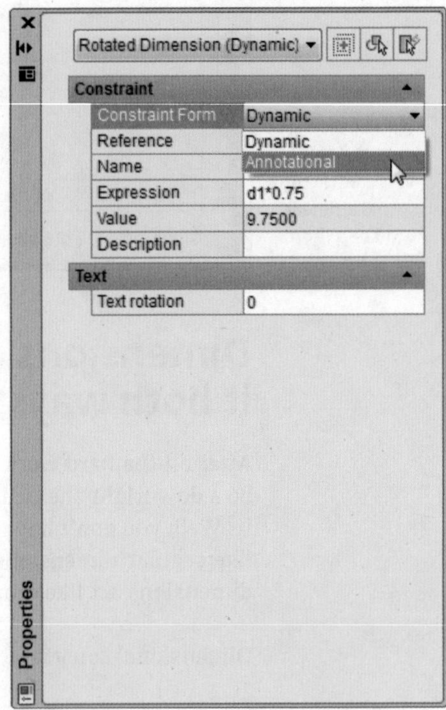

FIGURE 19-15: Turning dynamic constraints into annotational ones.

You can go the other way, too — from regular dimension objects to dimensional constraints to convert an existing dumb drawing to an intelligent parametric one.

4. **Add a linear, radius, diameter, aligned, or angular dimension (that is, a regular dimension, not a dimensional parameter) to your drawing geometry.**

   Nearly every type of dimension object has a parametric analog; the exceptions are arc length, jogged radius, jogged linear, and ordinate dimensions.

5. **Click Convert in the Dimensional panel, select a dimension type (refer to Step 4), and press Enter.**

   The Dimensional Constraint text box displays as soon as you click an associative dimension, and the dimension becomes a dynamic constraint as soon as you press Enter.

REMEMBER

The only clue that a dimension is an annotational constraint rather than a regular associative dimension is the padlock icon that appears next to the dimension value. You can turn off the display of the padlock from the Constraint Settings dialog box, but I recommend that you leave it on. It doesn't plot anyway, and you might decide to delete the dimension without realizing that it's controlling your object geometry.

Annotational dimensions can also be annotative so that they size themselves automatically to the drawing scale. I cover annotative dimensions in Chapter 14. Annotative and annotational dimensions are also associative, so you can have annotative annotational associative dimensions. Try saying that quickly after a few drinks.

REMEMBER

In this chapter, I present the two varieties of parametric constraints separately, but you get the most mileage from this feature when you incorporate both geometric *and* dimensional constraints with other precision techniques (described in Chapter 8). In fact, if you start a drawing by using Snap, Ortho, Osnap, and other precision techniques before you add dimensional and parametric constraints, you'll be well on your way to creating a library of intelligent drawings that maintain your design intent.

TIP

Geometric and dimensional constraints can be equivalent. For example, a 90-degree-angle dimensional constraint is equal to a perpendicular geometric constraint, but AutoCAD doesn't let you apply both.

All examples in this chapter involve a single drawing view. Geometric constraints apply "to infinity," so you can synchronize objects from one orthographic view with objects in another. For example, coincident and collinear constraints can link objects in the front view with their equivalents in the top and right-side views, and liberal use of the equal constraint means that you should never have to measure or dimension anything twice, you should never need more than minimal construction geometry, any changes in one view reflect through to the others, and your views always remain orthogonally aligned.

Lunchtime!

You may recall that in the opening paragraphs I laid out a scenario wherein your boss instructed you to make a number of changes, and I implied that you could do them almost instantaneously. If you haven't already done so, go to `www.dummies.com/go/autocadfd19` and download `afd19.zip`. Open the drawing named `afd19e.dwg`, which looks like Figure 19-16.

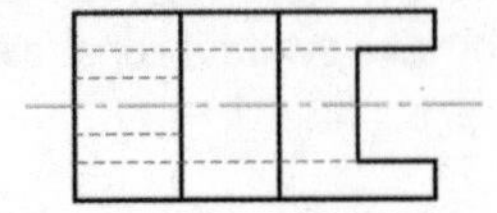

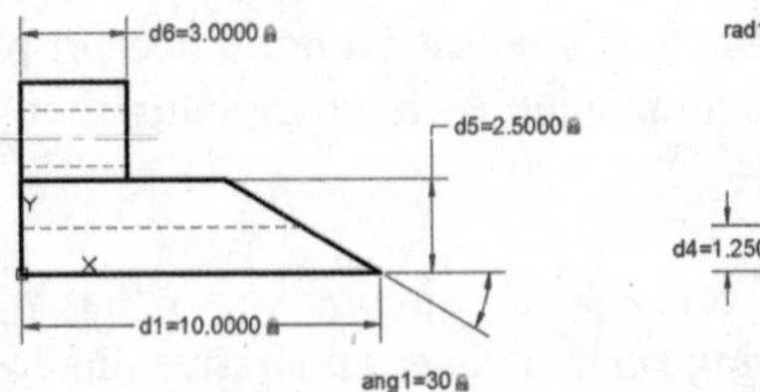

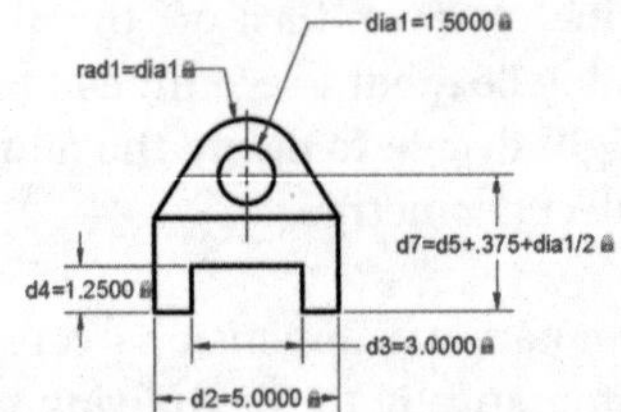

FIGURE 19-16: This part needs a few revisions.

Now double-click the `dia=1.5000` dimension and change it to `dia1=2.0000` and then press Enter. Surprise! Two more dimensions change and all three views update! While you're at it, change ang1=30 to ang1=60, change d5=2.5000 to d5=2.000, change d6=3.0000 to d6=2.5000, and change d1=10.000 to 8.1250. Then print your drawing and head to the door to meet your boss and the client with a drawing that looks like Figure 19-17.

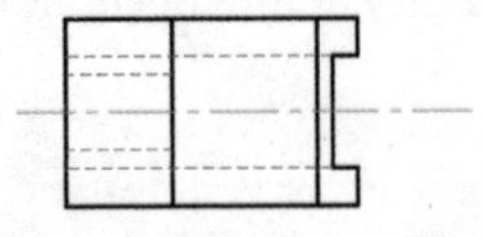

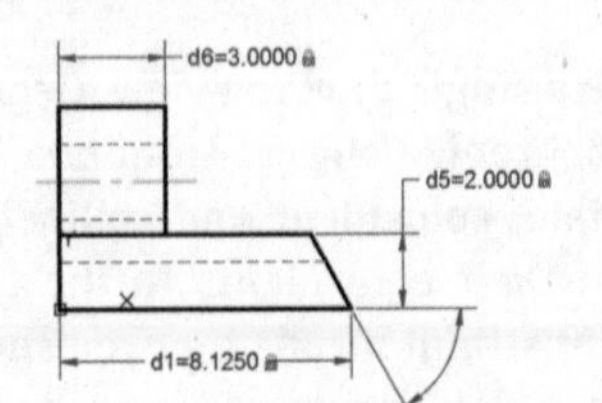

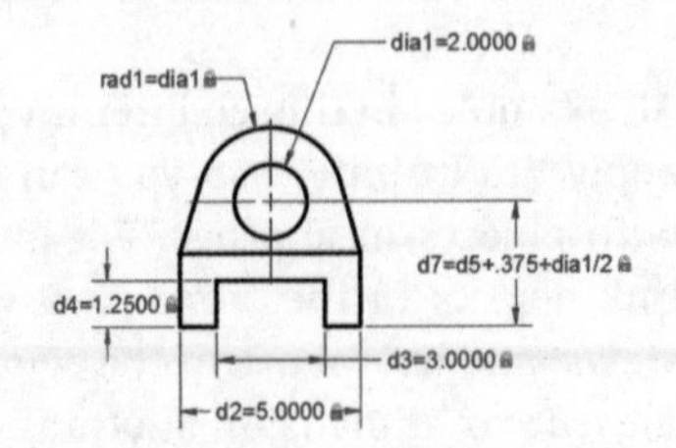

FIGURE 19-17: Thirty seconds later . . .

» **Fixing file dependencies with Reference Manager**

» **Protecting drawings with digital signatures**

» **Exchanging drawing files via the internet**

» **Sharing and collaborating over the internet**

Chapter 20
Drawing on the Internet

Unless you've been living under a rock for the past 35 years, you know that the internet is causing major changes in the way people work. Because of the 'Net, most of us communicate differently, exchange files more rapidly, and phone out for pizza less frequently. We still eat as much pizza, but ordering it over the internet is easy. However, they still have to work out a few of the details to deliver it that way. Years ago they tried FAX pizzas, but it didn't work out. 3D printing, on the other hand. . .

In this chapter, I show you how and when to use AutoCAD with the internet. I also cover how internet-related functions can be connected with traditional CAD tasks, such as file viewing and sharing. The emphasis of this chapter is on useful, no-nonsense ways to take advantage of the internet in your CAD work.

The Internet and AutoCAD: An Overview

The web is changing so quickly that it's almost impossible to write the definitive chapter that spells out exactly how everything works and what's best for you. My intention in this chapter is to steer you toward the features that I feel are of most interest.

WARNING

This is known as the CYA (Cover Your bAckside) paragraph. This chapter describes certain internet features as they are known to exist at the time of this writing, but things are subject to change without notice. There is no warranty, expressed or implied, that everything will be the same when you try to use it. For example, Autodesk changed the name of one function four times in six years.

You send me

Email and the cloud have largely replaced blueline prints and overnight delivery as the standard means of exchanging drawings. Snail mail is dead for envelopes but is otherwise rapidly growing because of online shopping.

Sending and receiving drawing files doesn't differ much from sending and receiving other kinds of files, except that

- **DWG files tend to be bigger than word processing documents and spreadsheets.** Consequently, you may run up against email attachment size limits.
- **You can easily forget to include all dependent files.** An AutoCAD drawing is not an island unto itself; it requires other files that go with it, such as XREFs and files that define fonts and linetypes. I tell you in the next section how to make sure that you send all the necessary files.
- **It's often not completely obvious how to plot what you receive.** Read Chapter 16 and the "Bad reception?" section, later in this chapter, to solve plotting puzzles.

TIP

Whenever you send DWG files, ask recipients to open the sent drawings as soon as they receive them, so that you both have time to respond should there be a problem.

Prepare it with eTransmit

Many people naively assume that an AutoCAD drawing is always contained in a single DWG file, but that's often not the case. Every drawing file created in AutoCAD can contain references to more than a dozen other kinds of files, the most important of which are listed in Table 20-1. Before you start exchanging drawings via the internet or even by USB thumbdrive, you need to assemble the drawings with all their dependent files.

TABLE 20-1 **Types of Files That DWG Files Commonly Reference**

Description	File Types	Consequence When Missing	Explained In
Custom font files	SHX, TTF	AutoCAD substitutes a different font that may make text illegible.	Chapter 13
Other drawings (xrefs)	DWG, DGN, DXF, DWF, PDF, and others	Stuff in the main drawing disappears.	Chapter 18
Raster graphics files	JPG, PNG, and others	Stuff in the drawing disappears.	Chapter 18
Plot style tables	CTB, STB	Lineweights and other plotted effects don't look right when the drawing is plotted.	Chapter 16
Linetype files	LIN, SHX	Linework looks continuous.	Chapter 9

TECHNICAL STUFF

Table 20-1 doesn't exhaust the types of files that your DWG files might refer to. Custom plotter settings (such as custom paper sizes) may reside in PC3 or PMP files. If you use sheet sets, DST files contain information about the sheet structure. (I don't cover sheet sets in this book; if you're interested, check the online Help system.) An FMP file controls some aspects of font mapping. Look up *sheet sets* and the *FONTALT* and *FONTMAP system variables* in the AutoCAD online Help system for detailed information.

Rapid eTransmit

Fortunately, AutoCAD's ETRANSMIT command does the heroic job of pulling together all files that the main DWG file depends on. Follow these steps to assemble a drawing with all its dependent files by using ETRANSMIT:

1. **Open the drawing on which you want to run ETRANSMIT.**

   If the drawing is already open, save it. You have to save the file just before using ETRANSMIT.

2. **Click the Application button and choose Publish ➪ eTransmit from the Application menu.**

   The Create Transmittal dialog box appears, as shown in Figure 20-1.

3. **On the Files Tree tab or the Files Table tab, remove the check mark next to any file that you want ETRANSMIT *not* to copy with the main drawing.**

TIP

   Fonts are initially not included in the list, because they may be protected by copyright. See the upcoming Warning paragraph about the Include Fonts setting. Unless you have assigned custom font mapping, you can omit the `Acad.fmp` file (in AutoCAD) or `acadlt.fmp` (in AutoCAD LT).

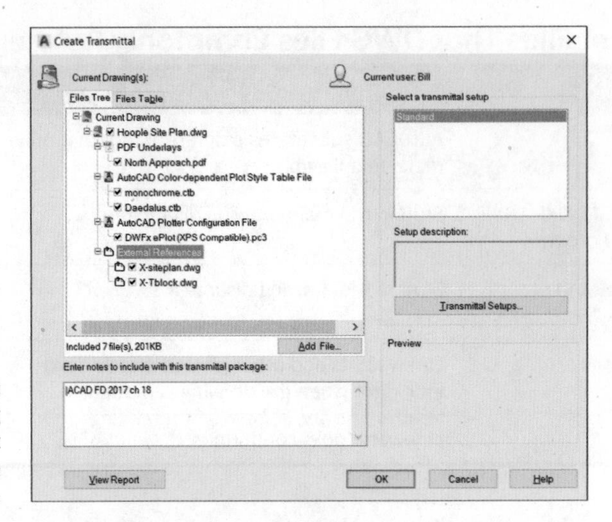

FIGURE 20-1: Using eTransmit to share drawings.

4. **Select a transmittal setup from the list.**

   Transmittal setups contain settings that control how ETRANSMIT processes the drawings and creates the transmittal package. Click the Transmittal Setups button to create new setups or modify existing setups. The default Standard transmittal setup works fine for most purposes. In any case, you should view the settings (click the Modify button) just to see what options you can change if you need to later.

WARNING

   If you want AutoCAD to include SHX and TTF font files, including any custom fonts that you're using, you must turn on the Include Fonts setting in the transmittal setup. As I discuss in Chapter 13, however, many custom fonts work like licensed software. Sending them to others is just like sharing your AutoCAD program installation media with others: illegal and unethical. Don't share licensed text fonts outside your organization.

5. **Click the View Report button.**

   You see a report listing the files that ETRANSMIT will copy, along with warnings about any files that it can't locate.

6. **Review the report and make sure that ETRANSMIT was able to find all the files.**

   If files are missing, find them and move or copy them to the location where ETRANSMIT expects to find them. See the upcoming section, "Help from Reference Manager."

7. **Click OK.**
8. **Specify the name and location of the transmittal package.**
9. **Click Save.**

ETRANSMIT creates the transmittal package, which by default is stored in a ZIP file. Zipping creates a single, tidy, compact package with all your DWGs, DGNs, DWFs, PDFs, raster images, plot style tables, and font files ready to be emailed or uploaded to a cloud storage provider.

TIP

Use eTransmit to create archives of your drawing projects as they progress. In this way, you can go back in time, in case questions arise about how the designs were developed. You may also want to check out the "Autodesk weather forecast: Increasing cloud" section, later in this chapter.

Increasing cloudiness

In the early decades of the internet, *File Transfer Protocol (FTP)* was the primary way to send and receive large files over the internet. Nowadays, cloud storage services, such as Google Drive and Dropbox, provide a more convenient interface than FTP. FTP and cloud storage services work on pretty much the same principle. You upload files to the cloud and optionally assign access rights. The intended recipients receive an invitation email that grants them access to specific files.

Dropbox Professional lets you view DWG files inside the Dropbox site, through an arrangement with Autodesk. Google Drive displays previews of DWG files after you authorize a third-party viewer, such as Softgateon CADviewer. Box opens DWG files in web-based apps, such as AutoCAD Web and Graebert Kudo, through the Integrations option. When a cloud service does not display DWG files, you need to download them and then view them with AutoCAD or a DWG viewer on your computer.

Bad reception?

If you are on the receiving end of incoming drawings (zipped, I hope!), as soon as you receive them, download the file to your hard drive or a network disk, and then unzip the files to their own folder. Fortunately, Windows and MacOS include the un-zipping function of what they call compressed folders. In both, double-click the ZIP file, and then specify the name of a destination folder.

TIP

Check at least a few drawings in the package to make sure that all xrefs, other reference files, fonts, and raster image files are included. Open each main drawing in the folder, press F2 to view the AutoCAD text window, and look for messages warning of missing fonts and xrefs, like this example:

```
Substituting [simplex.shx] for [helv.shx].
Resolve Xref "GRID": C:\Here\There\Nowhere\grid.dwg
Can't find C:\Here\There\Nowhere\grid.dwg
```

Write down the names of the missing files and then tell the sender to get on the ball (in a nice way, of course) and send you the missing pieces. While you're at it, tell that person to buy this book and read this chapter! Or buy it for that person yourself.

If you receive drawings with custom TrueType font files (files whose extensions are TTF), you must install those files in \Windows\Fonts folder (and not in one of AutoCAD's support folders) for Windows and AutoCAD to recognize them. All you need to do is to right-click each TTF filename and then choose Install; Windows does the rest.

Help from Reference Manager

If you move project folders around or transfer drawings to or from somewhere else, AutoCAD probably won't be able to locate linked raster image files and DWF/DWFx, DGN, PDF underlays (described in Chapter 18), or font files (described in Chapter 13). The ETRANSMIT command, described in the section "Rapid eTransmit," earlier in this chapter, does a good job of gathering dependent reference files, raster files, and font files, but it can't gather what AutoCAD can't locate.

The AutoCAD Reference Manager utility (not included with AutoCAD LT) is a real lifesaver if you're suffering from file-path perils, whether they occur in your own company or when sending files to, or receiving them from, others.

Reference Manager is a separate program, not a command inside AutoCAD. Follow these steps to launch the utility from the Windows desktop:

1. **Start Reference Manager.**

   As indicated in other chapters, starting this utility process depends on your version of Windows and how you have set it up. The bottom line is that Reference Manager is a separate program, so it must be started from the AutoCAD section of the Windows Start button.

   The Reference Manager program opens, as shown in Figure 20-2.

2. **Click the Add Drawings button to add one or more DWG files to the Drawings pane on the left.**

   Navigate to the folder that contains the drawings you want to send, select them, and then click Open.

3. **In the Reference Manager – Add Xrefs dialog box, choose Add All Xrefs Automatically Regardless of Nesting Level.**

   Reference Manager processes the drawings and displays all referenced objects in its right pane.

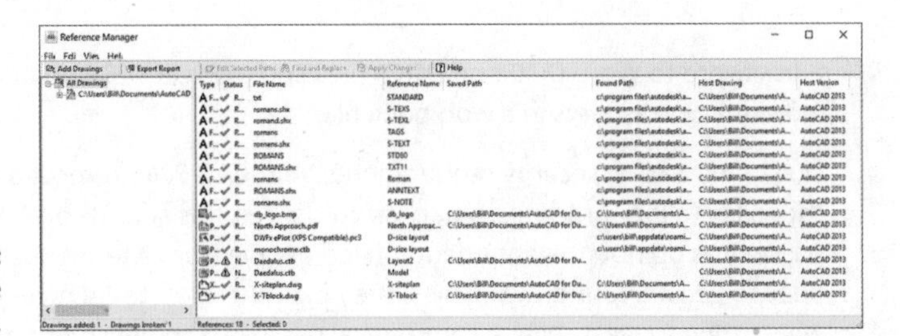

FIGURE 20-2: Reference Manager.

4. **Click Export Report to create a text report listing all dependent files and their paths, or click Edit Selected Paths to modify the paths of selected reference files.**

5. **If you choose to modify any selected paths, click Apply Changes.**

   When you're finished exporting reports and repathing reference files (or finding reference files that you may never have suspected you had if you hadn't run Reference Manager), close the Reference Manager window. Click the Help button in Reference Manager to find out more about the utility's capabilities.

TIP

If you always store parent and child DWG files in the same folder, which is the simplest approach to dealing with xref paths, you probably don't need to use Reference Manager.

TIP

AutoCAD can also save files in the ubiquitous PDF (Portable Document Format) from Adobe Systems. Files can be opened, viewed, and printed from virtually any type and model of computing or communicating device, including smartphones. This format is particularly useful when you want to show drawing information to people who aren't particularly CAD literate (that is, they aren't fellow geeks). I have more on this topic in Chapter 24.

The Drawing Protection Racket

Whether you're sending DWG or DWF files, you may be concerned about their misuse (that is, by the wrong people or for the wrong purposes), abuse (for example, modification without your consent), or reuse (on other projects or by other people without due compensation to you). Two basic procedures are available for securing your files when you send them to others:

» **Password protection:** This feature enables you to lock drawing files in older releases of AutoCAD. Only someone who correctly types the password you've specified can open it. Other programs, like Microsoft Office, also let you apply a password to Word DOC and DOCx files and to Excel XLS workbook files or to individual worksheets in a workbook file.

Due to increasing security requirements, AutoCAD 2016 removed the capability to lock drawings with passwords. Yes, this sounds backwards, but passwords do not prevent unauthorized people from altering drawings. If you need to secure your drawing files, see *passwords* in AutoCAD's help facility, where they suggest several commercial password and encryption products.

- **Digital signature:** Original paper drawings can't be modified without leaving evidence, and copies of paper drawings are obviously copies. Electronic files can be copied and modified without leaving much evidence, so a dispute can arise over which version is the original. ***Solution:*** AutoCAD supports digital signatures. Digital signatures don't prevent unauthorized people from opening the file, nor do they prevent anyone from modifying it. All they do is verify whether a file has or hasn't been modified since it was signed, if there's ever a dispute. The drawback is that you must first open an account with a digital-certificate provider before you can apply them. For more information, see *digital signatures* in the AutoCAD online Help system.

REMEMBER

Although electronic security features, such as the ones described in this section, can be useful as part of a strategy to protect your work from misuse, they don't substitute for communicating clearly, preferably in the form of written contracts, what constitutes appropriate use of drawings that you send to or receive from others.

TIP

The COMPARE command, introduced with AutoCAD 2019, compares two drawing files and points out the differences between them. This feature can be helpful in determining whether a drawing was tampered with, including altering it to make it look like someone else's drawing.

Outgoing!

Okay, you have your drawings bundled up and ready to go. Now what? When the package is small enough, say under 15MB, email is probably sufficient. On the other hand, if the package is large or you want discussion and collaboration, you need a cloud solution.

Autodesk weather forecast: Increasing cloud

Even if you just came out from under your rock in the past two hours, you have probably heard of the cloud, but you may not really understand what it means.

Okay, here's the truth: The *cloud* really just means the internet, which in turn really just means a huge network of interconnected computers. Some organizations have put together a large number of computers (a *cloud* of them, if you will) and connected them through and to the internet. You don't need to know the location of these computers or any specialized knowledge to access them. Collectively, they have tremendous computing and storage capabilities. So far, there are two fundamental uses for the cloud:

- **Computing:** Application software can live and run on the cloud, so your computer simply becomes a keyboard and monitor with a very long cord back to the cloud of computers. Computing can involve relatively simple applications, such as AutoCAD Web and Mobile (covered a little later in this chapter), or complex stress analysis operations, but the collective power of the cloud of computers makes it possible to do things that are impossible, or at least highly impractical, on stand-alone units.

  Probably the best example of this is the entertainment industry. Before you start producing your own blockbuster computer-generated movie, do the math. AutoCAD can produce high-quality, photorealistic images, but it can take anywhere from several minutes to a few hours to produce a single frame. Now you need 24 or 30 frames per second (fps; movies versus TV) for several hours. You soon hit geologic time frames. 3D movies, such as *Avatar* and *How to Tame Your Dragon,* were produced on rendering farms containing 20,000 to 30,000 (that's right; 20 to 30 *thousand*) computers. I've been in one: It was a warehouse full of rack after rack of main boards loaded with RAM and large hard drives. No monitors, no keyboards, no mice — just raw computing power. I have heard claims that if your personal computer costs one dollar per hour to run, several thousand cloud computers still cost you only about one dollar per hour.

- **File storage:** Cloud servers can collectively contain literally tens of thousands of hard drives, each with many terabytes (TB) of storage capacity. A secondary benefit of cloud storage is collaboration. If you set things up properly, you can share your files instantly, live, with anyone anywhere in the world who is also connected to the internet.

Both cloud computing and storage centers are located in multiple locations on most continents. Your computer automatically accesses the nearest one to minimize latency, the lag time it takes to get through the internet's wiring system.

The following sections take a look at a couple of available cloud computing solutions.

Your head planted firmly in the cloud

Besides file sharing, the other main use for the cloud is file storage. Many commercial sites store your files for you for a range of fees, including free sites such as Google Drive.

The pros

Here are some advantages of using the cloud for file storage:

- **Security:** Do you keep your money under your mattress, or do you put it in a bank? The *pro* side claims that your personal computer is the least-secure place in the world to keep your data files. Viruses sneak in, hard drives fail, computers get stolen, buildings burn down, and you may not (or, more likely, don't) have a proper backup procedure.

  A competent cloud server service is probably using RAID (Redundant Array of Independent Disks) technology. The users see it as a single disk drive, but, in reality, files are automatically copied to several physical drives. If one drive fails, the RAID operating system automatically disconnects that drive and continues working with the other drives. A technician swaps out the dead drive on the fly and plugs in the new one, and the RAID system automatically refreshes it from the good drives. Users don't even know that anything has happened.

- **Convenience:** Because the cloud server is part of the internet, you can access your files anytime from anywhere in the world. Some services, such as A360, can allow simultaneous access, in which several people scattered all over the globe can access, discuss, and edit files at the same time.

Here are some disadvantages of using the cloud for file storage:

- **Security:** Using a cloud service is based on promises over which you have no control. The cloud service might not actually be following proper RAID and backup procedures. Files might be stored in an easily visible native format instead of being encrypted. The site might not be secure against hackers and intruders. There may be a conflict over who really owns your files. Sites are known to raise their rates and then hold your files for ransom until you agree to pay the new rates. Providers can go bankrupt, and even suffer from sitewide fires.

  WARNING

  Anything you post to anywhere on the internet is there *forever*. You may think you have deleted a file after you uploaded it, but all you did was flag the occupied space as available. The raw data will survive until its disk space is needed and a new file overwrites it. Some cloud storage sites promise to not actually erase files you delete so that you can still recover them. By the way, this is also true for local hard drives. Many an embarrassing file has been recovered from discarded drives. In addition, RAID servers by definition hold multiple copies of the same file, or a competent cloud service makes regular backups, or both. Some cloud services keep the last ten versions of each file for up to a month.

The bottom line is that it's probably not a good idea to post things like the drawings of your unique nuclear reactor design, or those "interesting" photos of you and your partner.

- **Convenience:** I want to slightly amend one statement from the pros so that it now reads, ". . . you can *usually* access your files *almost* anytime from *almost* anywhere in the world." Autodesk is picky about which web browsers you can use. At the time of writing, the list specifies only the 64-bit versions of Google Chrome, Mozilla Firefox, and Chromium-based Microsoft Edge.

  As well, your device needs internet access, which cannot be always assumed. Here's the scenario: Your four-hour flight to visit an important client has just taken off. You have about an hour of work to do on one last critical drawing. You fire up your laptop, launch AutoCAD, and open — oh, poop! (or words to that effect). No internet connection. No cloud service. No files. You're doomed. Okay, the good news is that AutoCAD's web and mobile apps, outlined next, can overcome this problem with a little foresight and planning.

WARNING

As indicated earlier, Autodesk is continually increasing its online capabilities, and again as per my previous CYA statement, anything following this paragraph is subject to change. Because of this, I describe only general capabilities and don't go into detail about how to make things work. Don't you just hate it when a Help function says "You can do this and you can do that" but doesn't tell you how?

AutoCAD Web and Mobile

The bottom line as of this writing is that if you have a valid, current licensed copy of AutoCAD and an Autodesk account, you can use the Share function to do the following:

- Open a free storage at Autodesk.com.
- Save drawings to this account.
- Allow other people to have access to them for up to seven days.

- Open and edit these drawings from anywhere you have internet access, even by using a mobile device that doesn't have AutoCAD installed on it. (A version of AutoCAD runs on the Autodesk website so it does not need to be installed on the mobile device.)
- Allow other people to have access to the drawings, even with a mobile device that doesn't have AutoCAD installed on it.
- Set up a conference call for group review and editing sessions, even with a mobile device that doesn't have AutoCAD installed on it.
- Take photos with your mobile device and insert them directly into drawings.
- If you know you won't have an internet connection for a while, you can download files to a computer that has AutoCAD installed and edit them. When you are back online, you can sync the files.
- Mark up drawings with the Trace function (added to AutoCAD 2023).

My brain hurts.

5

On a 3D Spree

IN THIS PART . . .

Understand the basics of modeling in 3D.

Set up your 3D working environment.

Generate 2D drawing views and realistic renderings from your 3D models.

Access other CAD file formats.

» Accessing modeling tools

» Specifying coordinates in 3D

» Changing the current working plane interactively

» Managing user-defined coordinate systems

» Navigating through three dimensions

» Applying visual styles

Chapter 21

It's a 3D World After All

The addition of the Z coordinate releases your design work in AutoCAD from the planar world of two dimensions into a more lifelike three-dimensional space. AutoCAD's 3D capabilities have grown by leaps and bounds since AutoCAD 2007 was released with its souped-up 3D engine. Not only have the AutoCAD model creation and editing tools advanced, but it is now also a dab hand at visualization and rendering. You can view 3D models from any angle and slice through them to see what they look like on the inside, all while working with a pretty realistic visualization of the model. And finally, because the world of technical drawings is still a two-dimensional one, you can use AutoCAD's viewing options to generate 2D drawing views from 3D models semi-automatically.

If you're an AutoCAD LT user, just sit out most of this chapter and the next two. One of the major areas where AutoCAD LT differs from regular AutoCAD is in its extremely limited 3D functionality. Even viewing 3D models is more difficult than necessary in AutoCAD LT. Those users can acquire nearly all the 3D *viewing* capabilities of the full version of AutoCAD with a separate program. Just go to `www.autodesk.com/products/dwg/viewers` and access Autodesk Viewer online, or download the free DWG Trueview (click the Windows Viewer tab).

This chapter takes a look at some of the tools available in AutoCAD's 3D Modeling workspace and introduces you to many of the general concepts of creating 3D objects. You also discover how to look at your model from different viewpoints, and how to change the way it appears onscreen.

Three-dimensional modeling and visualization is a lot more demanding on computer hardware than 2D drafting; 3D models tend to be bigger than 2D drawings, so you may need more disk space. And you'll certainly be happier with more than the bare minimum of RAM required for 2D in AutoCAD. In the online extras for this book, I list AutoCAD system requirements and point out the increased resources needed for 3D work. In particular, note the desirability of using an engineering graphics card or a gaming card for heavy-duty 3D work.

The 3.5 Kinds of 3D Digital Models

Here are the three basic types of 3D computer models:

- **Wireframe:** Consists of edges and vertices only. Although they do occupy 3D space, they're rather unrealistic and difficult to read correctly because you can't tell which is the front of the model and which is the back. AutoCAD can help you create wireframe models, but nobody does so any more. You need only to view a complex 3D model in wireframe mode to understand why.
- **Surface:** Consists of infinitely thin skins that stretch from edge to edge of a model. Think of a balloon as the surface model of a bowling ball. AutoCAD can create two types of surfaces, and hence the 3.5 reference in the section heading.
  - *Mesh surfaces* are composed of thousands and thousands of three- or four-sided faces. Mesh surfaces tend to be imprecise. Think of the mirror disco ball in a nightclub. From a distance, it looks like a sphere, but as you get closer you can see that it is made up of many small, flat patches. A terrible punster would say that they make a mesh of the model, but I for one would never stoop so low. The type of mesh surfaces most often used for 3D animation are known as *sub-d,* short for sub-division modeling.
  - *NURBS surfaces:* NURBS (Non-Uniform Rational Bezier Spline, if you must know) surfaces are far more precise (as you'd expect with a name like that!) and are frequently used in consumer product design.

  Due to space limitations, I don't cover either type of surface modeling in this book.

» **Solid:** Has edges, surfaces, and mass. It's a true 3D model, so you can do interference checking between them to make sure that the parts of an assembly will fit together properly. You can also find the surface area, volume, center of gravity, and moments of inertia of 3D models. Solid models are also known as *b-reps,* short for boundary representation.

WARNING

Unfortunately, AutoCAD always assumes that the material has a density of 1. If your real-life part will be made from something other than water, you need to multiply the mass values by a suitable conversion factor, depending on the material being used. AutoCAD has several commands related to applying "materials" to solids, such as wood and marble, but these materials are ones that give the model a life-like appearance and do not specify the density.

Although surface and solid models are created and modified differently, the 3D objects that make up the models have the same elements that AutoCAD calls sub-objects. Following are the subobjects that make up any 3D object:

» **Vertex:** The corners of 3D models. A vertex is defined by a single X,Y,Z coordinate located at each corner of a 3D object with flat sides. Curved surfaces like spheres have no vertices.

» **Edge:** The boundary of a face between two vertices. Every flat side has at least three edges. Again, shapes like spheres have no edges.

» **Face:** The surface area bounded by three or more edges, except for a curved surface like a sphere, which is represented a single face over its entire surface.

In summary, every face has at least three edges, and every edge has at least two vertices — except for the face on a sphere. AutoCAD lets you select subobjects when you hold down the Ctrl key and then select an edge, a vertex, or a face. Figure 21-1 shows a simple 3D object (a six-sided cubic shape created with the AutoCAD BOX command) and its subobjects.

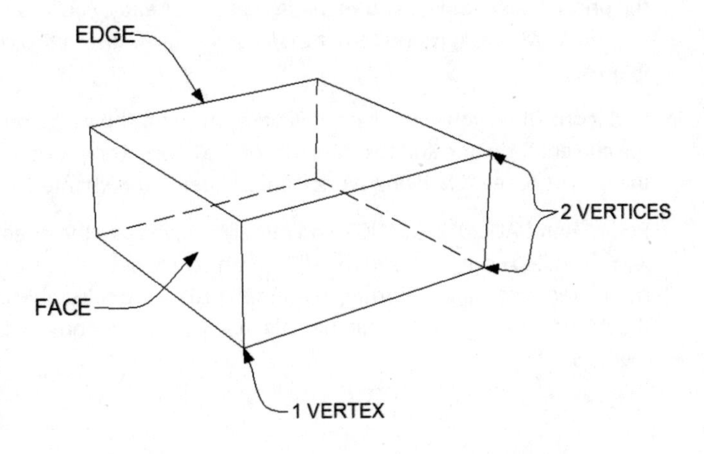

FIGURE 21-1: Subobjects of a 3D object.

Tools of the 3D Trade

AutoCAD includes a plethora of tools for creating, editing, viewing, and visualizing 3D models. You can find the most commonly used tools for 3D modeling on the 3D Modeling workspace's Ribbon.

AutoCAD comes with two workspaces for working in 3D, 3D Modeling and 3D Basics. I suppose the idea is that when you're new to 3D, you get your feet wet with the 3D Basics workspace. Then, after you gain some experience, you graduate to the 3D Modeling workspace. I think that the 3D Basics workspace is a waste of time; my preference is to have the whole enchilada available from the get-go. In this book, I work exclusively with the 3D Modeling workspace.

AutoCAD's 3D environment is a lot like its 2D version but with important differences. You should be aware of the following program settings and elements:

- **Hardware acceleration:** Graphics functions, such as changing visual styles and running 3DOrbit, can be processed on the main CPU — or, if hardware acceleration is enabled, on the graphics card. Turning on hardware acceleration, if your graphics card supports it (which I cover in the next section), nearly always improves 3D performance.
- **Workspaces:** AutoCAD comes with two 3D workspaces. I highly recommend that you use the 3D Modeling workspace. For more about the tools in the 3D Modeling workspace, see "Entering the third dimension," later in this chapter.
- **Drawing window:** The drawing window's background, cursor, and dynamic tooltip colors change when you switch between 2D and 3D views, and in 3D views when you switch between parallel and perspective projection. Click Options on the Application Menu, click the Display tab, and then click the Colors button to access color settings for the different interface elements. Note that I've tweaked the background color on my system so that it's clear on the printed page. Unless you've made changes already, you'll be looking at a very dark gray background rather than the light gray one in this chapter's figures.
- **UCS icon:** When you're working in 3D, the UCS icon shows the orientation of the current X,Y plane and the direction of the Z-axis. For information on using the UCS icon, see "Changing Work Planes," later in this chapter.

  Before AutoCAD 2012, the UCS icon merely showed you the orientation of the X-, Y-, and Z-axes. If you used the UCS command to set a new UCS, the icon reoriented accordingly. Starting in AutoCAD 2012, the selectable UCS icon is in the driver's seat in that you can directly manipulate the icon to change the UCS.

Warp speed ahead

When you start AutoCAD for the first time after installing it, the program assesses your computer's graphics card and — if it's up to the task — enables hardware acceleration.

The first time you run AutoCAD after installing it, a yellow notification balloon appears, reporting on the tests AutoCAD ran in the background on your graphics card. You can fine-tune the settings by entering the GRAPHICSCONFIG command. Autodesk tests a range of cards and drivers from AMD and Nvidia. To find out whether your graphics card is supported for AutoCAD, check out the graphics hardware list online at `http://usa.autodesk.com/adsk/servlet/syscert?siteID=123112&id=18844534`.

With hardware acceleration, you can change the view of your model — even models with lighting and applied materials — in real time, with no deterioration in quality. Hardware acceleration also boosts the overall visual quality of drawing objects as well as regeneration performance.

If your graphics card is unsupported, you may lose some or all of the shading and rendering options such as shadows, reflections, transparency, and refraction. On the other hand, current generic graphics hardware included with modern AMD and Intel CPUs is often acceptable, unless you are doing heavy-duty photorealistic renderings of things such as complex machinery or highly detailed architectural models.

In any case, if AutoCAD automatically turns on hardware acceleration, you are probably good to go.

Entering the third dimension

If you're new to the 3D game and you've been working in 2D until now, you need to do a couple of things before you can start a new 3D model in AutoCAD: You have to change the workspace, and then you have to open a new file by using a 3D template. The following steps explain how:

1. **Click the Workspace Switching button on the status bar, and then choose 3D Modeling.**

   Ribbon tabs flash on and off, and soon AutoCAD settles down to display the Ribbon, as configured for the 3D Modeling workspace with a few additional panels.

2. **Click the Application button and then choose New; then click Drawing.**

   The Select Template dialog box appears.

3. **Choose `acad3d.dwt` if you're working in Imperial units or `acadiso3d.dwt` if you're working in metric. Click Open.**

   A 3D modeling space appears (see Figure 21-2). You are looking at it at an angle from above, rather than straight down at the drawing area as you do in 2D.

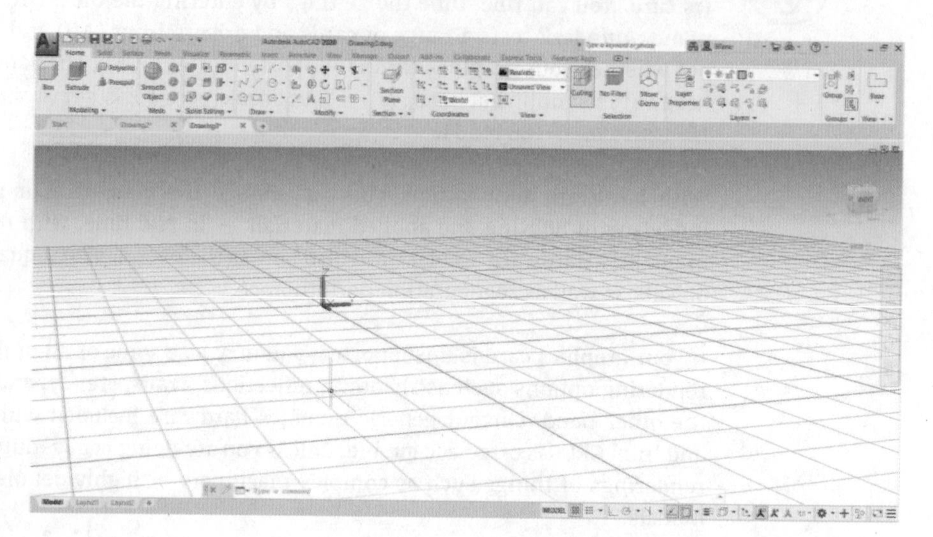

FIGURE 21-2: Ascending the Z-axis.

TIP

To switch from 3D to the 2D world, simply click the Workspace Switch button on the status bar and choose Drafting and Annotation.

Untying the Ribbon and opening some palettes

You can access most of AutoCAD's 3D modeling tools on the Ribbon and 3D-related palettes. The following list introduces you to the Ribbon tabs and palettes available in the 3D Modeling workspace. Note that not all items in this list are visible in Figure 21-2 and that this list omits some items that aren't unique to 3D.

» **Home:** Contains a subset of the most frequently used 3D tools for creating and editing solids, surfaces, and meshes as well as user coordinate system (UCS) command options, view tools, 2D drawing and editing commands, and layer and group tools.

- **Solid:** Contains tools for creating and modifying 3D solids. I show you some of the finer points of solid modeling in Chapter 22.
- **Surface:** Contains tools used to create, edit, and analyze NURBS surfaces.
- **Mesh:** Contains tools used to create and modify meshes, convert them to solids, and generate live sections.
- **Visualize:** Contains tools to add lighting and materials to a 3D model, and generate rendered output. Chapter 23 presents the basics of rendering in AutoCAD.
- **View:** Contains tools to navigate, visualize, and change the current view and UCS of a 3D model. I cover navigation tools, coordinate systems, and visual styles in this chapter.
- **Output:** Contains tools to output 3D models to DWF or PDF files. I cover plotting 2D drawings in Chapter 16, and outputting 3D designs to paper isn't significantly different. See the online Help system for specific information. I don't cover exporting to 3D DWF or 3D printing in this book.

The following list describes what you can do with the palettes found on the View tab of the Ribbon:

- **Materials Browser:** Apply visual materials to the objects in the current model, and manage materials in saved material libraries. These materials do not, sadly, impart mass or other physical properties to 3D models.
- **Materials Editor:** Apply, create, and modify visual materials in the current model. Chapter 23 introduces you to rendering materials.

WARNING

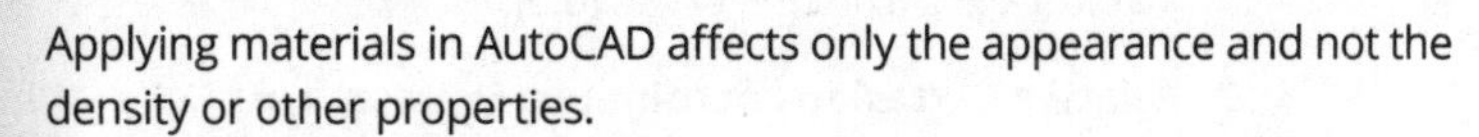
Applying materials in AutoCAD affects only the appearance and not the density or other properties.

- **Advanced Render Settings:** Manage rendering presets.
- **Visual Styles Manager:** Create, modify, and manage visual styles in the current model.

The following palettes are found on the Visualize tab of the Ribbon:

- **Lights in Model:** Use this palette to manage lights in the current model.
- **Sun Properties:** Use this one to specify the lighting settings for the Sunlight system in the current model.

Modeling from Above

So far in this book, I've explained how to work with X and Y coordinates only. X and Y coordinates by themselves define a two-dimensional plane known (logically enough) as the *X,Y plane*.

In this chapter and in Chapters 22 and 23, you supply a third value — a Z coordinate — to locate geometry above or below the X,Y plane. You use the same methods to specify Cartesian (X,Y,Z) coordinates as the ones I set out in Chapters 1 and 8. You just add a comma and another number, such as 1,2,3. The following techniques can be used to specify points in 3D:

- Coordinate and direct distance entry
- Point filter
- Object snap
- 3D object snap
- Object snap tracking

Using 3D coordinate input

I introduce you to 2D coordinate entry in Chapter 8. To recap, the input formats are

- **Absolute Cartesian coordinates:** Expressed as *X,Y* — the distances along the X- and Y-axes from the origin (0,0).
- **Relative Cartesian coordinates:** Expressed as *@X,Y* — the distances parallel to the X- and Y-axes from the last point.
- **Relative polar coordinates:** Expressed as *@d<a* — the distance and angle in the X,Y plane from the last point.

Do *not* enter spaces after the commas. AutoCAD treats the spacebar the same as Enter, and a space messes up the coordinates.

Although you can use these 2D coordinate input methods in 3D, you usually have to give AutoCAD just a little more information when you want to work in three dimensions. You can use absolute or relative Cartesian coordinates by simply adding a Z coordinate to the end. In addition, there are two 3D–only coordinate formats, both based on 2D polar coordinates. You can enter 3D coordinates with the following methods:

- **Absolute Cartesian:** Expressed as *X,Y,Z*. Working with absolute coordinates in 3D should look familiar because the input is like working in 2D except that you add a Z coordinate at the end.
- **Relative Cartesian:** Expressed as *@X,Y,Z*. Relative coordinates in 3D work just like they do in 2D except that you add the Z coordinate.
- **Cylindrical:** Expressed as *@d<a,z*. Cylindrical coordinates are similar to relative polar coordinates (*d* is the distance and *a* the angle from the last point *in* the X,Y plane) with the addition of a Cartesian Z coordinate to locate the point above or below the X,Y plane.
- **Spherical:** Expressed as *@d<a1<a2*. Spherical coordinates are also based on relative polar coordinates, but instead of a Z coordinate to specify a location above or below the X,Y plane, you specify an angle from the X,Y plane.

Using point filters

Coordinate entry (as described in the preceding section) is fine if you already know the coordinate values or distances in 3D space where you want to locate your points. Often, however, you may need to derive 3D points from existing geometry without knowing those exact values. Using point filters, you can locate new points based on existing points. You can use point filters, such as .yz and .xyz, to construct 2D and 3D coordinates.

When AutoCAD prompts you to specify a point, you can either enter point filter values at the command line or bring up the right-click menu and choose a filter option from the Point Filters submenu. After selecting a point filter option, specify a point to filter out part of the coordinate value and then enter the requested value. For example, enter the .xy filter to provide a Z coordinate value when specifying an X,Y point on the current working plane.

Object snaps and object snap tracking

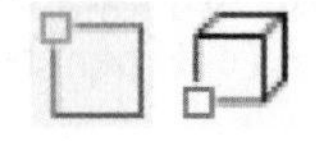

Object snaps allow you to accurately specify points on existing objects in a 3D model. You can use endpoints, midpoints, and center object snaps on 3D objects, and AutoCAD provides second set of object snaps specific to 3D modeling.

Three-dimensional object snaps allow you to specify points on 3D objects that regular object snaps don't recognize, such as at a vertex or at the center of a face. You enable 3D object snaps by clicking the 3D Object Snap button on the status bar. Right-click the 3D Object Snap button and choose the running 3D object snap modes that you want to use.

Object snap tracking allows you to calculate points that are not on an object when using object snap points on existing objects. While working in 3D, you can track points on not only the current X,Y plane but also along the Z-axis. While tracking points along the Z-axis, AutoCAD provides feedback in the form of a tooltip that lets you know you're moving along the Z-axis in the positive or negative direction.

The ELEVATION system variable allows you to specify a height above or below the current working plane. When you enter a 2D coordinate, AutoCAD uses the value assigned to the ELEVATION system variable (by default, it's 0) to create a 3D coordinate. You can set the OSNAPZ system variable to 1, to substitute the Z coordinate value of a point specified by using an object snap with the value of ELEVATION.

Changing Work Planes

Come fly with us while I explain changing planes in AutoCAD. Just step through this metal detector over here while I x-ray your bags.

Okay, it's not *that* kind of plane. Much of 3D modeling is done on 2D planes, specifically on faces. Faces can be at any angle in space, so it is important to align the drawing plane with a specific face quickly. This section covers the user-defined coordinate system (UCS) to specify the current working (or X,Y) plane. The UCS icon, which is usually displayed in the lower-left corner of the drawing window, shows you the orientation of the X, Y, Z axes of the current work plane. Sometimes, the UCS icon is found at the origin (0,0,0) or is turned off. I cover UCSs in Chapter 8.

Displaying the UCS icon

The UCS icon is displayed in both 2D and 3D views by default. While it's useful only on rare occasions in 2D, it's vital in 3D because you can lose your bearings easily when working in 3D. I and my technical editor strongly recommend that you keep the UCS icon in your sights at all times. To toggle it on or off, click the UCS Icon icon (yes, the UCS icon is labeled *UCS Icon*) in the Viewport Tools panel of the View tab on the Ribbon.

In older releases, the UCS icon merely pointed out the directions of the current coordinate system. When you set a new UCS in model space, AutoCAD adjusts the UCS icon to show you the new orientation of the X,Y plane and Z-axis.

You can select the UCS icon and see multifunction grips that let you change the UCS. Drag the square grip at the origin to move the icon; drag the round grips on the legs to reorient the coordinate system. There's even a right-click menu that lets you specify preset orientations.

In the next section, I show you how to create a new UCS by selecting and manipulating the UCS icon.

Adjusting the UCS

Every drawing you create uses the universal World Coordinate System (WCS), in which the origin is at 0,0,0 and the positive X axis goes to the right. This is the standard work plane for working in 2D and starting a new 3D model. When working in 3D, however, you'll need to create other work planes. For example, you might want to drill a hole on the angled face of a wedge. With the WCS, this is painfully difficult. But with a correctly placed user-defined coordinate system (UCS), it becomes laughably simple. UCSs are necessary to draw objects on work planes other than the WCS.

In the real world, think of the WCS as a latitude and longitude that gives you an absolute location on the planet. On the other hand, street names, house numbers, and hotel room numbers are local systems that are equivalent to UCSs. Many streets and hotels have the same house and room numbers, and likewise AutoCAD can have many UCSs that work in a specific part of 3D models.

Speedy modeling with Dynamic UCS

A couple of releases back, AutoCAD introduced Dynamic UCSs. What you get is a temporary coordinate system that changes as you move the mouse pointer over different planar faces of a 3D object.

To toggle Dynamic UCS on and off, click the Allow/Disallow Dynamic UCS button on the status bar, or just press the F6 key.

With Dynamic UCS, you focus on modeling, not on creating a UCS. The older, static named UCSs also allow you to work on different work planes, but they take an effort to set up and switch between. Those efforts can distract you when you're focused on 3D modeling.

Follow these steps to draw a circle on the side of a 3D solid box. Before you start, draw a 3D box with the BOX command.

1. **On the Home tab of the ribbon, click the BOX icon in the Modeling panel, or enter the BOX command at the keyboard.**
2. **Pick a point in the drawing to locate the lower corner of the box:**

```
Specify first corner or [Enter]:
```

3. **Pick another point to show the size of the base of the box:**

```
Specify other corner or [Cube Length]:
```

4. **Finally, pick a point to show the height of the box:**

```
Specify height or [2Point]:
```

5. **Press F6 to make sure that Dynamic UCS is enabled.**

   Dynamic UCS works with both 2D and 3D commands that let you draw and edit on faces of 3D models. It works only on flat faces, not on curved faces like spheres.

6. **Start the Circle command.**
7. **Pass the cursor over each face of the 3D solid box.**

   Note that AutoCAD highlights the face in blue and gives it a heavy blue border. This signals the face on which the dynamic UCS will be aligned.

8. **Click the face to fix the UCS with the face.**

   AutoCAD automatically aligns the UCS with the face, as shown in Figure 21-3.

9. **Finish the Circle command.**

   The circle is drawn on the face on the 3D solid, no matter how awkwardly angled. After you end the command, the previous UCS or the WCS restored.

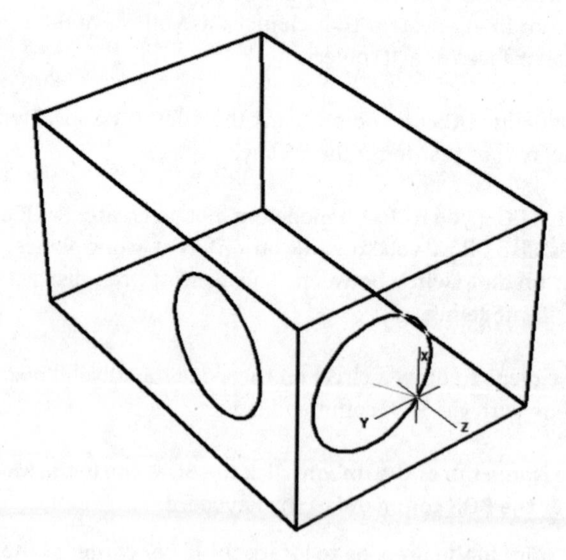

FIGURE 21-3: Dynamically create a UCS on a 3D solid.

Name that UCS

Sometimes it makes more sense to use the old UCS command to define named UCSs, just like you can use named views to return to specific views. The UCS command offers ten options to help you define a new UCS. These are the same options in the shortcut menu when you right-click the UCS icon. Access the UCS command options from the Coordinates panels on either the View tab or Home tab of the Ribbon.

After you select one of these UCS options on the Ribbon, follow the command prompts at the command line or the Dynamic Input prompt.

- **World:** Align the UCS to match the WCS.
- **Face:** Align the UCS to the face of a 3D solid.
- **View:** Align the UCS so that the X plane is perpendicular to your current viewing direction.
- **3 Point:** Specify a new origin for the UCS, and then the positive direction of the X- and Y-axes. Alternatively, use the multifunction grips on AutoCAD's UCS icon to create a new UCS by moving and realigning the icon.

For more information on all the UCS command's options, refer to AutoCAD's online Help system.

The more UCSs you make in your drawing, the more you need help managing them. AutoCAD offers a handy-dandy UCS dialog box for doing just that. Open it by clicking the dialog box launcher (the little arrow at the right end of the panel label) on the Coordinates panel of the View tab or Home tab, or simply type **UCSMAN** and press Enter. The three tabs in the UCS dialog box are

- **Named UCSs:** Lists world coordinate system and other types of user coordinate systems. Set a UCS current with the Set Current button, or right-click a UCS to rename or delete a named UCS. You can't rename or delete the World UCS.
- **Orthographic UCSs:** Lists the six default orthographic coordinate systems (front and back, left and right, top and bottom) relative to the WCS. These UCSs are automatically created by AutoCAD and can't be deleted or renamed.
- **Settings:** Controls properties of both the UCS icon and the UCS.

After you define a UCS that you think you might want to use again, you can save it in the UCS dialog box. I explain how in the following steps, which you begin by creating a solid box:

1. **Start a new 3D drawing by selecting `acad3d.dwt` (or `acadiso3d.dwt` for the metric crowd) for the template, and ensure that the 3D Modeling workspace is current.**

   Refer to the steps in the "Entering the third dimension" section, earlier in this chapter, if you need a refresher.

2. **On the Modeling panel of the Home tab, click Box.**

   AutoCAD prompts you:

   ```
   Specify first corner or [Center]:
   ```

3. **Type** 0,0,0 **and then press Enter.**

   AutoCAD anchors the first corner of the box at the origin of the WCS and prompts:

   ```
   Specify other corner or [Cube Length]:
   ```

4. **Drag the cursor away from the first corner and click a point to set the length and width of the box.**

   Exact distances don't matter in this example. AutoCAD prompts you:

   ```
   Specify height or [2Point]:
   ```

5. **Drag the cursor upward from the second corner and click to set the height of the box.**

   AutoCAD creates the 3D box and exits the command.

Then you define the UCS by following these steps:

1. **Move the cursor over the UCS icon.**

   The UCS icon shows the orientation of the world coordinate system. As you move the cursor over the icon, it turns a greenish-gold color, indicating that it can be selected.

2. **Click to select the UCS icon.**

   A square, multifunction grip appears at the origin, and round, multifunction grips appear at the ends of the icon's legs.

3. **Move the cursor over each multifunction grip and look at the grip menus.**

   Hovering over one of the round grips at the end of a leg lets you choose between realigning the selected axis and rotating the UCS around one of the unselected axes. Hovering the mouse pointer over the origin grip lets you move the UCS origin to a new location and either keep the current alignment of the X- and Y-axes or realign them. The third grip option, World, restores the WCS.

Now you use the UCS icon's multifunction grips to set a new UCS.

4. **Click the UCS icon to select it, and then move the cursor over the square, multifunction grip at the origin.**

   The cursor jumps to the origin, and the grip menu appears.

5. **From the grip menu, choose Move and Align.**

   AutoCAD prompts you:

   ```
   ** MOVE AND ALIGN **
   Specify origin point or align to face, surface, or mesh:
   ```

6. **Move the cursor to a different corner of the box, and when the UCS icon origin is over the corner, click to set the new origin.**

   If you want, you can drag the round grips on the axes to realign the new UCS.

Finally, you have to save it. Follow these steps:

1. **On the Coordinates panel of the Home tab, choose UCS, Named UCS.**

   AutoCAD displays the UCS dialog box.

2. **With the Named UCSs tab current, select Unnamed in the UCSs list.**

   The new, unnamed UCS is the current UCS in the drawing and is listed at the top of the list.

3. **Right-click Unnamed and choose Rename from the shortcut menu.**

   An in-place editor is displayed that allows you to rename the UCS.

4. **Type a name for the new UCS and press Enter.**

   The Unnamed UCS is renamed (see Figure 21-4).

5. **Click OK.**

   The UCS dialog box closes, and the new UCS is saved in the drawing.

To switch back and forth between the two UCSs, simply select the one you want to use from the Coordinates panel:

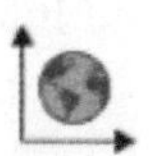

- **Use WCS:** On the Coordinates panel of the Home tab, choose UCS, World.
- **Use your custom UCS:** On the Coordinates panel of the Home tab, click in the Named UCS drop-down list and choose the name of the UCS you just saved.

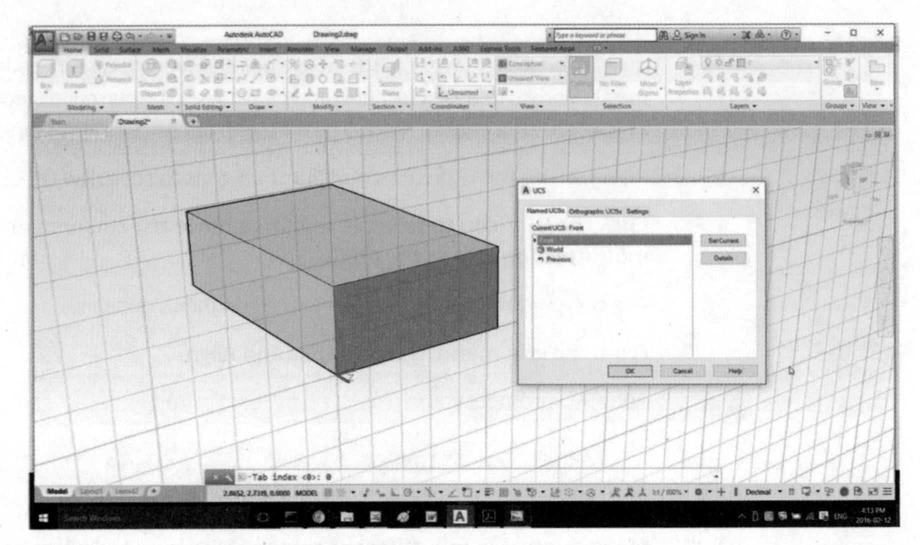

FIGURE 21-4: Naming a custom UCS.

TIP

AutoCAD likes giving you lots of choices. You can also restore a named UCS from the UCS drop-down list at the bottom of the ViewCube. I fill you in on the ViewCube in the "Taking a spin around the cube" section, later in this chapter. Or you can right-click the UCS icon, choose Named UCS, and then choose the UCS from the menu.

Navigating 3D waters

If you're new to 3D in AutoCAD, you may be wondering how to look at whatever it is you're modeling from whatever angle you desire.

The easiest way to switch to viewing from a different direction is to use ViewCube (shown in Figure 21-5) to switch to a standard orthographic 3D view or an isometric view.

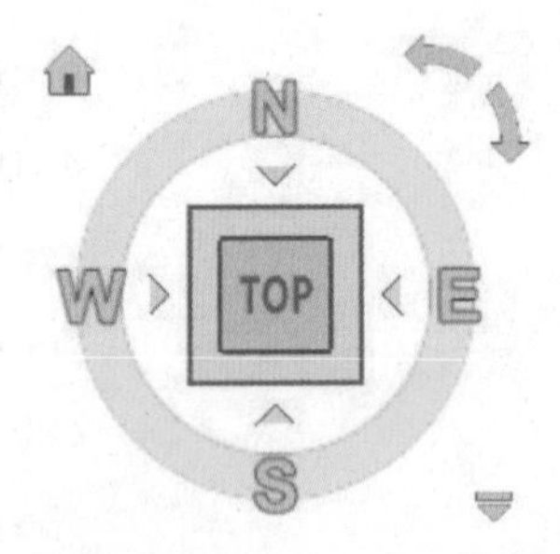

FIGURE 21-5: The 3D ViewCube.

- The six standard *orthographic* (straight-on) views are Top, Bottom, Left, Right, Front, and Back.
- The four standard isometric views are SW (left-front), SE (right-front), NE (right-back), and NW (left-back). An *isometric* view is one in which you see the object from above — or below.

Before I explain the ViewCube, however, here's a bit of background information for you.

TECHNICAL STUFF

An isometric view is an unrealistic 3D view because all lines that are parallel in reality are also parallel in the view. In a perspective view, lines that are parallel in reality appear to meet at a vanishing point, as parallel lines *appear* to do in reality. (Think of railroad tracks heading across the prairie.) Isometric views are parallel projection views, and you can view 3D models by using those views in either the full version of AutoCAD or AutoCAD LT. Perspective views are more commonly used in architectural applications, and parallel views are more commonly used in mechanical design.

WARNING

This warning is for AutoCAD LT users. Users of the full version of AutoCAD don't need to worry. AutoCAD LT can open drawings created in the full version. Those AutoCAD drawings can be saved with perspective projection current, and if you open such a drawing in AutoCAD LT, you can see it in perspective. You can switch from perspective to parallel projection by changing the PERSPECTIVE system variable's value from 1 to 0. *Be careful* if you do this, because you can't change it back again! You could use UNDO to reverse the change, but if you save and close the drawing, you're sunk.

The six orthographic and four isometric views work well for showing 3D models of common objects such as mechanical components and buildings. You can also change to *plan view,* which is a top-down view of either the world coordinate system or a UCS.

AutoCAD LT has limited 3D viewing capabilities. The same preset views are in both AutoCAD and AutoCAD LT, and you can also use the Viewpoint Presets dialog box (ddVPoint command) in which you set a viewing position by specifying angles in and from the X,Y plane. Finally, there's the really old command-line-only VPOINT command, left over from last-millennium releases. Man, that one's so ugly to use it's not even on the Ribbon!

Orbit à go-go

Preset views are fine for many 3D modeling tasks, but if you really want to have fun, 3DOrbit (not in AutoCAD LT) is your ticket to it. Orbiting a 3D model in AutoCAD is similar in concept to orbiting Earth in a satellite, only a lot cheaper. The three orbiter modes are Constrained, Free, and Continuous.

Free Orbit displays an *arcball* on the screen in the form of a circle representing a sphere around the object. You click various places inside, outside, and on the arcball and then drag to change the 3D view. The idea is that you're spinning an imaginary sphere containing the model. As you drag the cursor, AutoCAD updates the screen dynamically.

Constrained mode is pretty much like Free mode with training wheels in that you can orbit only horizontally and vertically.

Continuous mode lets you leave a mode spinning continuously. To change the orbit direction and speed, simply click-drag-release the mouse. The model orbits in the direction defined by the mouse motion, and the orbit speed is determined by how fast the mouse was moving when you released its button.

To use 3DOrbit during any command, hold down the Shift key with one hand, hold down the middle mouse button with your other hand, and then move the mouse. While you're using the 3DOrbit command, right-click and select Other Navigation Modes to switch between the three modes on the fly.

You might also want to experiment with different projection modes:

- *Parallel* projection is the default AutoCAD projection. Lines that are parallel in the 3D object remain parallel in the projected view on the screen.
- *Perspective* projection makes objects look more realistic (for example, train tracks appear to converge in the distance), but lines that are parallel in the model don't appear parallel in perspective projection.

If you manage to 3DOrbit out of control and can no longer see the model, right-click to display the 3DOrbit shortcut menu and then choose Zoom Extents. The Zoom, Pan, and Preset Views options offer other ways to get the model back in your sights.

When you start orbiting with no objects selected, AutoCAD tries to update the display of everything in the model in real time; with a complex 3D model, the process can take some time. To speed things up or to simply regain your bearings, select some objects before you start orbiting. Then AutoCAD updates the display of the selected objects only. When you exit Orbit mode, the entire model redisplays based on the new viewpoint.

Taking a spin around the cube

With so much talk about hybrid devices these days, Autodesk decided to create its own, in the form of the ViewCube (not in AutoCAD LT). The interactive ViewCube tool (see Figure 21-6) provides visual feedback about the current viewpoint, and allows you to set a preset view current or orbit the model, restore a named UCS, and define and restore the Home view of a model.

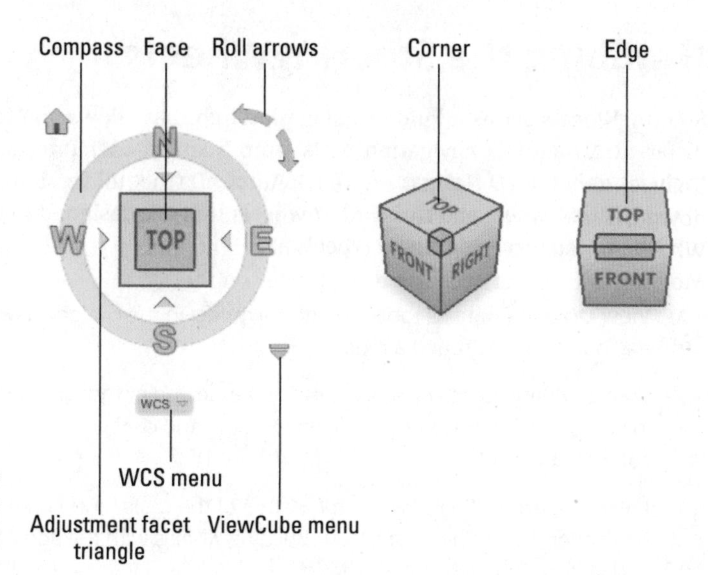

FIGURE 21-6: ViewCube, the multifunctional viewing device.

To change the view of the model by using the ViewCube, you can do one of the following:

- To align the current view with the same viewpoint represented on the ViewCube, click a corner, an edge, or a face.
- To orbit the view of the model, click and drag the main area of the ViewCube.
- To rotate the current view 90 degrees, click one of the roll arrows.
- To switch to the adjacent orthographic view indicated by the triangle, click one of the adjacent face triangles.
- To rotate the view around the center of the drawing, click one of the letters (N, E, S, W) or click and drag the compass.

Home view is a special view you can define so that you have a known reference view in the model. That way, if you lose your bearings, you always have a way to get back home. You can restore Home view from the ViewCube or from the SteeringWheels right-click menus.

Grabbing the SteeringWheels

SteeringWheels act as a kind of navigation hub that allows you to access several different 2D and 3D navigation tools from a single user interface. AutoCAD LT includes only the 2D Navigation wheel. AutoCAD (the full version) comes with the 2D Navigation wheel and three other wheels that are designed to be used with 3D modeling. The three additional wheels are

- **View Object:** Contains tools to center a model in the current view, zoom in and out, or orbit around a model
- **Tour Building:** Contains tools to move the viewpoint forward and backward, look around the model from a fixed location, and change the elevation of the current viewpoint
- **Full Navigation:** Contains a combination of the tools found on the 2D Navigation, View Object, and Tour Building wheels with the addition of tools to walk around or fly through a model

All wheels include the Rewind tool, which allows you to restore a previous view. The Rewind tool is similar to the Previous option of the Zoom command.

When a wheel is active, move the cursor over the wedge that contains the tool you want to use. For many of the tools, you either click over the tool or click and drag to use the tool. Some tools support alternative versions of themselves when you press and hold the Shift key. To get an idea of what each tool does, hover the cursor over the tool to display a tooltip message. To read more about each of the wheels and the tools on them, see AutoCAD's online Help system. To display the help that's specific to the wheels, right-click when a wheel is displayed and choose Help from the shortcut menu.

Visualizing 3D Objects

It really doesn't seem like that long ago when the only way to view 3D models was in Wireframe mode, but the last several releases have dramatically improved AutoCAD's visualization capabilities. It now has almost a dozen different display modes, called *visual styles*, that you can set with the click of a drop-down menu. Figure 21-7 shows the default visual styles, and you can add your own in the Visual Styles Manager palette.

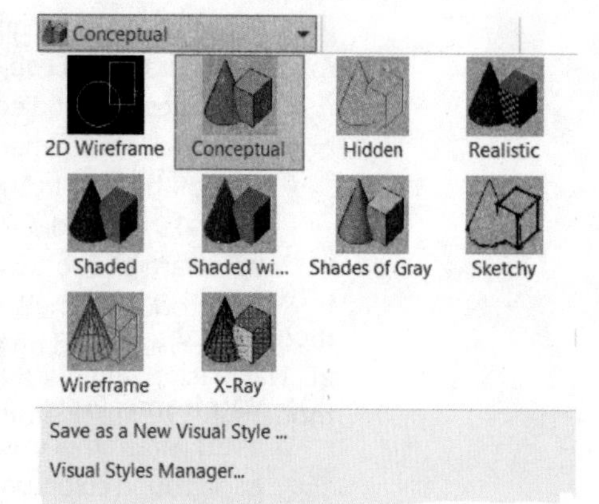

FIGURE 21-7: Some new styles to try on.

Visual styles are collections of settings that determine how a 3D model is displayed.

- **2D Wireframe:** AutoCAD's classic 2D viewing mode: full wireframe, dot-based grid, the 2D UCS icon, and no perspective. For better 3D performance, Autodesk recommends any visual style other than 2D Wireframe.
- **Conceptual:** An illustrative kind of shaded view. Colors are unrealistic and edges are heavy, but you get a good sense of the model's form with the guarantee that no surfaces will be too light or too dark. Refer to Figure 21-4 to see an example of the Conceptual visual style.
- **Hidden:** Looks slightly like Wireframe view (no surfaces are visible), but edges behind faces are hidden.
- **Realistic:** Fully shaded but not rendered visual style; edges are not displayed, and a default ambient lighting highlights the faces with different intensities of the object color. Materials and textures are visible after they've been applied. Have a look at Figure 21-8 for an example.
- **Shaded:** Similar to Realistic, but with more subdued lighting. Textures do not appear in this visual style. AutoCAD 2023 adds Shaded (Fast) for even faster shading, which is enabled automatically when your graphics board can handle it. The Shaded style might be named Shaded (Fast), which is a faster version that AutoCAD since 2023 enables when your computer's graphics hardware supports it.
- **Shaded with Edges:** Same as the Shaded visual style, except that edges and isolines are displayed.
- **Shades of Gray:** Like Shaded, except that all object colors are changed to different intensities of gray.

» **Sketchy:** Does not apply shading to the faces of 3D objects, but gives the edges of 3D objects a hand-sketched look. Somehow, I'm not sure of the logic in spending several thousand dollars on hardware and software so that you can produce something that looks like you sketched on a napkin.

» **Wireframe:** Pretty much the same as 2D wireframe, except for background color, optional perspective, and the 3D (rather than 2D) UCS icon. Figure 21-4 shows a simple model in the Wireframe visual style.

» **X-ray:** Similar to Shaded with Edges, except that Materials and Textures are applied like the Realistic visual style, and faces are set to 50 percent opacity for a see-through look.

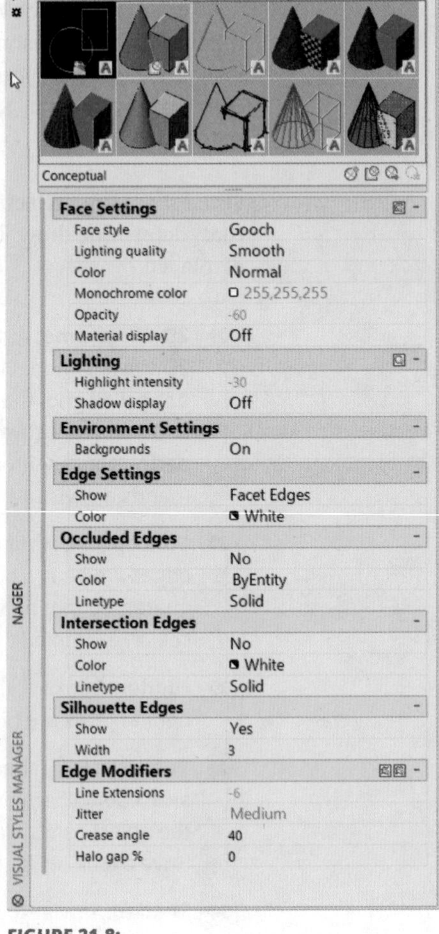

FIGURE 21-8: A manager for your visual styles.

AutoCAD's preconfigured visual styles are only the beginning. You can modify any of the styles and create new ones in the Visual Styles Manager palette (refer to Figure 21-8).

To display Visual Styles Manager, click the Visual Styles button on the Palettes panel of the View tab, or enter **VisualStyles** or **VSM** at the command line and press Enter.

On a Render Bender

The different visual styles presented in this chapter are good for helping you navigate while you're working on a 3D model. But when it comes time to impress a client, you need to use photo-realistic rendering, which I discuss in Chapter 23.

» Setting up a 3D working environment

» Creating 3D solid geometry

» Editing 3D models

Chapter 22
From Drawings to Models

For millennia, people have documented the design and construction of three-dimensional objects by drawing two-dimensional views of them. Most continued to use these classical methods in CAD drafting, because the methods are well understood and work reasonably well. After all, if 2D drawing was good enough for guys like Leonardo da Vinci and Andrea Palladio, why shouldn't it be good enough for us, right?

Nonetheless, the past decade or three has seen a trend toward creating 3D CAD models and letting the software generate the 2D views more or less automatically. This approach seems more logical, especially if the project documentation requires numerous views of the same complex object. Three-dimensional modeling is an absolute necessity when you want to create rendered views for presentation purposes.

And although AutoCAD 3D construction and visualization tools have improved dramatically over the years (trust me — you should have seen how poorly they behaved as recently as AutoCAD 2006!), 3D modeling is still a complex process that requires sophistication on the part of the AutoCAD user. Even through 3D modeling requires only one dimension more than 2D drafting, developing 3D CAD models can seem to be more complicated because it involves five more planes. Think of a sheet of paper versus the six faces of a cube. Users must master new techniques and contend with the 2D limitations inherent in most display screens and input devices. On the other hand, once you have a 3D model, generating traditional 2D drawing views — such as front (elevation), top (plan), and detail views — becomes almost trivial.

In Chapter 21, I show you how to move around 3D models that others have made, and I explain the principles of 3D coordinate systems that you need to understand to work in the AutoCAD 3D environment. This chapter introduces you to the concepts, tools, and techniques of AutoCAD 3D modeling; here, you can get your feet wet creating your own 3D objects.

Full 3D support is one main difference between full AutoCAD and AutoCAD LT. If you're using AutoCAD LT, you can look at and plot 3D models created in AutoCAD, but you can't do much 3D object creation or editing yourself. Also, viewing 3D models is less flexible in AutoCAD LT because it lacks nearly all the 3D navigation and rendering tools included in the full version of AutoCAD.

Is 3D for Me?

Traditional 2D drawings provide clues to help the viewer mentally construct a 3D model from the 2D image on paper. Multiple 2D views from different viewpoints in 3D space give experienced designers, drafters, builders, and tradespeople the information they need to make "3D sense" of 2D drawings. Design and drafting have succeeded pretty well by using 2D representations as guides to creating 3D objects. But at some point, nothing can replace a true 3D model, such as in helping someone understand how a building will look when it's constructed or how parts of an automobile fit together.

What does "using 3D in AutoCAD" mean? Fundamentally, it means that you create *models* instead of *drawings*. Rather than draft individual views of it from a limited number of viewpoints, you create three-dimensional models of objects. This 3D depiction of objects includes all the information necessary for AutoCAD to display views from any point of view. Using a properly constructed 3D model, AutoCAD can output commands to machines to create the actual 3D objects, whether they're 3D printed or created by computer-controlled machine tools.

As I explain in Chapter 21, AutoCAD lets you work with three types of 3D models: wireframes, surfaces, and solids. In most practical applications of 3D, you select only one type. While AutoCAD doesn't prevent you from mixing all three types of 3D objects in the same drawing, it does a poor job of converting between them, should the need arise. So, it's best to choose one for all the objects in the model, based on ease of construction and the intended use of the model.

After you determine the type of 3D representation to use, you decide on the appropriate level of detail and then construct the model using the commands and techniques introduced in this chapter. If you need to, you can go on from there to create any required 2D or rendered views (or both) for plotting or viewing onscreen.

Getting Your 3D Bearings

The first challenge faced by the pioneers of 3D modeling was figuring out how to see a three-dimensional model on our two-dimensional computer screens. In AutoCAD, the normal view of model space shows a single, projected, 2D view of the model — the top-down, plan view by default.

AutoCAD provides two model space capabilities that enable you to escape this visual flatland:

- **View*port*:** Carve the model space drawing area into smaller rectangular areas, each of which shows a different side of the model.
- **View*point*:** Change the point in 3D space from which you look at the model. By setting a different viewpoint in each viewport, you can look at several sides of the model on one monitor. It's like looking at one of Picasso's cubist paintings, except that what you see is more orderly.

No matter how much or how little 3D modeling you're thinking about doing, it's well worth your while to set up a template. I fill you in on drawing templates in Chapter 4. If you've ever started a new drawing, you're probably aware that AutoCAD already comes with a template for 3D modeling named `acad3d.dwt` (or `acadiso3D.dwt`, for the metrically inclined). This is fine as far as it goes, but it shows only a single view of the model. The next section explains how to improve on this template.

Creating a better 3D template

Model space viewports enable you to see several views of the model at one time, each from a different viewpoint. For this reason, model space viewports are especially useful when you're creating and editing objects in 3D. As you draw and edit, the different views help ensure that you're picking points that are located correctly in 3D space.

Chapter 12 discusses viewports in paper space, which are useful for creating layouts for use in plots and presentations in both 2D and 3D. (Paper space, by definition, shows 3D models flattened as in 2D mode.) Model space viewports, which are cousins of paper viewports, are less flexible but simpler, and they're a great help in constructing 3D models.

Model space viewports divide the screen into two or more rectangles that do not overlap. Unlike when you use paper space viewports, you can't move model space viewports or have any shape other than a rectangle. Starting with AutoCAD 2015, however, you can drag a viewport boundary to resize the rectangle. When you do this, adjacent viewports resize themselves to maintain the "no gaps and no overlaps" rule. You can't plot multiple viewports on the Model tab, but you can with layout tabs, which is what they are for. And, unlike the situation in layouts, a layer that's visible in one model space viewport is also visible in all of them.

TECHNICAL STUFF

You may hear or read references to *tiled viewports,* which is just another name for model space viewports. *Tiled* refers to the way in which model space viewports always fill the drawing area, with no gaps and no overlapping allowed. Conversely, paper space viewports are sometimes called *floating* viewports, because you can move them around, leave gaps between them, and overlap them.

One of the best 3D viewing features goes by the somewhat obscure title of the *in-canvas viewport control.* This little text label that you may have noticed at the upper-left corner of the graphic area (see Figure 22-1, for example) is a clickable control that lets you set the visual style and the view without needing to enter commands. What's especially nifty is that double-clicking the plus or minus sign toggles the drawing area between multiple tiled viewports already configured for 3D viewing and a single, maximized viewport. Out of the box, double-clicking the minus (–) sign switches to four viewports of equal size showing different views of the geometry. In the figure, you see four viewports, four different viewpoints, and four different visual styles. That's extreme for everyday work, but it gives you an idea of the possibilities.

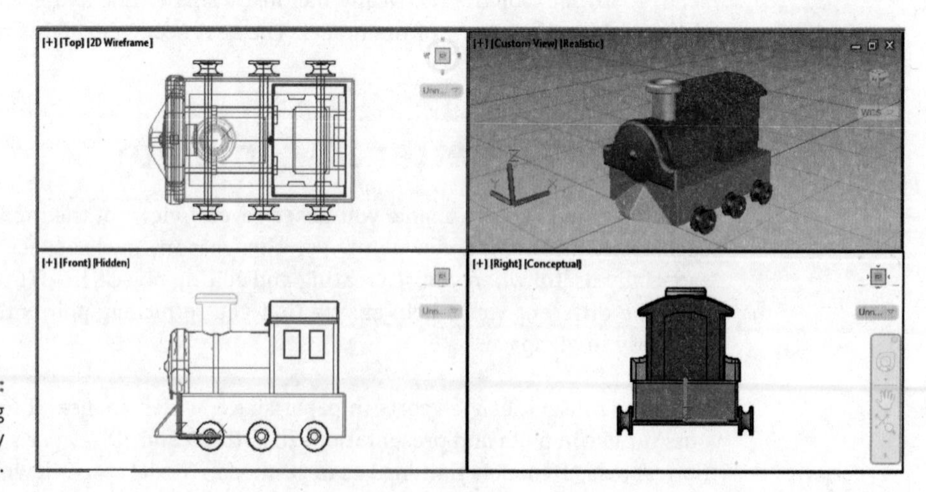

FIGURE 22-1: 3D viewing from every which way.

I highly recommend working with multiple viewports when you're modeling in 3D. That way, you get to see exactly what you're doing in all three dimensions simultaneously, in real time. My preference is to work mostly in the isometric viewport, so I make that one larger than the other three. The in-canvas viewport toggle switches back and forth between the last multiple viewport you set up and a maximized viewport. Follow these steps to set up the tiled-viewport configuration shown in Figure 22-2:

1. **Switch to the 3D Modeling workspace if you're not already in it.**

   Click the Workspace Switching button on the status bar and select 3D Modeling. Or in AutoCAD 2016 and earlier, you can also click the Workspace drop-down list on the Quick Access toolbar and choose 3D Modeling. If the Materials Browser palette opens, close it.

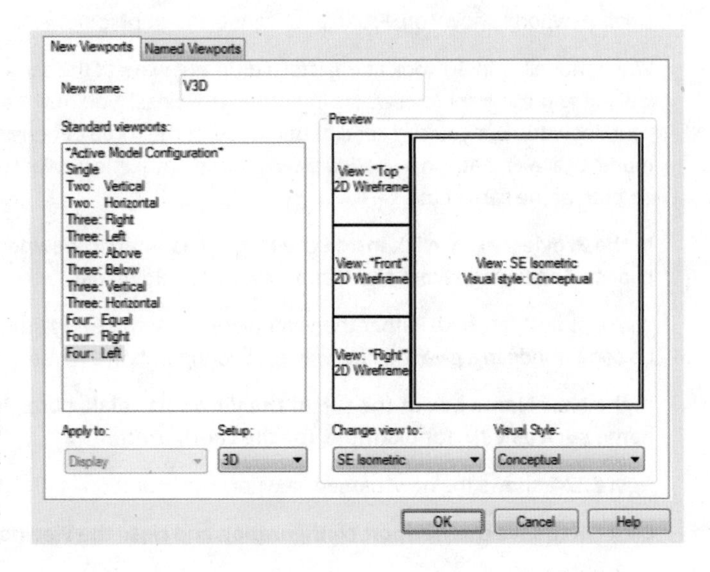

FIGURE 22-2: Setting up a 3D work environment in the Viewports dialog box.

2. **Click New on the Quick Access toolbar to open the Select Template dialog box.**

   If a new, blank drawing appears and you don't see the Select Template dialog box, someone has assigned a default template to this button in the Options dialog box. In that case, click the Application button (the Big Red A) and choose New, and then choose Drawing from the submenu.

3. **Choose `acad.dwt` (choose `acadiso.dwt` if metric is your preference), and click Open.**

   Yes, I know about the ready-made 3D templates (`acad3d.dwt` and `acadiso3d.dwt`), but trust me — starting this setup is easier from a 2D template.

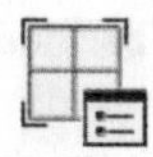

4. **From the Model Viewports panel on the Visualize tab of the Ribbon, choose Named.**

   The VPORTS command starts and the Viewports dialog box appears.

5. **Click the New Viewports tab to make it current; then choose Four: Left (my preference) or Four: Right from the Standard Viewports list box.**

   The Preview panel shows a large, squarish viewport occupying most of the work area, with three small, squarish viewports stacked on the left (or right). The default 2D setup shows the visual style of all four viewports as 2D Wireframe. I explain visual styles in Chapter 21.

6. **From the Setup drop-down list, choose 3D.**

   The visual styles remain as 2D Wireframe, but the view direction (listed as Current in 2D Setup) is now SE Isometric in the large viewport, and the three small viewports show Top, Front, and Right orthographic views.

TIP

When modeling in 3D, look at objects in different ways at the same time. When looking at orthographic views (as in drafted drawings), you may want to see all your linework. But you can get a better sense of the three dimensionality of the model by looking at it in a shaded view. By using multiple viewports, you can do both at the same time.

7. **In the Preview area, click inside the large SE Isometric viewport and then choose Conceptual from the Visual Style drop-down list.**

   You're almost finished. Rather than complete this setup every time you want to do some modeling, give the new viewport configuration a name.

8. **In the New Name box at the top of the Viewports dialog box, enter a name such as V3D, for example, for the configuration.**

   Figure 22-2 shows the new, named viewport configuration.

9. **Click OK to save the viewport configuration and close the Viewports dialog box.**

   In Figure 22-3, I turned off the grid in the three small orthographic viewports, but whether you leave it on or turn it off is a matter of personal preference.

WARNING

The Home button in the ViewCube can trip you up by changing to a view you don't expect. I introduce the ViewCube in Chapter 21. If you want to keep the current SE Isometric viewpoint, be sure to reset the Home view. To do this, simply right-click anywhere in the ViewCube and choose Set Current View as Home from the menu that appears. And because you've gone to the trouble of setting up a shaded 3D viewport, you might want to make it a little more realistic and turn on perspective mode at the same time.

TIP

Personal preference rules here, but architects tend to use perspective, and mechanical designers use parallel views.

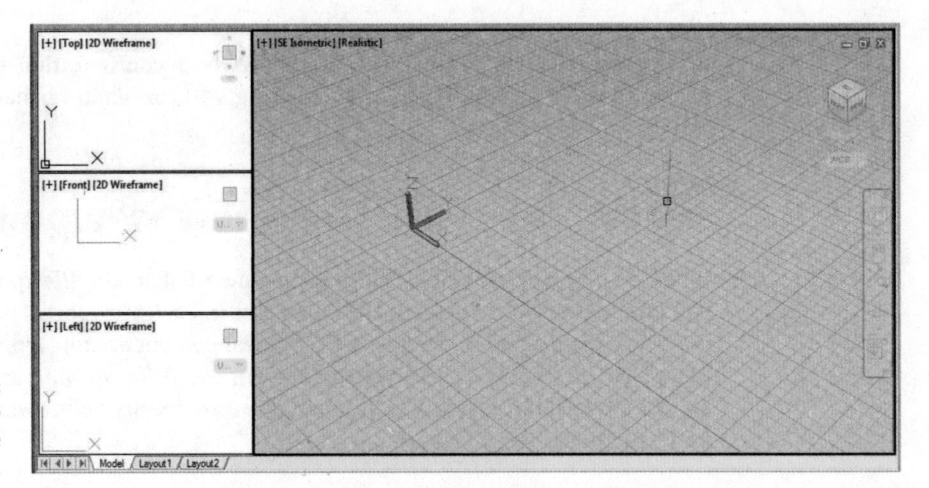

FIGURE 22-3: It's 2D and 3D all at the same time.

10. **In the large 3D viewport, right-click the ViewCube to display its shortcut menu and then choose Perspective with Ortho Faces (see Figure 22-4).**

Selecting Perspective shows the orthographic views in Perspective mode, and usually you don't want that. Selecting Perspective with Ortho Faces resets the projection mode from Perspective to Parallel when you switch to an orthographic view.

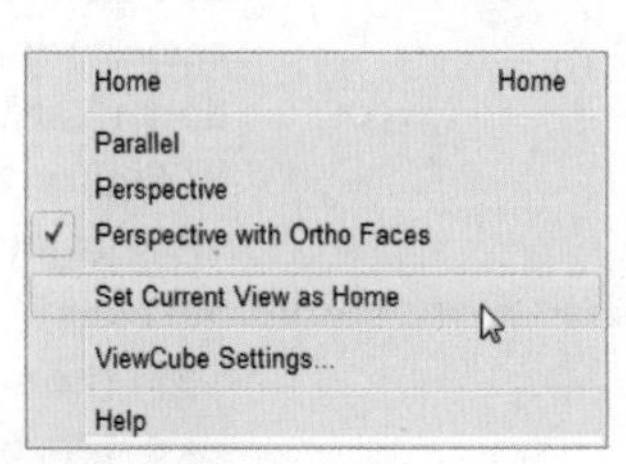

FIGURE 22-4: Changing settings in the right-click menu in the ViewCube.

11. **Repeat Step 10, this time selecting Set Current View as Home.**

Now you can readily return to this viewport configuration, keeping the same viewpoint in all viewports, perspective projection in the 3D viewport, and parallel projection in the orthographic viewports.

The last thing to do is save the configuration you just set up as a template (`.dwt`) file.

12. **Click Save on the Quick Access toolbar. From the Files of Type drop-down menu, choose AutoCAD Drawing Template (*.dwt) and enter a filename. Click Save.**

AutoCAD saves the new 3D template file in the same location as your other templates so that you can always select it when you want to do some serious 3D modeling.

To return to a single viewport later, simply click the plus (+) sign on the in-canvas viewport control.

Conversely, you can restore your four-viewport configuration at any time by clicking the Named Viewports tab, choosing V3D (or whatever name you used in Step 8), and clicking OK.

Seeing the world from new viewpoints

When you choose 3D in the Setup drop-down list in the Viewports dialog box, you direct AutoCAD to change the viewpoint in each viewport. The default viewpoints when you choose a four-viewport arrangement are top, front, right, and SE (southeast) isometric. These viewpoints work well for viewing and constructing simple models, but eventually you may want to specify your own, custom viewpoint in a particular viewport.

In both AutoCAD and AutoCAD LT, the best (but not the easiest) way to change viewpoints is to use the drop-down list in the Views panel of the View tab, as shown in Figure 22-5, to switch to a standard orthographic 3D view or an isometric view:

- **The six standard orthographic (straight-on) views:** Top, Bottom, Left, Right, Front, and Back.
- **The four standard isometric views:** SW (left-front), SE (right-front), NE (right-back), and NW (left-back). An isometric view is one in which you see the object from above, but not from too high above, as though you were hovering in a low-flying helicopter or receiving an image from a remote-operated drone, should they still be legal by the time you read this.

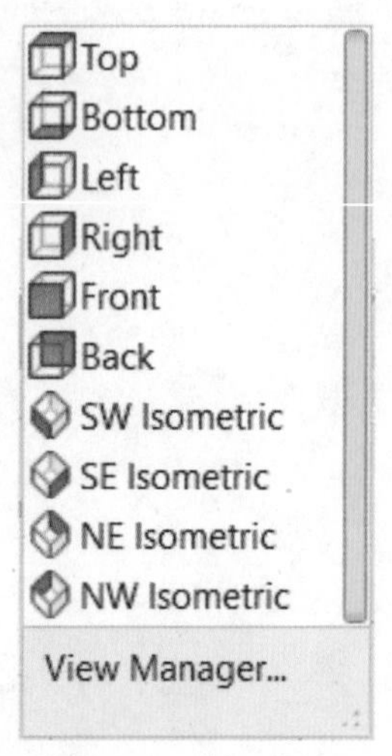

FIGURE 22-5: The preset views drop-down menu.

WARNING

The easiest (but not the best) way to change viewpoints is to click a named face on the ViewCube (not available in AutoCAD LT). I don't recommend this strategy because the ViewCube is only a *viewing* aid. You usually change to an orthographic view because you want to do some drawing or modeling there, and for that task, the user coordinate system (UCS) ought to orient itself with the view. See Chapter 21 for the skinny on UCSs. Clicking Front on the ViewCube shows you a nice front view of the model, but it leaves the UCS unchanged, and that can make drawing anything difficult. Clicking Front in the Views panel on the View tab *does* change the UCS to Front so that it matches the view and lets you draw without difficulty.

These ten views are standard because they're often used in manual drafting and rendering work. They work well for showing 3D models of common objects such as mechanical components and buildings. You can also change to plan view, which is a top-down view of either the world coordinate system or a user coordinate system. I describe coordinate systems in Chapter 21.

The full version of AutoCAD has the ViewCube, the 3DOrbit command, and the Shift+middle mouse button methods of changing 3D viewpoints. All are described in Chapter 21.

The options in AutoCAD LT are much more limited — you can set nonstandard viewpoints by typing **DDVPOINT** and pressing Enter. In the Viewpoint Presets dialog box that appears, specify these settings:

- **A viewing angle *in* the X,Y plane:** Imagine moving a camera on a dolly around an object while keeping the camera at the same elevation.
- **An angle *from* the X,Y plane:** Imagine using a boom to swoop the camera to a different height so that you're looking at the object from increasingly steep angles.

AutoCAD users can set viewpoints the same way, but it's grossly inefficient compared with the other viewing options.

TIP

By default, AutoCAD shows 3D models in 2D Wireframe mode, even if you've created surface or solid objects. If you want to better visualize which objects are in front of which other objects, especially in an isometric or other non-orthogonal view, you have a couple of options:

- Select a shaded visual style such as Realistic, Conceptual, X-Ray, or Shades of Gray, from the Visual Styles drop-down list on the Visual Styles panel of the View tab. See Chapter 21 for the lowdown on visual styles.
- Render the model, as described in Chapter 23.

AutoCAD LT doesn't include visual styles because they're not much use in 2D drafting. A drawing saved in a visual style in the full version of AutoCAD *does* display that style when it's opened in AutoCAD LT. However, you have no way to change it to 2D Wireframe (or anything else) so that you can work on it.

From Drawing to Modeling in 3D

This section introduces three techniques for creating 3D objects: drawing 3D lines and polylines, creating 3D objects from 2D geometry, and creating solids. In AutoCAD LT, you can use only the first two techniques. AutoCAD is also quite capable at surface modeling, offering both freeform mesh and NURBS surfaces. If you're interested in either of these, check out the online Help system; in the Search field, enter *creating meshes* or *surface models.* Topics include creating surfaces and meshes from scratch and creating solids and surfaces from 2D objects.

When you draw 3D objects (as when you draw 2D objects), place them on appropriate layers and use precision techniques to specify each point and distance. See Chapter 8 for more information about precision techniques.

Drawing basic 3D objects

The most basic forms of 3D geometry are wireframe-like objects created by picking points or entering X,Y,Z coordinates. These objects have no surfaces, so they look the same in 2D Wireframe mode or photorealistic renderings. They're most useful as paths for sweeps and lofts, or as edges for surface creation. Basic 3D objects include the ones described in this list:

- **Line:** The line is a 2D object, but you can draw it in 3D space by specifying a different Z coordinate for each end. You can also use lines for constructing objects in 3D space.
- **3D Polyline:** Created with the 3DPOLY command, the 3D polyline is similar to the 2D one (which I describe in Chapter 6), except that the vertices of a 3D polyline can have different Z coordinates and a 2D polyline must be planar. Three-dimensional polylines are useful as paths for sweeps or for fly-throughs. I don't cover sweeps or fly-throughs in this book.
- **Spline:** This free-form curve is created by using the SPLine command. I describe it in the 2D context in Chapter 7. Splines are 3D objects as their vertices can have different Z values. The spline is a better option than the 3D polyline for sweeps because it has smoother curves.
- **Helix:** A helix can be either 2D (think of a mosquito coil or the element on an electric range) or 3D (think of Mr. Slinky). It's made by using the HELIX command and is especially useful as a path for threaded objects.

You can find the Line, 3DPOLY, and SPLine commands in the Draw panel on the Home tab of the Ribbon; the HELIX command is on that panel's slideout.

The 3DPOLY command is similar to the PLine (plain old 2D polyline) command. Though both commands draw a series of connected line segments, they have different capabilities:

» The 3DPOLY command accepts 3D points for the line segments' vertices. The PLine command requires that all vertices be on the same plane.

» 3DPOLY is limited to straight line segments. The PLine command can draw arc segments and create segments with a uniform or tapered width.

» Segments created using 3DPOLY can't display dash-dot linetypes, so 3D polyline segments always display as continuous lines.

The command sequence for drawing 3D segments with the Line or 3DPOLY command is the same as for drawing 2D segments with the Line or PLine command; see Chapter 6 for information on drawing lines. The only difference is that you specify 3D coordinates instead of 2D ones. Figure 22-6 shows an example.

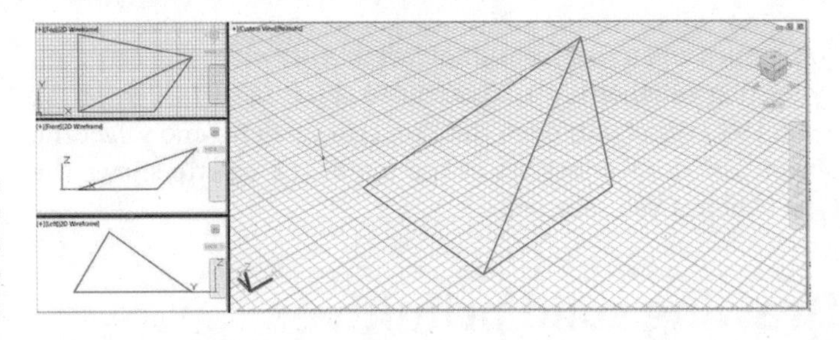

FIGURE 22-6: Entering 3D coordinates to draw a 3D polyline.

Creating 3D representations in this way is straightforward, although tedious, for all but the simplest objects. More important, a wireframe model becomes increasingly difficult to decipher as the complexity of the model increases. You see a mass of lines representing the edges, and you have difficulty telling which parts are in front of others. To reduce this visual confusion, you need to graduate to surface or solid modeling commands. I introduce you to solid modeling next in this chapter.

Gaining a solid foundation

Solid modeling is in many ways the culmination of 3D CAD. Solids more accurately represent most real-world objects than do wireframes or surfaces. And even when representational accuracy isn't the main issue, it's easier to construct many kinds of models with solids.

Many special-purpose solid modeling programs use a combination of solid and surface modeling techniques for maximum flexibility in constructing and editing 3D models. These kinds of programs — and solid modeling in general — are especially popular in mechanical design.

Constructing the basic building blocks, or solid primitives, for a solid model in AutoCAD isn't difficult. Just follow these steps:

1. **Define a suitable user coordinate system (UCS).**

   The UCS controls the construction plane and basic 3D orientation of the solid. Read about changing planes in Chapter 21.

2. **Click the down arrow below the leftmost button in the Modeling panel on the Home tab, and then choose a solid primitive from the drop-down list.**

   As shown in Figure 22-7, your choices are Box, Cylinder, Cone, Sphere, Pyramid, Wedge, and Torus. The Ribbon remembers the last one used.

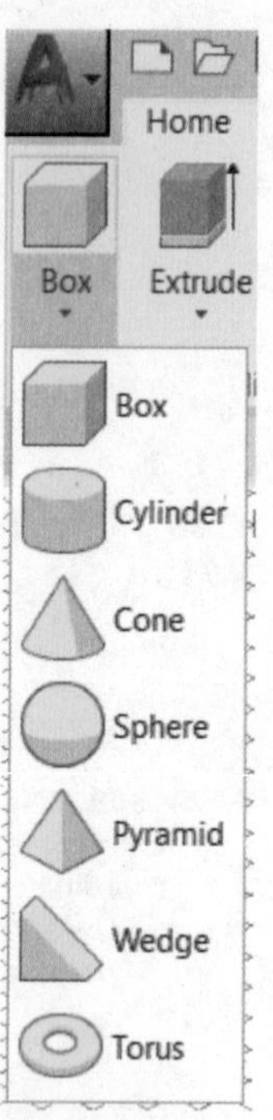

FIGURE 22-7: Everything you need for a solid foundation.

TIP

Sometimes when you see a 3D object in a drawing, you can't tell by looking whether it's a 2D extruded object, a surface mesh, or a solid — to find out, open the Properties palette and select the object. The drop-down list at the top of the palette shows the type of object you selected.

Drawing solid primitives

Solids are the easiest kinds of object to work with if you're new to 3D. The two types of 3D solid object are

- **Primitive:** The most basic of 3D building blocks. Primitive solids are based on simple geometric shapes, such as boxes, cones, and spheres.
- **Complex (or compound):** Made by combining primitive solids and (optionally) editing them with Boolean operations. The following sections focus on creating and modifying 3D solids.

TECHNICAL STUFF

Boolean editing operations (named for the Victorian English mathematician George Boole) use arithmetic-like functions to combine solid objects or remove parts of them. The three Boolean commands in AutoCAD are UNIon, SUbtract, and INtersect. See "Boolean operations," later in this chapter, for more information.

Adding the Third Dimension to 2D Objects

A common tactic in 3D modeling is to create 3D objects from 2D objects using a number of techniques. For instance, you can add thickness (extrude) an open 2D object to create a surface, or to a closed 2D object to create a solid. You can take multiple 2D cross sections and create a lofted shape that adapts to the shape and size of each cross section that's selected, like an airplane wing. 2D objects that are the source of the 3D model are sometimes called *sketches.*

Everything mentioned in this main section applies only to AutoCAD (not to AutoCAD LT), with the exception of the thickness property.

Adding thickness to a 2D object

Most 2D objects, such as lines and circles, have a thickness property. Changing the thickness property of a 2D object doesn't create a true 3D object, but it does create a pseudo-surface that looks like a 3D surface. This was the original way to make 3D models in AutoCAD and was called 2-1/2D because the tops were always flat. These pseudo-surfaces can hide objects behind them, but they have limitations. If you add thickness to a circle, an open cylinder is created without caps on the top and bottom. If you change the thickness of a rectangle (which is a polyline), you end up with four walls but no top or bottom.

Extruding open and closed objects

Extruding an object (think of squeezing toothpaste out of a tube) is a better approach than adding thickness to objects. You can extrude open or closed 2D objects to create 3D objects. Extruding a *closed* 2D object (such as a closed polyline, spline, ellipse, circle, or region) creates a 3D solid or 3D surface — AutoCAD gives you the choice. But extruding an *open* 2D object (such as an open polyline, spline, line, or arc) always creates a 3D surface. Figure 22-8 shows the result of extruding open and closed objects.

If you want to produce a 3D solid from a boundary composed of multiple objects, read about PRESSPULL in the later section "Pressing and pulling closed boundaries."

To extrude an open or a closed 2D object, click Extrude in the Solid panel on the Solid tab, or click the Extrude split button in the Modeling panel on the Home tab. (If you see a button labeled Loft, Revolve, or Sweep, click the lower part of the split button and choose Extrude from the drop-down menu.) Use the MOde option to control whether you create a surface or 3D solid from closed objects, select the objects you want to extrude and specify an extrusion height or drag the mouse to create the 3D solid.

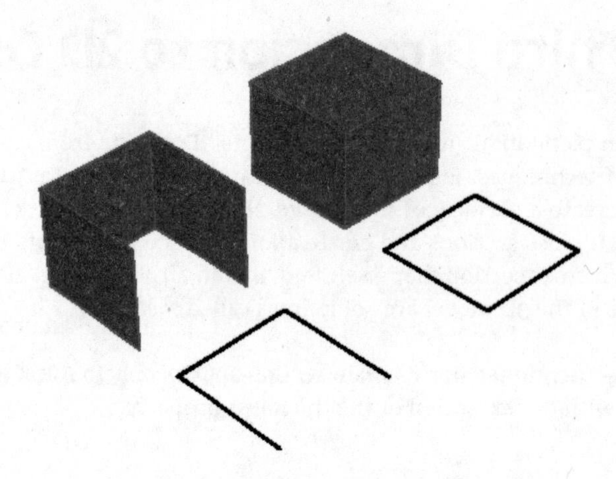

FIGURE 22-8: Converting 2D to 3D by extrusion is as easy as 1-2-3.

WARNING

The DELOBJ system variable controls whether the 2D objects you use to create 3D solids are retained or erased following any command that makes 3D objects out of 2D ones. Most of the time, you'll want to keep the source geometry, but the default DELOBJ setting erases it. If you want to keep the original 2D geometry, type **DELOBJ** and set its value to 0. See Chapter 23 for more about system variables. DELOBJ also controls whether the source objects for associative arrays (introduced in Chapter 18) are retained or deleted.

Pressing and pulling closed boundaries

The PRESSPULL command allows you to create 3D geometry by extruding a closed boundary. PRESSPULL differs from EXTrude in that the EXTrude command *requires* a single 2D object, whereas PRESSPULL can use closed boundaries formed by other 3D solids, such as faces, or by several objects that form a closed region. PRESSPULL searches for a suitable boundary in exactly the same manner as the Hatch command, which I cover in Chapter 15.

To press or pull in a closed boundary to create a 3D object, click Presspull in the Solid panel on the Solid tab, click inside an enclosed boundary, and then drag or specify the distance on the extrusion you want to create. For instance, you can draw a circle on a face of a 3D object, and then, with the PRESSPULL command, use the mouse to drag the circle into a cylinder.

Lofting open and closed objects

Lofting lets you select a two or more 2D cross sections to create surfaces or 3D solids of varying cross sections, such as aircraft wings or a boat hull. The selected cross sections define the outer surface that's generated. You must select a minimum of two cross sections. If you select open objects, the resulting loft is a 3D surface; if you select closed objects, the resulting loft can be either a surface or 3D solid, depending on the current value of the MOde option for the LOFT command.

You can loft objects along a path with guide curves, or only between selected cross sections, as shown in Figure 22-9. To loft open or closed 2D objects, open the Sweep/Loft drop-down menu in the Solid panel on the Solid tab. Use the MOde option to control whether you create a surface or 3D solid from a closed object cross section, and specify how the loft object should be calculated.

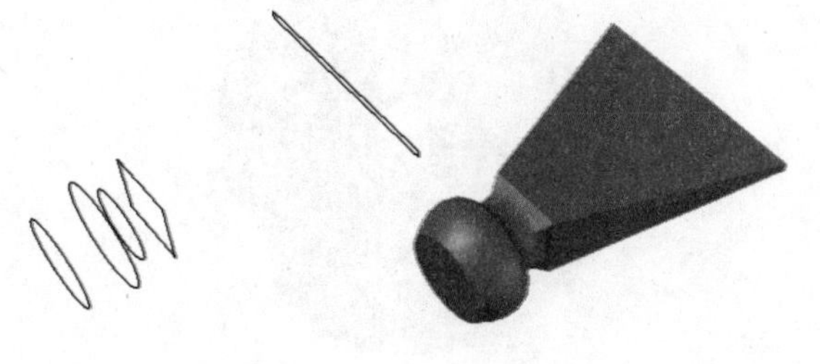

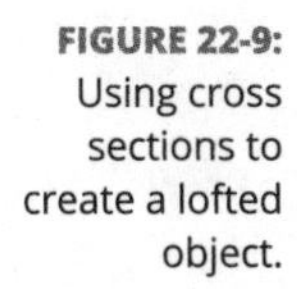

FIGURE 22-9: Using cross sections to create a lofted object.

Sweeping open and closed objects along a path

Sweeping is similar to extruding 2D objects, except that you specify a *path* rather than a height or distance. The object used as the path can be either an open or a closed 2D object, and must be only a single object, as shown in Figure 22-10. If you sweep an open object, the resulting swept object is a 3D surface; if you sweep a closed object, the resulting swept object is a 3D surface or 3D solid, based on the current value of the MOde option of the SWEEP command. To sweep an open or a closed 2D object, open the Sweep/Loft drop-down menu in the Solid panel on the Solid tab. Use the MOde option to control whether you create a surface or 3D solid from a closed object, select the objects to sweep, and specify a path for the sweep.

FIGURE 22-10: Sweeping a closed object along a path.

Revolving open or closed objects around an axis

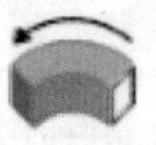

Revolving lets you create a 3D surface or 3D solid by turning selected 2D objects around an axis, as shown in Figure 22-11. If open objects are selected, the resulting revolved object is a 3D surface; if you select closed objects, the resulting revolved object is a 3D solid.

FIGURE 22-11: Revolving a closed object around an axis.

To revolve an open or a closed 2D object, click Revolve in the Solid panel on the Solid tab, or click the Revolve split button in the Modeling panel on the Home tab. (If you see a button labeled Loft, Extrude, or Sweep, click the lower part of the split button and choose Revolve from the drop-down list.) Use the MOde option to control whether you create a 3D surface or 3D solid from a closed object, select the objects to revolve, and specify the axis to use.

The good news is that you can use parametrics to control and manipulate the 2D objects that define your 3D object. I cover parametrics in Chapter 19. The bad news is that as soon as you use the 2D objects to create a 3D object, all parametrics disappear. If you want the 2D parametrics to control the 3D objects, you need to switch to Autodesk Inventor.

Modifying 3D Objects

Many modification techniques and commands that you use in 2D drafting can be applied to 3D modeling. In addition, a specialized set of 3D editing commands is available. All these commands are in the Modify panel on the Home tab when the 3D Modeling workspace is set.

Selecting subobjects

Three-dimensional models themselves can be complex enough to be made up of several hundred, or perhaps even thousands, of objects. Although these objects never stray from their parents, you can access them individually or in groups via *subobject selection:* selecting a vertex, an edge, or a face of a 3D object. After you select a subobject, you can use grip editing and the 3D gizmos (discussed in the later section "Working with gizmos") to manipulate it.

To more easily select subobjects, you can enable subobject filtering from the Ribbon. Vertex, edge, and face filters are located on the Subobject Selection Filter split button in the Selection panel on the Home and Solid tabs.

TECHNICAL STUFF

The CULLINGOBJ and CULLINGOBJSELECTION system variables help limit object selection to the faces that are visible in the current view by ignoring faces around the back of the objects being viewed. By default, both these variables are turned off, so you can select objects in front and behind. If you have a complex model, turn on culling by clicking Culling in the Selection panel on the Home tab. To turn off culling, click the Culling button again. The button is shaded blue when the feature is enabled.

TIP

If you need to select only a face on a 3D object, press and hold the Ctrl key and select the face you want to select. This method can be faster than turning subobject filters on and off.

Working with gizmos

Although you can use the Move, ROtate, and SCale commands to modify 3D objects, they can sometimes give unexpected results in 3D. Enter the 3DMOVE, 3DROTATE, and 3DSCALE commands, which all use a gizmo or grip tool when a non-orthographic view is current (see Figure 22-12).

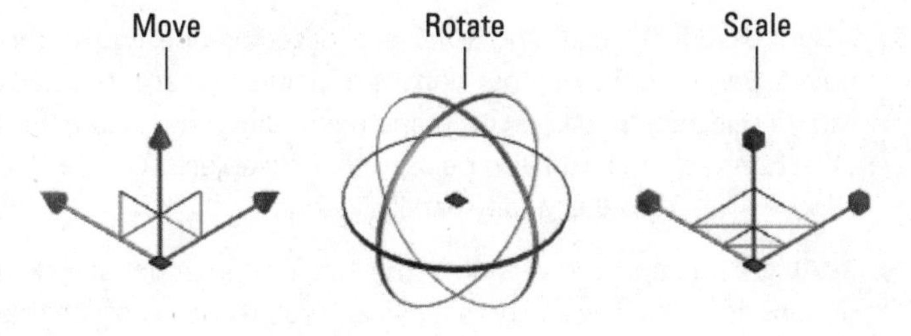

FIGURE 22-12: Using gizmos to modify objects in 3D.

A gizmo restricts or constrains movement along the X,Y plane and the Z-axis. To use a gizmo, click an axis on the tool to restrict movement to that axis. You can access the three gizmos by using these methods:

- **3DMOVE:** Click 3D Move in the Modify panel on the Home tab, or type **3DMOVE** at the command line.
- **3DROTATE:** Click 3D Rotate in the Modify panel on the Home tab, or type **3DROTATE** at the command line.
- **3DSCALE:** Click 3D Scale in the Modify panel, or type **3DSCALE** at the command line.

You can also access the Move, Rotate, and Scale gizmos in 3D by selecting an object in the drawing when no command is active. You can set the default gizmo that is displayed when you use grips in 3D by opening the Gizmo drop-down menu in the Selection panel on the Home tab and selecting the gizmo that you want to be active. Right-clicking a gizmo lets you switch between different gizmos and constraints.

More 3D variants of 2D commands

Moving, rotating, and scaling objects are certainly the Big Three operations of 3D editing, but a number of other 3D variations on 2D editing commands are hiding in the wings, awaiting their turns in the spotlight. For more about spotlights, see Chapter 23.

Getting your 3D ducks in a row

Making objects align with each other in 3D can be a challenge at times, especially if you need to not only move an object in 3D but also rotate and scale it based on the specified alignment. AutoCAD has two commands you can use to align objects:

- **ALign:** Used to align 2D and 3D objects based on one, two, or three pairs of points. Based on the number of pairs of points that are specified and how they're selected, the ALign command might move and rotate the selected objects into place. It can also be used to scale objects. On the Home tab, choose Align from the Modify panel slideout.
- **3DALign:** An improved version of the ALign command that includes additional options as well as the ability to move and rotate a copy of the selected objects and use Dynamic UCS with the command. Click 3D Align in the Modify panel on the Home tab.

Holding up a mirror

The MIrror command is limited to working on the X,Y plane. If you want to mirror objects in 3D, you use the MIRROR3D command. Click 3D Mirror in the Modify panel on the Home tab. The MIRROR3D command is similar to the MIrror command, but you can control the plane on which the mirroring is performed.

The generic 2D MIrror command works also in 3D, but you must employ a trick. As indicated in the preceding paragraph, MIrror works only in the X,Y plane — but it doesn't have to be the World X,Y plane. The command works equally well in the X,Y plane of the current user coordinate system (UCS).

Associative arrays, which I discuss in Chapter 18, work in 3D, as well as in 2D. AutoCAD has long had the 3DARRAY command; it's similar to the old-style ARray command (refer to Chapter 11 to read about that one) in that it doesn't create an associative array object. For information on creating rectangular, polar, and path array objects, refer to the online Help system.

Editing solids

You can edit 3D solids in a variety of ways that you can't edit other objects. You can use grip editing to change the shape of 3D solids, or use Boolean operations on a 3D solid to create complex models. You can fillet and chamfer the edges of a 3D solid by using the FILLETEDGE and CHAMFEREDGE commands.

Using grips to edit solids

Grip-editing is one of the most direct ways to modify 3D objects, such as making them shorter or taller. To edit a 3D solid by using grips, select the 3D solid when no command is running, and then select the grip that you want to use to edit the solid. Pay close attention to the grip you select; triangle grips give you control over changing the size of parts of a solid, such as the face or top radius of a cone. Square ones move the solid or change the size of the entire solid. Figure 22-13 shows a pyramid with its top radius being edited with a triangle grip.

Boolean operations

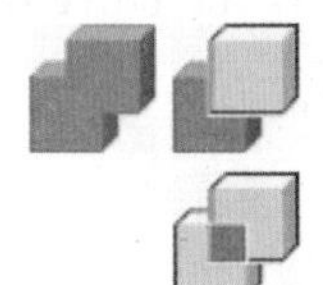

As I mention in the earlier "Drawing solid primitives" section, you can join 3D solids by using the UNIon command to create a new 3D solid. You can also subtract volume from a 3D solid by using another, intersecting 3D solid to determine what should be removed with the SUbtract command. The INtersect command can be used to calculate a new 3D solid based on the volume that is common to two or more intersecting 3D solids. See Figure 22-14 for examples. You can find these three commands in the Boolean panel on the Solid tab.

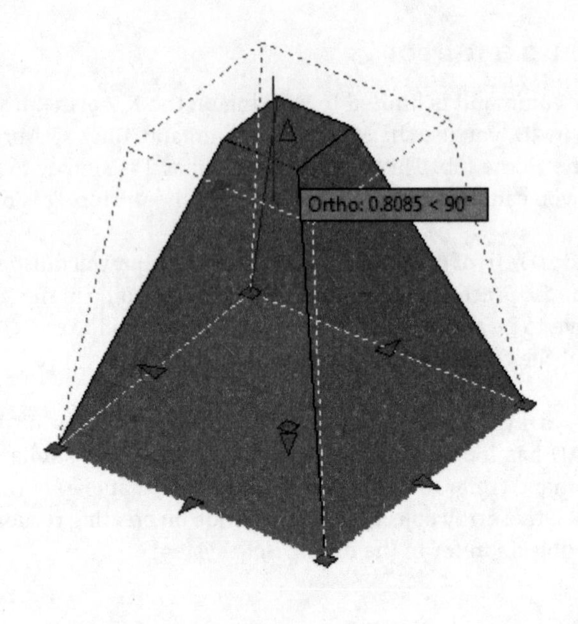

FIGURE 22-13: Grip-editing a pyramid.

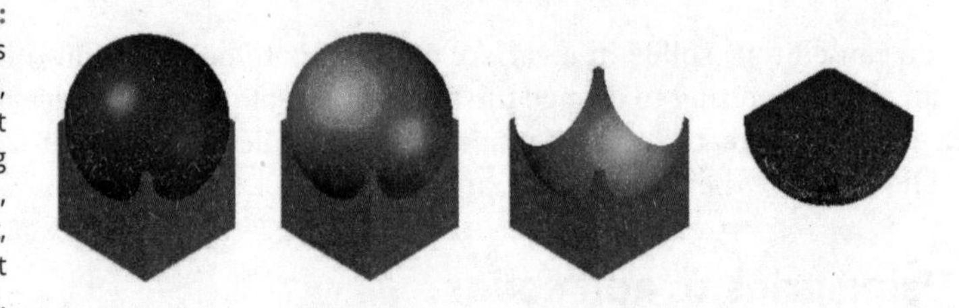

FIGURE 22-14: Solid primitives on the left, and the result of using the UNIon, SUbtract, and INtersect commands.

Boolean operations are probably the most-used of the 3D commands. Yes, primitive solids such as a box or a wedge are easy to use, but most real-world 3D objects are more complex than the simple primitives. Something as apparently simple as a connecting rod on a car engine might require a dozen or so Boolean operations to add and subtract simpler solids until you arrive at the final design.

You will often need to change to a different user coordinate system (UCS) before placing the next subobject to be combined with the base object. The easiest way to do this is to turn on UCSDETECT (F6) on the status bar. Then, as you move the cursor around while a create command is active, AutoCAD will automatically snap the UCS to any planar surface that it passes over.

Filleting and chamfering

Because fillets and chamfers are common real-world features, it's logical that tools for creating them are available. You can fillet or bevel the edges of a 3D solid by clicking the Fillet Edge (or Chamfer Edge) split button in the Solid Editing panel on the Solid tab.

Both commands allow you to select multiple edges to fillet or bevel. When you select an edge, AutoCAD gives you feedback on how the selected edge will be affected. Figure 22-15 shows an L-shaped 3D solid that has been filleted and chamfered.

FIGURE 22-15: Before and after filleting and chamfering a 3D solid.

TIP

To remove a fillet or chamfer, start the Erase command. Then, at the `Select Objects` prompt, hold down the Ctrl key and select the fillet or chamfer to be removed. For fillets, you may need to also remove nearby filleted corners that might have been created during the application of the fillet.

REMEMBER

The FILLETEDGE and CHAMFEREDGE commands were new in AutoCAD 2011. You can still use the Fillet and CHAmfer commands on 3D solids, but the new commands are much more efficient.

Slice

The SLice command lets you cut a 3D solid along a plane. You can slice a 3D solid by using a planar curve, such as a circle, 2D polyline, or surface, among many others. When you slice a 3D solid, you can choose which part of the 3D solid is retained, or you can keep both. Figure 22-16 shows a solid model that has been sliced in half.

FIGURE 22-16: Carving up a solid model with the SLice command.

To start the SLice command, choose Slice from the Solid Editing panel on the Solid tab. After the command is started, specify a 3D solid to slice, an axis or object to define the cutting plane, and then, finally, which new 3D solids to keep.

» **Visualizing 3D models**

» **Lighting your models**

» **Applying materials**

» **Assigning a background**

» **Applying render presets to 3D models**

Chapter 23
It's Showtime!

The word *design* has the same Latin origin as *designate*: "to point out to or to show to others." If you have an idea and produce it yourself, you aren't a designer — you're an artist or a craftsman (nothing wrong with that!). You don't become a designer until you tell other people about your idea so that they can do all the dirty, heavy work of producing it.

Two-dimensional orthographic drawings date to Roman times, but only in the past few years have I seen a significant change in how things are designed — and how designers communicate their visions to others. No, I don't mean 2D CAD on a computer, because that's simply a more efficient way of producing Roman-era drawings. The world we live in is 3D, but paper is 2D. Orthographic engineering drawings, which show the left, front, top, and so on sides, were developed for the sole purpose of transferring the designer's 3D idea into 2D format. The recipient of the drawing then transfers the idea back into 3D, first as an image in the mind and then as the physical object.

The significant change occurred when personal computers became inexpensive enough and powerful enough and software become sophisticated enough that designers could work in 3D on their own computers, including laptops. Around the same time, 3D printers became available to take the 3D model on your computer and produce a real-world object. This type of printer was developed with mechanical designs in mind, but architectural designers use it to print scale models of their buildings. More recently, the scale factor for printing buildings has become 1:1. That's right: Full-scale concrete multi-story apartment blocks are being printed in 3D directly from the computer model! However, the real world still needs (or at least wants) 2D drawings.

Combining relatively low-cost computers with powerful software made it possible for designers to produce photorealistic rendered images of their 3D models. Clients and customers are keen to see the product before it becomes real.

I start this chapter by showing you how to generate 2D drawings from 3D models, and then move on to discuss rendering of 3D models.

Get the 2D Out of Here!

Would you believe that it took less than five minutes to produce Figure 23-1?

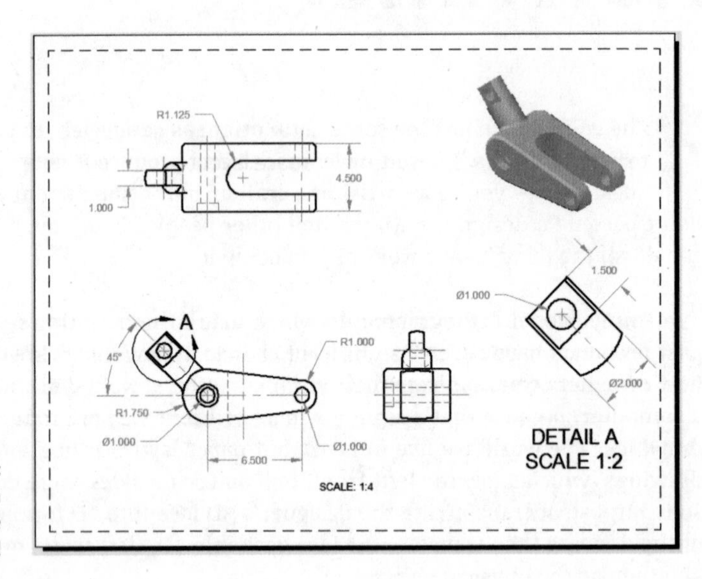

FIGURE 23-1: A drawing, in less than five minutes!

Follow these steps to produce a 2D view from a 3D model. There are several ways to approach this task; I show you the quickest:

1. **Click the Workspace Switching button on the status bar, and then choose 3D Modeling.**

   Ribbon tabs flash on and off, and soon AutoCAD settles down to display the Ribbon, as configured for the 3D Modeling workspace with a few additional panels.

2. **Set up a 3D model.**

   Create a new model by using the techniques I discuss in Chapters 21 and 22, or open an existing file that contains a 3D model.

   You can find the files I use in this step list at this book's companion website: Go to `www.dummies.com/go/autocadfd19` and download `afd23.zip`. The drawing named `afd23a.dwg` contains the model I use in the following steps.

3. **Switch to paper space.**

   Click the Layout 1 tab near the lower-left corner of the screen.

4. **Delete the existing viewport by clicking the viewport object (the frame of the viewport) and then pressing Delete.**

   By default, new drawings created from a standard template file contain a single viewport. If you find that you create new drawings like this one frequently, set up a template file with the viewport already deleted. I cover templates in Chapter 4 and paper space layout viewports in Chapter 12.

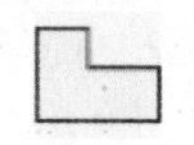

5. **Click the Base button from the Create View panel on the blue Layout contextual tab of the Ribbon, and then choose From Model Space in the drop-down list.**

TIP

   The VIEWBASE command creates two new layers automatically, one for visible lines and the other for lines that could be shown with the hidden linetype. By default, they're the opposite of the screen color (black or white), but they always print in black. You can change these layers to any color you want (see Chapter 9).

6. **Position the base view.**

   Pick a suitable place in the upper-left quadrant of the layout sheet. AutoCAD selects what it thinks is an appropriate scale automatically. The size of the views is based on the assumption that you'll want to fit four views of the 3D model onto this sheet — the three standard orthographic views and one pictorial (isometric) view. You can change the scale later if you want.

   The blue Drawing View Creation contextual tab appears on the Ribbon (as shown in Figure 23-2), a drop-down list of view options appears at the cursor, and an option list appears on the command line.

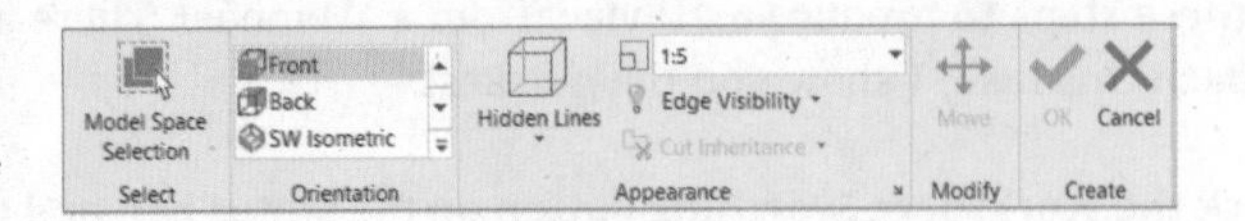

FIGURE 23-2: The Drawing View Creation context tab of the Ribbon.

7. **Define the base view.**

   The base view is the initial 2D drawing made from the 3D model; the other 2D drawings are based on it.

   Using any one of the user interfaces — at the Ribbon, with the dynamic input menu, or through the command line prompts — set up the base view as follows:

   - *Orientation:* The view shows what appears to be the bottom view of the part, because AutoCAD defines the top, bottom, and so on relative to the world X,Y coordinates. Select Orientation and then Top to place the top view of the 3D model.
   - *Hidden lines:* The preview images of the 3D model always appear in shaded mode as you place the views, regardless of the visual style of the model in model space. Change the Hidden Lines option to Visible and Hidden. The view won't change yet, but don't worry — it will after you complete these steps.
   - *Scale:* This setting defaults to 1:4, which is suitable for your purposes if you started from our sample drawing.
   - *Visibility*, or *Edge Visibility:* This setting specifies how to display the edges that are formed where tangent surfaces meet. The normal practice is not to display them, but it sometimes causes features to disappear (or partially disappear). If you want to change this setting, hover the cursor over the Edge Visibility button in the Appearance panel of the Ribbon and pause for a few seconds; a much more extensive tooltip list then explains each option.
   - *Move:* Specify a new location for the view before it is finally created. This isn't such a big deal, though, because views can always be easily moved later.
   - *Exit:* Or press Enter.

TIP

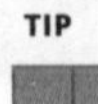

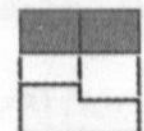

   You can edit all these view options later, view by view.

8. **Place more drawing views below and to the sides of the first one.**

   When you finish placing and defining the initial base view, AutoCAD automatically starts the VIEWPROJ command, and all you need are three quick clicks to place the top, isometric, and right-side views. Press Enter to have AutoCAD generate the hidden-line views.

9. **Edit the isometric projection.**

   Isometric projections don't normally show hidden lines. Double-click anywhere in the isometric projection to bring up the Drawing View Editor tab on the Ribbon, as shown in Figure 23-3.

   Click Hidden Lines on the Appearance tab and choose Shaded with Visible Lines from the drop-down list.

REMEMBER

   If a Ribbon button has a drop-down list, the Ribbon displays the last button that was used. Any one of four different buttons may be in this particular location.

   Figure 23-4 shows three ortho views and a shaded isomeric view, which I created in 37.6 seconds.

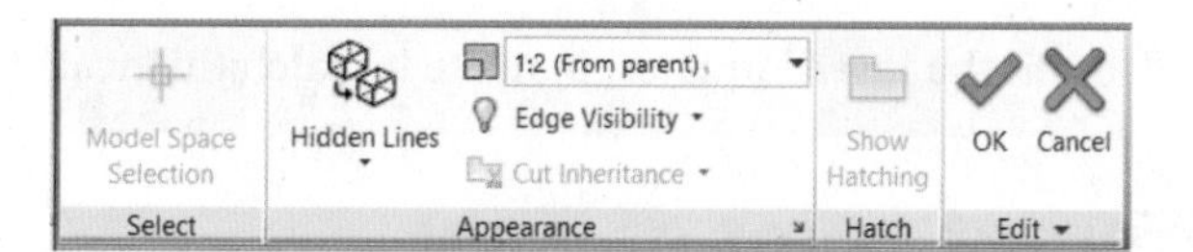

FIGURE 23-3: The Drawing View Editor context tab on the Ribbon.

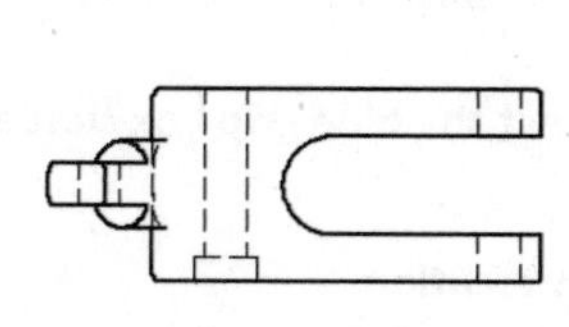

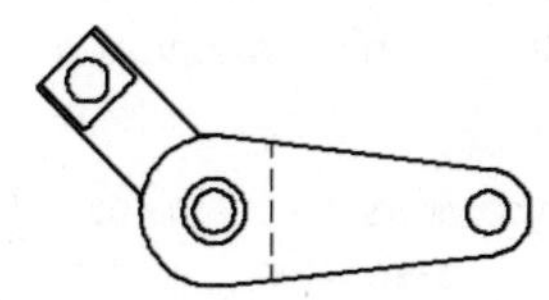

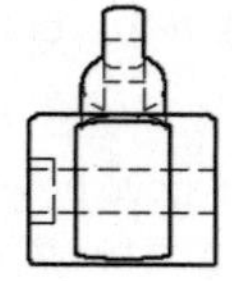

FIGURE 23-4: Average creation time: 9.4 seconds per view.

10. **Add annotations.**

   Add dimensions and text notes in the paper space layout. Dimensions are associative to their matching geometry if you use object snaps to the geometry when you place them. I cover object snaps in Chapter 8, text in Chapter 13, and dimensions in Chapter 14. While you're at it, perhaps you can try using VIEWDETAIL to create a detail view at a different scale.

   When you submit a drawing to your boss, he or she will be impressed that you managed to create such a complex drawing, including the shaded isometric projection, in only three days.

Always place text and dimensions on their own layers. I discuss layers in Chapter 9.

An isometric *view* and an isometric *projection* are different creatures. An isometric *view* is normally drawn so that lines that are parallel to the three principal axes appear in their true length, and an isometric *projection* foreshortens them due to the tilting and rotating of the viewing angle of the object. Traditional paper-and-pencil drawings use isometric *views*, whereas AutoCAD creates isometric *projections*.

If you truly want an isometric *view*, the solution is simply to ignore the usual rule about drawing and inserting at full size. When creating an isometric projection, use this approximate scale factor to produce an isometric view:

```
1.2247441227836356744839797834917
```

You can also edit the insertion later, to make it match this scale factor.

A different point of view

You can edit 2D views that were generated from 3D models in two ways. The easy way is to edit the view specifications themselves:

1. **Select the base view, and then select the blue grip box that appears in the center of the view.**
2. **Drag and drop the view into a new location.**

   Interesting! If you move the base view, all the ortho views projected from it follow along, with some limitations. The ortho views don't move in perfect unison as a single group, but they maintain their orthographic relationship to the base view.

   Similarly, you can move projected ortho views in only the direction that still maintains their ortho relationship to the base view.

   Better yet, all attached dimensions (you hope) also follow along.

The second way is to double-click a view and then change the specifications that were used to create it. Refer to Step 9 in the step list in the preceding section.

To experience the magic of AutoCAD automatically associating 2D views with the 3D model, return to model space and edit the model. For example, add a second hole (hint: subtract a cylinder) and extend the length of the peg. Return to the paper space layout. All your views and their dimensions are updated, as shown in Figure 23-5.

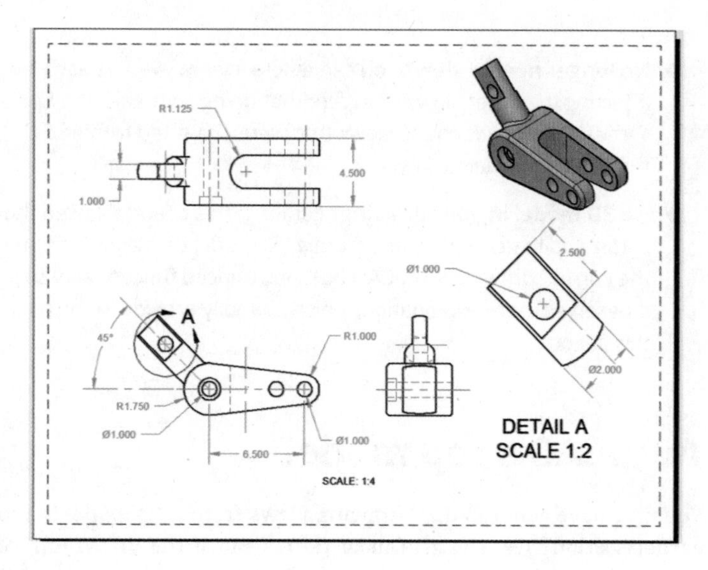

FIGURE 23-5: When anything changes, everything changes.

Behind the scenes, AutoCAD creates the views as a series of *anonymous* blocks. They behave much like regular blocks, but because they don't have normal names, you can't access them directly to edit or explode them. I discuss blocks in Chapter 17.

Additional 3D tricks

In earlier sections of this chapter, I only touch on the 3D-to-2D capabilities of AutoCAD — although, if you're reading the entire chapter, I may have whetted your appetite for more. To fully cover the 3D capabilities of AutoCAD would easily require a full book on its own (*AutoCAD 3D For Dummies*, perhaps?), but meanwhile, here are a few high points:

- **Don't want four views?** If you don't want four standard views, create only the base view and the one or two additional views you want. Then change their scale factor to better suit the sheet size.
- **Need additional base views?** If necessary, you can have more than one base view in a single layout. For example, one large drawing might show an assembly and its component parts. Just rerun the VIEWBASE command.
- **Didn't create enough views?** Use the VIEWPROJ command to add more projected views later. They don't have to project from the original base view, but can project from an existing projected view.

- **No longer need a view?** You can delete a view, even a base view, without affecting the other views — except that doing so breaks the horizontal and vertical links between the views that were projected from it. Select a view and then press the Delete key.
- **No 3D model in your drawing?** Earlier in this chapter, I used the VIEWBASE command to generate views from a 3D model that lived in the model space of the current drawing. AutoCAD 2013 introduced functionality so significant that it deserves more explanation, which I happily provide in the next section of this chapter.

AutoCAD's top model

You may have generated 2D drawing views from a 3D model in model space in the earlier section "Get the 2D Out of Here." But if the VIEWBASE command cannot find a 3D model in the current drawing file, it opens a standard file dialog box so that you can browse for an Autodesk Inventor part or assembly file.

TECHNICAL STUFF

Autodesk Inventor is the 3D parametric modeling software from Autodesk primarily intended for the mechanical design field. Inventor is fully parametric: Dimensional constraints drive 2D profiles that define the solid features that consist of the parts that comprise the assembly model that drives the 2D drawing views that Jack built. If you change a dimension on a 2D drawing of a part, everything updates, all the way down the line.

AutoCAD accepts Autodesk Inventor files during the VIEWBASE command. The Inventor file isn't inserted into the AutoCAD file but is attached like an xref. I cover xrefs and DWG files in Chapter 18. VIEWBASE creates a 2D drawing view based on it, and additional views can be projected from the base view.

Here's the magic part: The AutoCAD drawing views are linked to the Autodesk Inventor file so that any changes made to the Inventor file are reflected in the AutoCAD file, updating it. On the other hand, whenever the AutoCAD DWG file has access to the Inventor file, the AutoCAD drawing views update and remain in step with any changes made to the Inventor model. (The AutoCAD file contains only anonymous blocks for the 2D views and has nothing in model space.)

Better yet, you can send the AutoCAD DWG file to a client or vendor without having to send the source Inventor file.

OLDER 3D COMMANDS

AutoCAD begat several generations of tools to produce 2D views from 3D objects, and each generation generally improves. I list these older 3D commands here because they all still work, and you will probably encounter older drawings that were made by using them:

- **FLATSHOT:** This one produces a quick-and-dirty 2D view in model space. It basically sets the Z-axis coordinate of all objects to 0 (zero). It creates suitable new layers, and then makes a block definition of the flattened model, and inserts it as a 2D model. The benefit is that if you edit the model, you can run the command again and redefine the block; the drawback is that FLATSHOT can't create section views, and it takes some complex fiddling to make multiple views face and align correctly.
- **SOLVIEW** (Solid View), **SOLDRAW** (Solid Draw), **SOLPROF** (Solid Profile), **MVSETUP** (Multi-View Setup): This series of commands was used in the old days to create suitable layout viewports and then to generate appropriate 2D views into them. The views are created as anonymous blocks. The benefit is that they can produce simple cross-section views, and they're produced directly in layout viewports. The drawback is that the commands must be used in a specific order, and the views aren't associative back to the model. If you edit the model, you have to delete the views and start over.
- **SECTIONPLANE**, **LIVESECTION**, **SECTIONPLANEJOG**, **SECTIONPLANETOBLOCK:** This family of commands is used to create a series of blocks containing 2D views, much like FLATSHOT but with much more versatility and control. The benefit is that they can produce more-sophisticated cross sections, such as jogged section planes. If you change the model, the block definitions can be re-created to show the updated version. The drawback is that it takes some complex fiddling to get multiple views to face and align correctly.

Here are more tips for working with AutoCAD and Inventor files:

» **Mix and match 2D drawing views.** You can have more than one base view in an AutoCAD drawing, so you can mix and match 2D drawing views. One or more base views can come from an internal AutoCAD 3D model, and others can be linked to external Inventor files. If VIEWBASE finds an AutoCAD solid in your part, you can tell it to ignore it and attach it to an Inventor file instead. For example, you might have a mechanism assembly drawing consisting of several 3D parts, of which one is a simplified conceptual part. When you create 2D drawing views, you can tell AutoCAD to ignore the simplified part and instead link it to the more complex and detailed Inventor model.

» **Select additional solids.** When model space contains more than one solid, VIEWBASE allows you to switch back to model space to select and deselect solids to appear in the base view. For example, a model of a gearbox assembly might consist of many components. Separate views can be created, perhaps on several different layouts: one showing an outside view of the entire gearbox (which doesn't need to include the internals such as gears and bearings); another showing only the input shaft, gear, bearings, and seals; and another showing the output shaft and its related components.

» **Choose a different scale.** The VIEWDETAIL command generates detail views at scales different from the parent view, which matches standard drafting practice.

» **Use different section views.** The VIEWSECTION command has five options for creating section views: Full, Half, Offset, Aligned, and From Object. This command creates section views based on existing views in the layout. The cutting plane line that it generates can be manipulated like any regular polyline, and the section view then updates accordingly.

Visualizing the Digital World

Rendering allows you to see how a final product might look before it's manufactured or built in the real world (see Figure 23-6); it definitely helps to sell concepts to clients. Although you can show a client hidden line and shaded views of 3D models, showing a series of photorealistic renders helps make it feel that much more real.

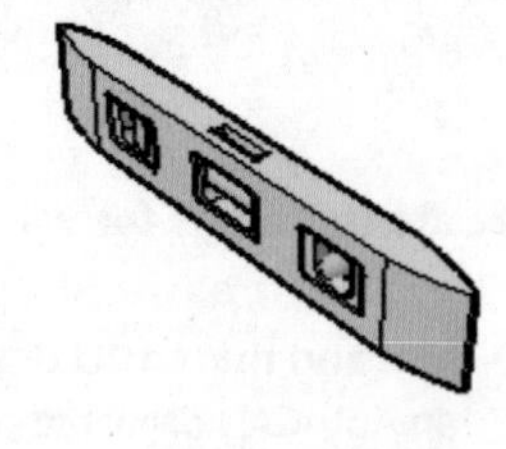

FIGURE 23-6: Rendering compared with the Conceptual visual style.

Before rendering a 3D model, you should

» Add lights to highlight features and define shadow types.

» Apply materials to 3D objects by face, object, or layer.

» Set a background to your 3D model

All these actions bring realism to a 3D model. This chapter focuses next on each of these tasks and then explains the steps to create a rendering.

Adding Lights

One key ingredient that makes renderings look good is lighting. *Lighting* helps give a model depth by applying highlights and shadows. As in the real world, objects that are closest to the light source appear the brightest, and objects the farthest distance away appear darker.

AutoCAD uses both *default* and *user-defined* lighting. Default lighting, as its name suggests, is on and available in every drawing. It uses two lights to give basic form to your 3D model when you click the Render button. User-defined lights can cast shadows.

Most tools that you use to create and edit lights are located on the Lights panel and the Sun & Location panel on the Render tab.

Default lighting

Over the years, AutoCAD's default lighting has improved in quality. Before AutoCAD 2007, default lighting consisted of a single, distant light source, always directed toward the target of the current view from behind your back. AutoCAD 2007 added a second default light to help increase the lighting level, and to balance the lighting in a viewport.

Default lighting can't cast shadows, so I don't recommend using it for final renderings. It does, however, work pretty well for quick conceptual renderings. You have control over the brightness, contrast, and midtone levels for default lighting by using the slider controls on the slideout of the Lights panel on the Visualize tab of the Ribbon (see Figure 23-7). AutoCAD 2014 and earlier have these controls on the Render tab of the Ribbon.

FIGURE 23-7:
AutoCAD to lights; extra brightness is a go.

Use the DEFAULTLIGHTING system variable to enable and disable the use of default lighting in the current viewport. Set DEFAULTLIGHTING to 0 when you want user-defined lighting to render your 3D model. If you're using default lighting, the DEFAULTLIGHTINGTYPE system variable controls whether one or two default distant lights are used: When set to 0, one default light is used; when set to 1, two default lights are used.

User-defined lights

Default lights are fine for quick renderings, but they don't bring renderings to life in the way that user-defined lights can. User-defined lights are lights that you create (with one exception) and modify in your 3D model. The only user-definable light type that you can enable and modify, but not create, is the sunlight system — someone else already created it.

The first time you use a user-defined light in a drawing, the Lighting – Viewport Lighting Mode alert box is displayed, advising you to turn off default lighting in order to see light from user-defined sources. You can have only one type — default or non-default — so click Turn Off the Default Lighting to disable the default lighting. (To turn default lights back on, open the Lights panel slideout on the Render tab and click Default Lighting.)

You can create two types of user-defined lights:

- **Generic:** This type provides many controls over how the light should be emitted.
- **Photometric:** This type is more sophisticated because it mimics the way actual lights illuminate the physical world.

The LIGHTINGUNITS system variable controls whether user-defined lights in a drawing are represented as generic or photometric. When LIGHTINGUNITS is set to 0, generic lights are used; when LIGHTINGUNITS is set to 1 or 2, photometric lights are used — 1 indicates American lighting units, and 2 indicates international lighting units. Lights don't have to be removed and added to switch between generic and photometric lights; you only need to switch the system variable.

To add a user-defined light to a drawing, click the lower half of the Create Light split button on the Lights panel of the Render tab. You can choose from four distinct types of lights on the flyout (you can see the differences between light types in Figure 23-8):

* **Point light:** Emits light uniformly in all directions, similar to a candle or a light bulb. The light emitted falls off as it moves farther from the source.

* **Spotlight:** Emits light in a one direction, like the ones used at concerts. As light travels away from the source, it spreads out in the shape of a cone. You can define the *hotspot* (the brightest part in the center of the emitted light) and a fall-off for the light.

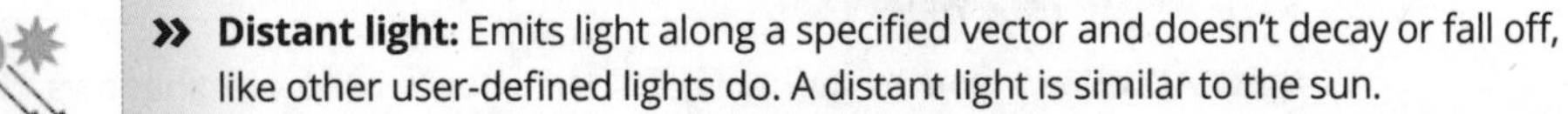

* **Distant light:** Emits light along a specified vector and doesn't decay or fall off, like other user-defined lights do. A distant light is similar to the sun.

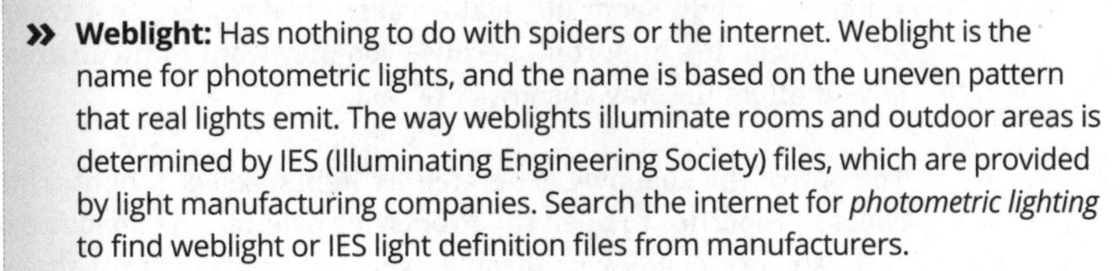

* **Weblight:** Has nothing to do with spiders or the internet. Weblight is the name for photometric lights, and the name is based on the uneven pattern that real lights emit. The way weblights illuminate rooms and outdoor areas is determined by IES (Illuminating Engineering Society) files, which are provided by light manufacturing companies. Search the internet for *photometric lighting* to find weblight or IES light definition files from manufacturers.

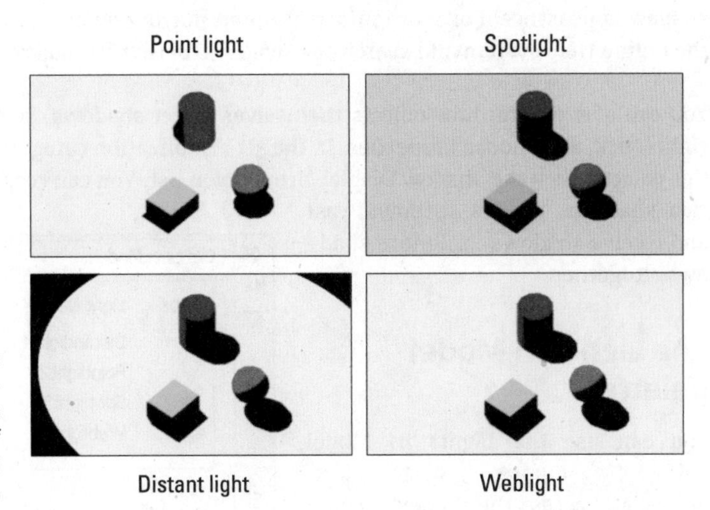

FIGURE 23-8: Lighting primitives with the four types of user-defined lights.

WARNING

AutoCAD — and I — strongly advise against using distant lights when using photometric lighting. If you decide to ignore this warning, turn down the intensity factor of the distant light to avoid washing out the rendering.

TECHNICAL STUFF

Point lights, spotlights, and weblights can be selected and edited directly in a drawing because a glyph is displayed to show where the light is located. Distant lights have no associated glyph. A *glyph* is a nonprinting object displayed in a drawing that enables you to select an object that isn't part of the actual model.

TIP

To avoid some of the frustration of applying lights, visit a portrait studio, have your portrait taken, and notice how the lights are set up. A brighter light is typically placed (about 45 degrees) above and (about 45 degrees) to one side of the camera as you face it, and a second, dimmer light is placed lower (about 30 degrees) and on the other side of the camera.

The shadow knows. . .

AutoCAD allows you to define lights that don't generate shadows. Though this statement might seem illogical because shadows are cast when light is obscured by an object, it's important because you may want to fill an area with light but not have it affect the way shadows are cast.

To control the shadows generated by lights, select a light, right-click, and then choose Properties to open the Properties palette. The Shadows option in the General category enables or disables shadow creation from the selected light. The Rendered Shadow Details category gives you a large degree of creative control over shadow appearance. For more information on putting objects in the shade, open the online Help system and search for *Render 3D objects for shadows.*

TIP

You can also control how objects themselves affect shadows. Select a 3D object, right-click, and choose Properties. In the 3D Visualization category of the Properties palette, click the Shadow Display drop-down list. You can configure objects to cast shadows, receive shadows, cast *and* receive shadows, or ignore shadows altogether.

The Lights in Model palette

You can use the Lights in Model palette (see Figure 23-9) to select, delete, and access the Properties palette for selected lights. The palette is especially handy for the distant light, which doesn't display a glyph for you to edit because it's located at infinity. To display this palette, click the Lights panel launcher (the little arrow at the right end of the Lights panel label).

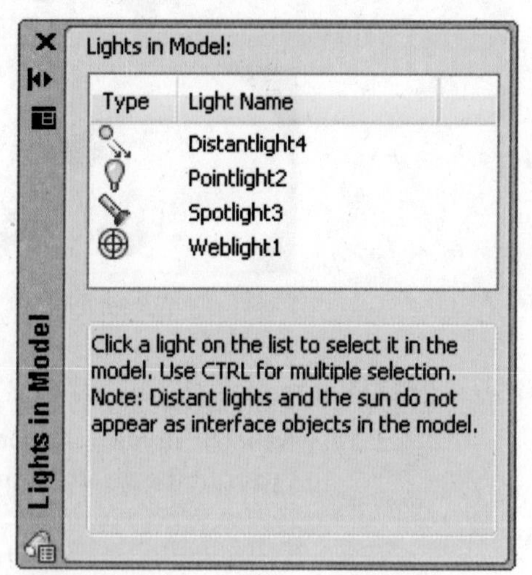

FIGURE 23-9: Ah, yes, you're on the lights list.

Sunlight

You control the way sunlight affects your 3D model through the Sun Properties palette (see Figure 23-10). Adding sunlight to a drawing and then specifying the date, time, and geographic location allows you to perform sunlight studies. Sunlight adds realism to both indoor and outdoor renderings.

FIGURE 23-10:
The Sun Properties palette gives you the power to control day and night.

You use the Sun Properties palette to locate the sun's position based on the nearest large city or on latitude and longitude values. You can also use KML or KMZ files exported from Google Earth to specify a location. In addition to location, you can designate the northern direction to find the correct daily movement of the sunlight in your drawing. To adjust the sunlight in a drawing, open the Sun Properties palette by clicking the Sun & Location panel launcher (the little arrow on the panel label).

Creating and Applying Materials

Materials, even more so than lights, bring your models to life. Materials can be as simple as glossy paint, or as complex as rough bricks, or anything in between. You can apply representations of realistic, real-world materials like stone, marble, glass, wood, metal, and fabric — the list is almost endless. The materials can be opaque, transparent, reflective, or nonreflective. AutoCAD comes supplied with a material library of literally thousands of material types. Assigning them to objects can be as simple as dragging a material from a palette and dropping it on an object. Or you can define your own materials as complex as you care to make them.

Before AutoCAD 2011, you created and stored materials in individual drawings, which made them difficult to manage. AutoCAD 2011 introduced Materials Browser, which makes managing materials much easier. Material definitions, which live in a central repository, are attached to each drawing as an external reference.

REMEMBER

All Autodesk products that use material definitions use the same central library, so a model rendered in AutoCAD looks the same as one done in other Autodesk programs like Inventor, Maya, and 3ds Max.

The partner to Materials Browser is Materials Editor, which edits the properties of materials to define their look. It's accessible from a button in Materials Browser.

The material libraries introduced in AutoCAD 2011 occupied vast amounts of hard drive space, which was considered wasteful for people who never did renderings. So, as of AutoCAD 2012, most material libraries are online. The first time you click the Render button, AutoCAD asks whether you want to go online to install the Medium Material Library, about 700MB. (*Medium* refers to the resolution of the image files used for materials, typically 1024 x 1024 pixels.) If rendering looks interesting to you, go ahead and install the library. The download is an MSI installer file, which you run to install the library. The library is specific to each release of AutoCAD.

Use the MATbrowser command to display the Materials Browser palette (see Figure 23-11), from which you can create, edit, and manage in the current drawing or material libraries. (Before you can use Material Browser, you must install the Medium Material library.) You can add preconfigured materials to drawings or create custom materials.

Materials Browser palette

Materials Editor palette

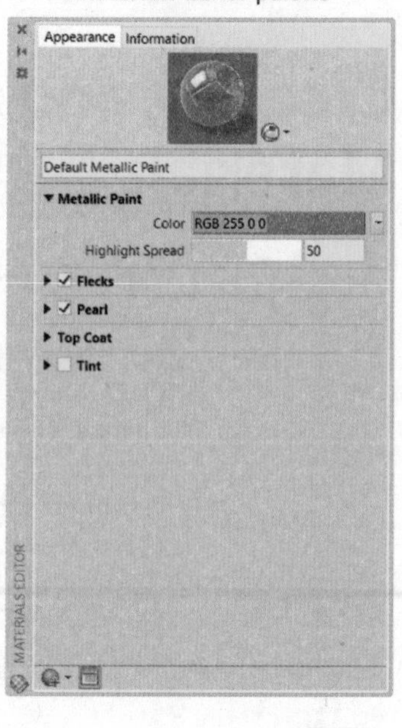

FIGURE 23-11: Adding color and texture to a 3D model with the Materials Browser palette (left) and the Materials Editor palette (right).

Follow these steps to create and manage a new material:

1. **On the Visualize tab, choose Materials Browser from the Materials panel.**
2. **On the Materials Browser palette that opens, click Create a Material (in the lower-left corner) and then choose the material template you want to start with. For example, choose Metallic Paint.**

   The Materials Editor palette is displayed (see Figure 23-11).

3. **On the Materials Editor palette, click in the Name field, located below the preview of the material, and enter a name.**
4. **Enter new values in the appropriate attributes for the material, and close the palette when you're done.**

   The attributes that you can edit vary, based on the type of material template that you choose to start with. When you finish editing the material, you can close or hide the Materials Editor palette. The material is then automatically added to the current drawing.

5. **(Optional) Save the material in a library to use it in more than one drawing. In the Materials Browser palette, at the bottom, click Manage, and then select Create New Library.**

   The Create Library dialog box is displayed.

6. **Enter a name and location for the new library. Click Save.**
7. **In the Materials Browser palette, drag the custom material that you created onto the name of the new library.**

   Now you can access the material from any drawing. You can create categories if you want to manage multiple materials within a library. In Materials Browser, right-click the library name, choose Create Category, and then enter a name for the new category. After the category is created, simply select the material from the right side of Materials Browser and drag it to the new category.

TIP

You can open Materials Editor directly from the Ribbon by clicking the Materials panel launcher (the little arrow at the right end of the Materials panel label).

Which materials you use in a model depends on what you're trying to represent. For example, you might choose to make a material semitransparent, to communicate an idea rather than a true material selection. After you create a material, you can apply it to the objects in your 3D model. You can apply materials to objects by

» **Layer:** Assign materials to all objects on a particular drawing layer. To assign materials by layer, open the Materials slideout on the Render tab and then choose Attach by Layer.

» **Object:** Assign materials to an object by selecting the object and then, in the Materials Browser palette, right-click the material you want to assign. From the menu that appears, choose Assign to Selection. You can also drag and drop a material from the Materials Browser palette onto an object in a drawing. You can change an object's Material property by using the Properties palette.

» **Face:** Assign materials to individual faces of a 3D solid. Select a face using Subobject filtering by holding down the Ctrl key and selecting the face to which you want to apply a material. Right-click the material in the Materials Browser palette, and then choose Assign to Selection from the menu that appears.

Defining a Background

Although your 3D model looks more realistic with some user-defined lighting and custom materials assigned to it, its setting might look a little, well, empty. To fill that empty space beyond the 3D model, you can assign backgrounds to fill your scene. A background is a property of named views, so you set them up in the View Manager dialog box. I cover the View command and View Manager dialog box in Chapter 5.

You can define the background as a solid color, a gradient (two or three colors merging into one another), a raster image like a photograph, or a simulated sky with a sun (this last one is available only with photometric lighting).

Follow these steps to create a new named view and assign a background to it:

1. **On the Views panel of the Visualize tab, click View Manager.**

   The View Manager dialog box appears.

2. **Click New to display the New View/Shot Properties dialog box.**

   For more on creating named views, see Chapter 22.

3. **In the Background area, click the Default drop-down, choose an option to override the default background, and then click OK.**

   Choose Solid or Gradient if you want a studio-type scene, or choose Image if you have a suitable raster image in which to immerse your model. Any of these options displays the Background dialog box from which you make your settings (see Figure 23-12). Choose Sun & Sky if you want to place your model in the great outdoors. (It's always fair weather in AutoCAD! Just be aware that sunny, fair days also tend to slow down the program.) Choosing the last option displays the Adjust Sun & Sky Background dialog box, as shown in Figure 23-13.

4. **After the background is defined, select the view from the Views tree and click Set Current.**

   When View Manager closes, the new background is displayed.

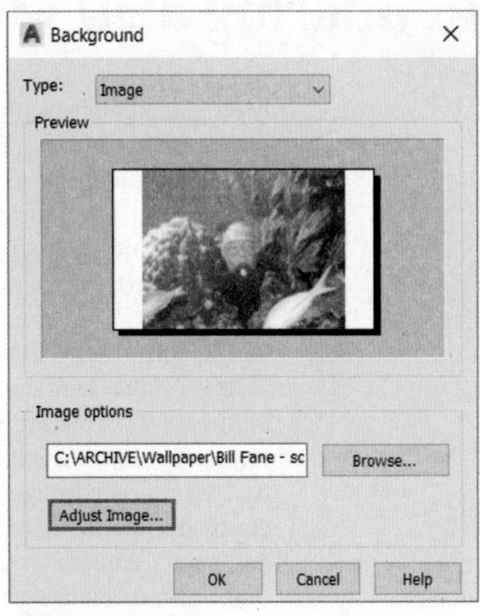

FIGURE 23-12: Painting the canvas of the drawing window.

For more information on View Manager, see Chapter 5.

TIP

If you don't want to create a named view first, enter the BACKGROUND command to display the Background dialog box directly.

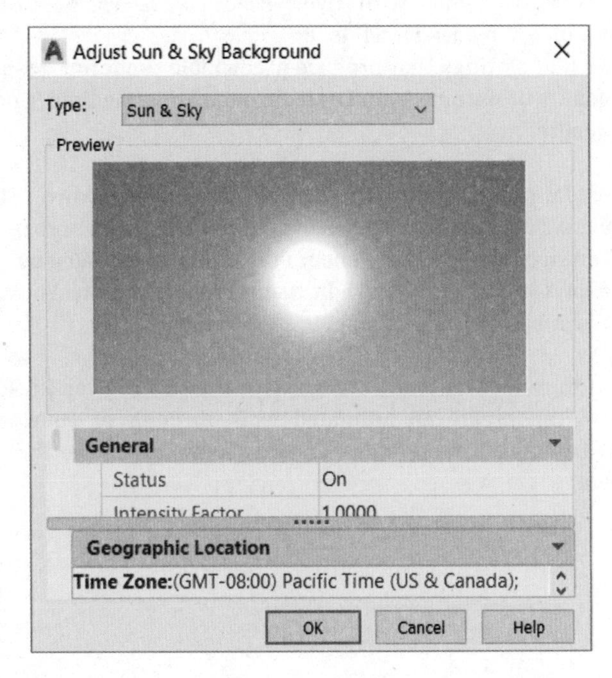

FIGURE 23-13: Here comes the sun!

Rendering a 3D Model

After you add lights, apply materials, and define a background for your 3D model, you're ready to press the magic button! Clicking Render tells AutoCAD to round up all the materials and apply them to the objects and faces to which they're assigned. After materials are assigned, the background is applied to the current viewport, and then, finally, AutoCAD calculates light and shadows based on all those property settings I show you how to make earlier in the chapter.

By default, rendering is performed in the render window, as shown in Figure 23-14. To start rendering your virtual fat into digital lard — how's that for a grisly image? — click Render to Size on the Visualize tab's Render panel. And to anyone who says that stuff in Figure 23-14 looks like red paint, not red wine, I would reply "You haven't tasted my homemade red wine, have you?"

FIGURE 23-14: Rendering a rendering in the render window.

To help make rendering as easy as possible, AutoCAD comes with five preset renders. A *render preset* is a configuration of settings that produce predictable rendering results. The presets range in quality of output, from Draft through Low, Medium, and High to Presentation Quality.

To set a render preset, choose the Render Presets drop-down list on the Render tab's Render panel. Choose Manage Render Presets at the bottom of the Render Presets drop-down list to open the Render Presets Manager dialog box. Render Presets Manager allows you to create and edit custom render presets. You can start with an existing render preset when you create a custom one.

In addition to rendering a still image with the RENDER or RENDERCROP commands, you can create an animated walk-through of your model by using ANIPATH. To find out more about creating animations, see the AutoCAD online Help system.

» Importing other file formats into AutoCAD

» Dealing with translation issues

Chapter 24
AutoCAD Plays Well with Others

So your drawing masterpiece is ready to go. Chapter 16 covers plotting to paper, and Chapter 20 discusses the details of transmitting drawing files using the internet. So far, so good. But what if you want to include a drawing as an illustration in a written report? Or send a drawing to someone using a CAD program other than AutoCAD? No problem. This chapter starts by revealing several processes for extracting drawing information in a variety of file formats.

Next, we come to the flip side. What if you receive a drawing file that isn't in DWG format? Okay, maybe that's not much of a problem because AutoCAD can import and even edit several other file formats.

This chapter concludes with a section on some of the possible pitfalls of dealing with translations between file formats.

Get Out of Here!

AutoCAD has several processes for exporting drawings in formats other than DWG. In fact, it often has more than one process for the same format.

In Chapter 1, I discuss the difference between the two fundamental file formats for storing graphic images. All CAD programs use a vector format, in which the file contains definitions and properties for each object type, such as a line, a circle, or an arc. Most other graphic programs use some variant of a raster (bitmap) format, wherein the file simply contains a mapping of the color of each of the tiny dots (pixels) that produce the image on screen. Some, like illustration programs, combine both vector and raster data in their files.

The bad news is that when you want to include CAD drawings in non-CAD applications, well, many cannot handle the vector format that's native to CAD software. The good news is that AutoCAD can produce raster and other kinds of vector files that can then be inserted into things like a Word document file. For instance, Word can import several raster formats plus three vector ones, of which AutoCAD exports two: PDF and WMF. So, it's a matter of comparing lists of file formats: Those AutoCAD can export with those the other software can import.

Making a splash with PNG

The most common compressed raster file formats are JPG ("jay-peg"), which is used by digital cameras, PNG ("ping"), which is used on websites, and TIF ("tiff"), which is used in desktop publishing.

As I noted earlier, raster image files consist mostly of a whole bunch of numbers that indicate a color for each pixel in the image. One problem with raster files is that as the resolution (the number of rows and columns) goes up, the file size goes up exponentially. When you, say, double the resolution, the file size quadruples.

The file size problem is greatly reduced by applying file compression techniques. The basic principle of file compression is that the software looks for repetitions in the data, and then describes them appropriately. For example, if green is color number 3, a horizontal green line in a raster file could be stored as "3 3 3 3 3 3 3 3 3 3 3 3." A compressed file could store the same information as "12x3" — 12 pixels of green in a row — which obviously requires less space. The bad news is that some compression formats, such as that used by the popular JPEG format, achieve very high compression rates by eliminating finer details, so that "3 3 3 3 3 3 3 4 3 3 3 3 3" might also be stored as "12x3." The missing 4 is false information and is called an *artifact.* And artifact may not be a problem in photographs but becomes a problem when details are lost in CAD images.

Because JPEG is overly aggressive in applying compression to images and because storage media have become much cheaper and more plentiful, PNG was developed to eliminate the artifact problem yet apply reasonably good compression. The format is now common on websites and in desktop publishing. All the images in this book, for instance, were saved in PNG format.

The good news is that AutoCAD offers a number of ways to extract raster images from drawings. The bad news is that it offers a number of ways to extract raster images, ranging from convenient to flexible. No one central command can do them all:

- **Convenient: Press Alt+PrtSc on your keyboard.** This method captures the screen view of the current application to the Windows clipboard. You can now paste it directly (with Ctrl+V) into many other applications, such as Word or a graphics program such as Paint for further editing, cropping, and saving to a variety of raster formats. This keystroke combination is convenient because there are no options to choose, and it works with every Windows application.
- **In between: Use AutoCAD's EXPort command.** This method saves the current drawing to a file on the hard drive, converting it to mostly a variety CAD and vector formats. For non-CAD applications, the two formats to consider are WMF (Windows metafile) and BMP (bitmap). WMF is a mixed raster-vector file format, which means you can zoom in or out without losing details. Oh, small stuff may not be visible when you zoom out too far, but it hasn't gone away when you zoom back in. After you insert a WMF file into a Word document, you can resize it at will to fit the available space without losing clarity or detail.

  This command prompts you to choose an export format and then select the entities to export; press Enter to select all visible ones. Other than that, there are no options to consider.
- **Flexible: Use AutoCAD's PLOT command.** Refer to Chapter 16, where I discuss plotting. The Printer/Plotter Name drop-down list includes the two most common raster formats through the Publish to Web JPG and Publish to Web PNG options. When you select either, the Paper Size drop-down list displays a variety of image resolutions. Choose a high resolution for the clearest image. To specify a very high resolution, such as 4K: In the Plot dialog box, choose Plot to PNG, click Properties, choose Custom Paper Sizes, click Add, and then follow the steps in the Custom Paper Size Wizard.

  You may need to do a bit of playing with the Plot Area and Plot Scale values to get what you want. Extents and Fit to Paper are usually good starting points.

WARNING

Doing a screen capture or a plot to a raster file is strictly a case of WYGIWYS (what you get is what you see). If a detail is small enough that it isn't visible onscreen at the time the image was captured, it isn't in the raster file and, contra CSI Las Vegas, no amount of zooming in will bring it back. The only way a raster image can be resized is by dropping or adding pixels. Resizing smaller results in a loss of detail, which eventually messes up the drawing image. Resizing bigger adds no detail but makes the image blockier, unless you're in that CSI lab, in which case enlarging an image reflected from an eyeball retains full resolution.

I strongly recommend that you play with the capture settings so that the image can be inserted 1:1 in the target document, without the need to resize it.

The three panels of Figure 24-1 show possible side effects of resizing a raster image.

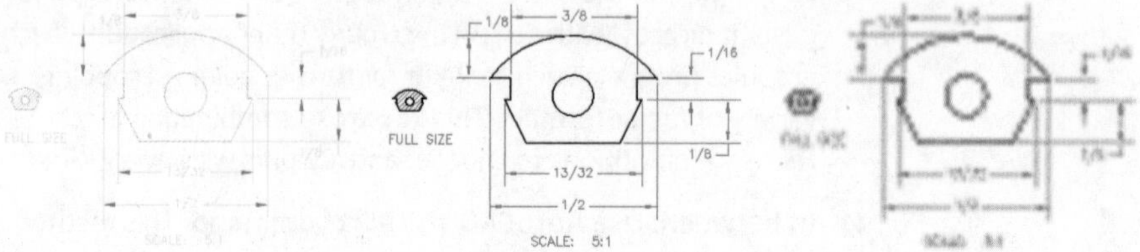

FIGURE 24-1: How to mess up a raster image.

The first major issue with exporting a drawing to a raster file is that the objects to be exported have to be visible onscreen and they must all reside in model space or paper space. For example, if your border and title block reside in a paper space layout and a viewport shows model space, exporting a raster file from paper space will capture only the border, title block, and viewport boundary — and nothing from model space.

Fortunately, there is a simple solution. By the way, have you noticed that I almost never bring up a problem unless I also have a solution? Anyway, all you need to do is to be in the desired paper space layout. Now go to the Application menu, choose Save As, and then click Save Layout as a Drawing. From here, the steps are self-explanatory. All visible objects from both universes are collected, and then everything is copied into the model space of a new drawing. Now you can export the new drawing as a WMF file.

The second issue involves framing. When you create a raster file, its boundary is determined by the visible area of the screen and not by the extents of the objects you select. This means that when you zoom out to the full extents of your drawing but select just a few objects, the resultant image when inserted into a Word document will show a large white area with your selected objects displayed as a very small detail. The solution is to zoom and frame your objects so that they just fill the screen before you select them. This process can even include stretching and shrinking the height and width of the drawing window.

PDF to the rescue

The PDF file format by Adobe Systems has become the de facto standard for many types of information exchange. PDF stands for Portable Document Format, with the emphasis on *portable*. In the early days of computing, a document created on the PC version of Word couldn't be opened by the Mac version and vice versa. On the other hand, PDF files were designed be opened and displayed by any Windows or Mac or Linux computer or any iOS or Android or ChomeOS device, such as a tablet or a Chromebook. All you need to view a PDF file is the appropriate free reader app from Adobe and other software firms.

PDF has several advantages over other export file formats. It's a vector format, so it can be zoomed and resized without loss of detail. It retains drawing information like layers and hyperlinks, and stores fonts. It can be viewed by nearly any device, including in web browsers, and can be opened by nearly any software program. It has become the default file format for distributing AutoCAD drawings to people without AutoCAD. The one drawback to be aware of is that PDF files do not have the universe-spanning precision of AutoCAD, and so the PDXF export precision maxes out at 0.0000001 in metric and 1/128 in imperial, which is good enough in many cases.

To generate a PDF version of the model tab in an AutoCAD DWG file, go to the Application menu and choose Export ➪ PDF (EXPORTPDF command). The Save As PDF dialog box provides options for how you want the drawing saved in PDF format. The Open in Viewer when Done option is handy for immediately checking that the export worked as you expected.

To export layouts, however, you have to first switch to a layout *before* starting the command; when the drawing holds multiple layouts, you have the option to export the current one or all of them as pages in the PDF file. Although PDF has long had the capability to embed 3D models, AutoCAD exports only flattened 2D representations of 3D models. When you need an AutoCAD 3D model in a PDF document, consider a third-party PDF exporting program. Autodesk recommends the Bentley View program; a free program that converts 3D models in DWG to 3D PDF is ODA Viewer, at `www.opendesign.com/guestfiles/oda_viewer`.

What the DWF?

The AutoCAD DWG format works well for storing drawing information on local and network drives, but the high precision and large number of object properties that AutoCAD uses make for comparatively large files.

To overcome this size problem and to encourage people to publish drawings on the web, Autodesk developed an lightweight vector format for representing AutoCAD drawings: Design Web Format (DWF, pronounced "dwiff"). A DWF file is a compact representation of a DWG file. DWF uses less space because it's less precise and doesn't have all the information stored in the DWG file. Therefore, it takes less transfer time over the web and by email.

Use the EXPORTDWF command to create a DWF file from a drawing in the Model tab; DWF can't export from layout tabs. Click the Application menu and then click the Export button to use the EXPORTDWF command.

DWFx is a version of regular DWF that's compliant with Microsoft's XML Paper Specification (XPS). Had your fill of alphabet soup yet? Everything about DWF also applies to DWFx. The DWFFORMAT command determines whether the PUBLISH, 3DDWF, and EXPort commands output to DWF or DWFx. If this sounds like a topic that might be important in your work, look up *DWFFORMAT* in the online Help system. Windows Vista through Windows 7 include an XPS Viewer, and the long and the short of this feature is that drawings plotted to DWFx can be viewed automatically in these Windows versions without AutoCAD or any special viewing software.

On the other hand, Autodesk seems to have come to the realization that PDF is the dominant force in this area and no longer promotes DWF.

3D print

Okay, all you old *Star Trek* fans! The Matter Replicator exists! Well, not really, but 3D printing is getting pretty close. For example, materials that can be used for 3D printing include a wide range of plastics, concrete, metals, and even human cells for replacement organs, and the size can range up to multi-story buildings. The good news is that this section of the book is short.

AutoCAD exports 3D solids and watertight 3D meshes in the industry-standard STereoLithography (STL) file format, which can be used by virtually any type, brand, and model of 3D printer. I cover 3D in Chapter 21.

To export an STL file, follow these steps:

1. **On the Application menu, choose Print ➪ 3D Print.**

   The 3D Printing — Prepare Model for Printing dialog box appears.

2. **Choose whether to learn more about preparing a 3D model for printing or click Continue.**
3. **Select one or more 3D object and then press Enter.**
4. **Confirm the options in the 3D Print Options dialog box and then click OK.**
5. **Enter a suitable file name and then click Save.**

You can also send your file to a 3D printing service bureau, which will 3D print the model and mail it to you — for a fee, of course. The export process is the same as saving a model as an STL file. Alternatively, you can choose Publish ➪ Send to 3D Print Service from the Application menu. The result is the same: an STL file, which you can email or deliver on a USB drive. At one time, Autodesk had ambitions to develop its own Spark 3D printing preparation software and even showed off its own line of Ember 3D printers, but they abandoned the idea.

But wait! There's more!

Between them, AutoCAD's print, export, save as, and publish functions can export drawing files into about 30 different file formats. These formats include the ones already discussed in this chapter, plus several other CAD formats. Plus AutoCAD can roll newer AutoCAD files back to earlier releases.

Open Up and Let Me In!

Among other things, Chapter 18 discusses how and why to attach raster images, PDF files, and external references (XREFs) to an AutoCAD drawing file.

Editing other drawing file formats

AutoCAD can also create, open, edit, and save a number of other file formats, including from several other brands of CAD software. I won't go into all the ins and outs here, but I will give you a hint: On the Application menu, look at the options under Open and under Import, and then click the Files of Type drop-down list to see the list of supported formats.

PDF editing

AutoCAD 2017 added the capability of editing PDF files. From a practical point of view, this is viable only when the PDF contains vector data, such as when the PDF is exported from a drawing in AutoCAD or another CAD program. Yes, AutoCAD can edit a PDF that came from a Word document, for example, but trust me, you don't want to do that.

There are two ways to bring in a PDF file to edit. One is to place the PDF file as a reference, as I describe in Chapter 18, and then convert it. The other is to open the PDF file from disk. Both methods are handled by one command, PDFIMPORT.

1. **Open a new or existing drawing file.**
2. **On the Application menu, choose Import ➪ PDF.**
3. **Press Enter.**
4. **Select a PDF underlay already in the drawing or enter the File option to browse for and select the desired PDF file.**
5. **Click Open.**

   The Import PDF dialog box shown in Figure 24-2 appears. For your first try, I suggest you go with the defaults.
6. **Click OK.**

   You might have to wait a while as AutoCAD peruses the PDF file, converting elements into DWG equivalents.

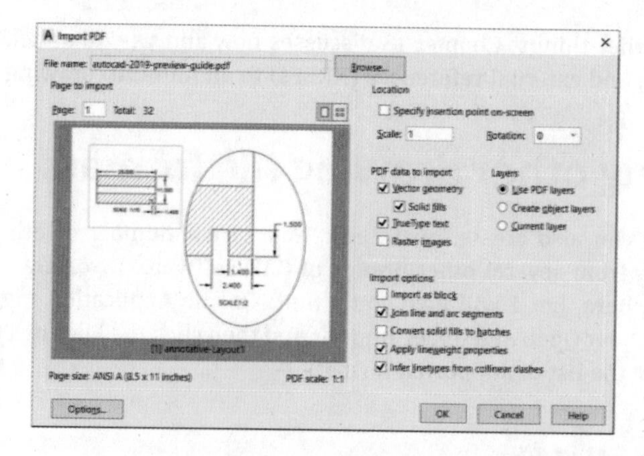

FIGURE 24-2: The Import PDF dialog box.

Magic! The imported PDF looks exactly like the original drawing that was used to create the PDF! Even the text, dimensions, colors, layers, and linetypes look correct. Layer names are prefixed with PDF_ to distinguish them from AutoCAD layer names.

Yes, AutoCAD can now edit a PDF file, but (and it's a big, uppercase, bold **BUT** — not to be confused with a big butt) before you get all excited at the prospect, you may want to consider the following discussion of some of the limitations. It's possible that these may be inherent in the PDF format or with AutoCAD's capability to analyze the file or both.

First, all complex object types have been exploded to their basic constituents. For example, all hatch patterns and dimensions are reduced to a bunch of individual polylines. That's right, every line segment in a complex hatch pattern has become an individual, independent polyline segment. Similarly, dimensions get exploded to polylines and text, and are no longer associated with the original definition points.

Worse yet, some text is totally exploded, while some is not. Text made with TrueType fonts tends to convert to mtext, but the individual characters in AutoCAD's SHX fonts get redefined as a bunch of polyline segments. Autodesk addressed this problem with a text recognition command, as described later in the bullet named "AutoCAD SHX fonts."

I'm not finished! Curves such as splines and arcs also end up as a series of arcs, polyline arc segments, splines, or a combination of all three. As for blocks, AutoCAD makes an attempt to identify them; some, like door swing symbols or staircase outlines, are converted into single polylines that mimic the look of the original blocks, but they are not blocks.

Ah, but I saved the worst for last. In Chapter 8 in particular, I harp on the need for accuracy in AutoCAD drawings. Unfortunately, the PDF file didn't read that chapter, and so most objects seem to come in with sizes that vary from the original drawing in the third and fourth decimal place. This may not sound like much of a problem, but you will now find that a line that is supposed to be tangent to an arc or two lines that are supposed to touch at their ends possibly don't. Even when you import a PDF exported by AutoCAD, the results are less than stellar.

Other than that, PDF editing is perfect.

Okay, the good news is that later releases have greatly improved on importing PDF files. I won't go into all the gory details, but two significant ones follow:

- **Non-continuous lines:** If you select the Infer Linetypes from Collinear Dashes box in the Import PDF dialog box, AutoCAD will try to guess which standard linetype variant was used to create the original non-continuous linetypes. The name of the linetype is suffixed with PDF_IMPORT. This feature can significantly increase the processing time when importing the PDF file, especially when the file contains a lot of crosshatching that uses noncontinuous linetypes.
- **AutoCAD SHX fonts:** AutoCAD created their proprietary font file, known as SHX fonts, in the days before standard TrueType fonts. Most current drawings probably use TrueType fonts, but older drawings or current drawings started from an older template will use SHX fonts. If you insert a PDF file and encounter text that consists of many small line segments instead of individual characters, you have SHX characters.

AutoCAD 2018 added the capability to convert all those little SHX line segments into text. Click the Recognize SHX Text button on the Insert tab of the Ribbon. You will be invited to select objects that you believe to be text. AutoCAD will then analyze them and try to match them to one of four of its most probable SHX fonts. If it succeeds, it will produce a single editable text object of all the characters. If it fails, you need to use the SEttings option of the command to select one or more fonts out of 78 possible ones or to reduce the recognition threshold. The good news is that AutoCAD remembers any selected fonts for future conversion attempts. Be sure to check the converted text for accuracy!

Translation, Please!

As indicated earlier in this chapter, AutoCAD can both import and export drawing files from several brands of CAD software.

WARNING

No translation is perfect, whether between computer file formats or human languages. Neither of them have exact equivalencies for certain items or terms. The first test of any translation is to do a round trip out and back into the original format, and then look for the differences. For example, legend has it that programmers did a round-trip test when they were writing the first human-language translators. What went in was "The spirit is willing but the flesh is weak." What came back was "The booze is okay but the meat's gone bad." Even rolling an AutoCAD DWG file back to an earlier release can cause loss of data, such as parametrics or objects that didn't exist in the earlier release.

You have probably guessed by now that AutoCAD isn't the only CAD program on the market, nor is it even the only one that can process files with the DWG format. A number of programs, ranging from free to mid-priced, claim to be AutoCAD-compatible in file format, command structure, and user interface. As AutoCAD is more expensive, some design firms substitute lower-cost workalikes to do everyday 2D drafting, saving AutoCAD for complex drawings and when compatibility is crucial.

Varying degrees of "compatible" exist, and compatibility isn't necessarily a function of price. Degrees of compatibility range from AutoCAD object types that don't exist in the compatible programs (but can still be displayed) to objects that can be damaged or destroyed by a round trip.

Examples of some of the varying degrees of compatibility include the following:

- **3D:** Other CAD programs may not be able to create or edit 3D solids and surfaces, may not be able to display them, and might even strip them out of an incoming DWG file.

- **Self-scaling annotations:** Other CAD programs may not be able to create or edit self-scaling annotations or display more than the first scaled variant applied to an object. Some programs might strip out the other scales, so they don't survive a round trip.
- **AutoCAD-specific objects:** Other CAD programs are unable to create and edit objects Autodesk covers with patents, such as dynamic blocks.
- **Parametrics:** CAD programs that support parametrics probably don't recognize the ones created by Autodesk software, and so strip them out or convert them to regular dimensions. They do not survive the round trip.
- **Tables:** I discuss tables in Chapter 13. When spreadsheets are inserted in AutoCAD drawings as tables, they can then do such things as read values from AutoCAD objects and link between individual cells in Excel spreadsheets. Earlier releases of AutoCAD, and some other CAD programs, don't support these features. At best, tables will be translated into simple line elements and text objects that display the current cell value. Connectivity will be lost. At worst, the table may disappear entirely.

The bottom line is that you should conduct round-trip cycles on test files before committing hard data.

The Importance of Being DWG

To take full advantage of AutoCAD in your work environment, be aware of the release number of the *DWG* file format — the format in which AutoCAD saves drawings. From time to time, Autodesk makes changes to it. The good news is that DWG is unchanged since 2018. Here are some DWG facts to keep in mind:

- **In many cases, an older release of AutoCAD can't open a DWG file that's been saved by a newer AutoCAD release.** Table 24-1 shows the relationship between AutoCAD versions and their corresponding file formats.
- **A newer release of AutoCAD can *always* open files saved by older versions.** I have sample files dating back to 1984 that open in AutoCAD 2023.
- ***Some* previous AutoCAD releases can open files saved by a subsequent versions.** As Table 24-1 shows, Autodesk began changing the DWG file format every three years or so starting in 2000 but since then has leveled off recently so that drawings created in or saved by AutoCAD 2023 can be opened with AutoCAD 2018 and newer.

- **You can use the Save As option to save a file from a newer release to an older DWG format.** In fact, AutoCAD 2022 can Save As all the way back to AutoCAD Release 14, in the last millennium (1997). In addition, you can save a file as a simple text-based DXF format as far back as Release 12 (1992). Table 24-1 shows which versions use which DWG file formats.

WARNING

Earlier formats may not support all the features of later formats. AutoCAD does its best at translating, but some items may be lost or may not fully survive the round trip to an older release and back to the newer one.

TABLE 24-1 **AutoCAD Versions and DWG File Formats**

AutoCAD Version	AutoCAD LT Version	DWG File Format
AutoCAD 2018–2023	AutoCAD LT 2018–2023	AutoCAD 2018
AutoCAD 2013–2017	AutoCAD LT 2013–2017	AutoCAD 2013
AutoCAD 2010–2012	AutoCAD LT 2010–2012	AutoCAD 2010
AutoCAD 2007–2009	AutoCAD LT 2007–2009	AutoCAD 2007
AutoCAD 2004–2006	AutoCAD LT 2004–2006	AutoCAD 2004
AutoCAD 2000, 2000i, 2002	AutoCAD LT 2000, 2000i, 2002	AutoCAD 2000
AutoCAD R14	AutoCAD LT 1998, 1997	AutoCAD R14
AutoCAD R13	AutoCAD LT 1995	AutoCAD R13
AutoCAD R11, R12	AutoCAD LT R2	AutoCAD R11

So, do you *have* to use AutoCAD to produce DWG files? The answer is no. Several low-cost to no-cost programs claim to be compatible with AutoCAD's DWG file format. They usually come with a few crunchy bits, however.

- Because these programs have to reverse-engineer the file format, they are often a half-release behind AutoCAD's current version.
- These programs usually don't fully support all AutoCAD features because of copyright, patent, or perceived market size limitations. In particular, annotative objects (Chapters 13-15), parametrics (Chapter 19), dynamic blocks (Chapter 19), and 3D models (Chapters 21-23) may not survive unscathed from a round trip to the other brand and back to AutoCAD — or even exist in the other brand.
- Yes, a line is a line and a circle is a circle in a DWG file, but the commands that you use in other programs to put them in the file may not operate in the same way.

6
The Part of Tens

IN THIS PART . . .

Find additional AutoCAD resources.

Uncover system variables to make your life easier (or maybe just a little more fun).

Discover ten more tricks to working in AutoCAD.

- World Wide (CAD) Web
- Your local user group
- Autodesk User Groups International

Chapter 25
Ten AutoCAD Resources

As much as I want to make a claim to the contrary, to reach the highest peaks of AutoCAD mastery, you probably need more information than this book, a copy of the software, and the software's Help system provides. This chapter lists ten aids likely to help you find out more about your favorite software program.

Autodesk Discussion Groups

Although independent, newsreader-based discussion groups are still out there, the majority of the AutoCAD action nowadays happens in moderated discussion groups at Autodesk's website. They're user-to-user groups, but from time to time Autodesk employees do jump in to answer questions in their areas of expertise. To see what's there, point your browser to `https://forums.autodesk.com/` to find the list of 80 or so product-based discussion groups for AutoCAD, AutoCAD LT, and so on.

Autodesk's Own Blogs

Several Autodesk employees run their own blogs via the company servers, and they're chockablock with tips, techniques, and (occasionally) highly entertaining digressions. For a list of the nearly 50 Autodesk blogs, go to `www.autodesk.com/blogs`.

Autodesk University

Autodesk University (AU) is an annual event sponsored by Autodesk. For the past decade, the event has settled in Las Vegas, Nevada. With the arrival of coronavirus, however, AU went online and became free. Whether in Las Vegas or online, nearly all sessions are recorded so you can view them at your leisure.

Attending the in-person AU can be expensive, so when your boss balks at sending you, quote a few figures: At AU about 850 instructors present about 840 classes and labs to about 10,000 attendees on virtually every conceivable topic related to virtually every Autodesk product. Another 9,000 attend regional sessions in other parts of the world, and 53,000 take virtual classes online.

Add over 300 exhibitors showing their wares in about 200 booths — plus, the AU support staff and techie types — and you end up sitting down for lunch with about 13,000 of your best friends. The meals and the evening social sessions are great times for networking with your peers, and you can often learn as much as you do in the classes. Check out `www.autodesk.com/autodesk-university/` to find out how to schmooze like a pro and become a guru!

Autodesk Channel on YouTube

If you search the web for solutions to particular AutoCAD issues, you've probably already discovered the YouTube Autodesk channel. You can find dozens and dozens of video clips that cover not only AutoCAD and AutoCAD LT but also other Autodesk products such as Inventor and Revit. You can subscribe and receive email notifications of new videos every few weeks. Check it out at `www.youtube.com/autodesk`.

Independent CAD users and AutoCAD retailers also offer tutorial videos on YouTube. Search for *AutoCAD tutorial*.

World Wide (CAD) Web

I'm reluctant to list specific sites because they tend to come and go, but *Cadalyst* magazine (`www.cadalyst.com`) has been in existence for nearly 40 years. In particular, click the CAD Software Tutorials link to see the series of tutorials *Circles and Lines*, by Lynn Allen, and *The Learning Curve*, by the technical editor of this edition. I have an active blog at `www.worldcadaccess.com` and a weekly emailed newsletter at `www.upfrontezine.com`.

Your favorite search engine can also be a good friend. Apart from Autodesk's own web presence, thousands upon thousands of other sites are scattered around the world. Search by entering *AutoCAD tutorials* for online exercises, or try *AutoCAD blogs* for independent views and opinions as well as tips and tricks for using AutoCAD. You can also try specific Google or Bing searches. For example, entering *AutoCAD line command* yields about 6,180,000 hits, and entering *Bill Fane* gets you 4,380,000 hits, but they aren't all for this book's technical editor. For example, he's not the villain in an old Grade B British movie!

TIP

Click the triangle next to the shopping cart icon in the upper-right corner of the screen for direct links to a number of web resources in AutoCAD.

Your Local Authorized Training Center

Autodesk Authorized Training Centers (ATCs) are located around the world in both private institutions and public colleges and institutes. At an ATC, you attend scheduled, instructor-led classes where you learn to use AutoCAD from the ground up. Courses are designed for rank beginners or experienced users who want to learn the latest customization techniques. To find the location of your nearest ATC, browse to `www.autodesk.com/atc`.

Your Local User Group

AutoCAD has inspired an incredibly loyal following in the 40 years since its initial release. One primary reason is the especially enthusiastic individuals who arrange to meet one evening per month to talk about what they've discovered they can do with AutoCAD. These folks are still out there, and although some now have grayer hair and more wrinkles, they still love to get together and talk about AutoCAD. And most are especially welcoming to newcomers!

To find a nearby group, type **AutoCAD user group** and the name of your city in your browser's Search box. The bad news is that online searching has greatly reduced the appeal of live, in-person user groups. VAUS (Vancouver AutoCAD Users Society) was the world's first AutoCAD user group but folded a couple of years ago.

Autodesk User Groups International

Autodesk User Group International (AUGI) is the umbrella organization of the global user group community. You don't even have to belong to a local user group to participate. Individuals can join too, and basic membership is free. Browse to `www.augi.com` to find out more.

AUGI is probably best known for the annual wish list it presents to Autodesk; the list is compiled from requests from members for changes or new features in AutoCAD. It's sometimes the case that top wish-list items find their way into new releases of AutoCAD, so it's a place where you can influence future releases of AutoCAD. AUGI also supports a series of online and live training sessions, and your membership usually earns you a discounted rate to attend Autodesk University.

Books

Although I'd like to fill you in on *all* the wonders of AutoCAD, I can do only so much in a book of this size. I simply don't have the space to cover such topics as data linking and customization.

For more information, visit `www.wiley.com` and search for *AutoCAD* to see a list of nearly 100 titles. And I have more than 100 specialized e-books at `www.worldcadaccess.com/ebooksonline`.

Autodesk Feedback Community

If you're interested in helping shape future releases of AutoCAD, sign in to the Autodesk Feedback Community portal (`https://feedback.autodesk.com`) using your Autodesk account, and then apply by clicking the Create Account link. If you're accepted, you will, in return for small (or sometimes considerable) time commitments, contribute your ideas to the AutoCAD Futures group or apply to beta-test the next release of AutoCAD.

- » DIMASSOC
- » MENUBAR
- » MIRRTEXT
- » OSNAPZ
- » PICKBOX
- » REMEMBERFOLDERS
- » ROLLOVERTIPS and TOOLTIPS
- » TASKBAR
- » VISRETAIN

Chapter 26
Ten System Variables to Make Your AutoCAD Life Easier

System variables are the settings that AutoCAD checks before it decides how to do something. Nearly a thousand system variables control all aspects of AutoCAD's operations. For example, if you leave the system variable SAVETIME at its default value of 10, AutoCAD saves the drawing file automatically every 10 minutes; if you change SAVETIME to 5, the time between automatic saves is 5 minutes.

You can set many system variables through the OPtions command, which groups them into categories such as Files, Drafting, and 3D Modeling. You can also change the value of every system variable by typing its name at the AutoCAD command prompt and pressing Enter. AutoCAD displays the current value of the system variable setting and prompts you for a new value. If you don't want to change it, press Enter to confirm the existing setting; otherwise, type a new value and press Enter to change it. The change takes effect the next time you use a command that depends on the system variable.

Some system variables affect only the current drawing session while others apply to all drawings. AutoCAD stores them in three places, depending on where they should have an effect:

- **Saved in the Windows Registry:** Affects all drawings when you open them on your system but not on other computers.

- **Saved in the drawing:** Affects only the current drawing, but on every computer that opens the drawing. The Drawing icon in the Options dialog box indicates which variables are saved in a drawing. The icon is useful when making template drawings because it points out which variables are saved for all subsequent drawing. When you change system variables and then save the drawing as a template, all subsequent drawings that you start from this template inherit these values. I cover templates in Chapter 4.
- **Not saved anywhere:** Lasts only for the current drawing session.

A few system variables only report a value, such as Date and Platform. These are called read-only variables, and report the status of AutoCAD and Windows. You can't change their values.

The System Variables section in the Resource area of the online Help system indicates each system variable's type. Several hundred system variables exist, but understanding the ones described in this chapter can help make you a happy CADster.

APERTURE

APERTURE controls how closely the cursor must be to an object before an object snap marker appears. I discuss object snaps in Chapter 11. You can set APERTURE from the command line to any value between 1 and 50 pixels, or you can set it in the Options dialog box by adjusting the slider to a maximum of 20 pixels. The initial value is 10, which in most cases is good enough.

DIMASSOC

DIMASSOC controls how AutoCAD dimension objects are created. See Chapter 14 for the lowdown on dimensioning. When DIMASSOC is 0, AutoCAD creates exploded dimensions of separate lines, 2D solids, and text. You *really* don't want to work on DIMASSOC=0 drawings.

When DIMASSOC=1, dimensions are referred to as *nonassociative* — they're single objects, but they don't update if you change the size of an object.

When DIMASSOC=2 (the default setting), AutoCAD creates fully associative dimensions. You can set this value from the command line or select the Make New Dimensions Associative check box after choosing Options ➪ User Preferences.

If you work with older drawings, you'll likely find that the value of DIMASSOC is 1, which is the best that releases earlier than AutoCAD 2002 could manage. The DIMASSOC setting is stored in the drawing and carries forward from the template.

MENUBAR

When MENUBAR is set to 1, AutoCAD displays its traditional classic text-only menu bar below the application title bar. If MENUBAR is set to 0 (the default), the menu bar is hidden. The menu bar can also be turned on and off by using one of the settings revealed by the down arrow at the right end of the Quick Access toolbar.

TIP

If you've been away from AutoCAD for a few years and you feel that you could benefit from training wheels while you become accustomed to using the Ribbon, display the classic menu bar. Experienced users like the author and his technical editor can find most commands more quickly by using the menu bar.

MIRRTEXT

You've got to love seeing a clear indication of a system variable's purpose from its name. ROLLOVERTIPS, anyone? You'll frequently build up a drawing by mirroring already-drawn components of it. That's what the MIrror command does. Luckily, AutoCAD realizes that you rarely want to produce a mirror image of any text that may be included among the set of objects you're mirroring. Having to use a mirror to read the text on your drawing is counterproductive.

On the other hand, suppose that you want to draw the mold for a cast part, and the finished part has some text molded in it, such as a company name or a part number. The text features in the mold need to be mirror images so that the final cast text is read correctly. No problem: Simply set the MIRRTEXT system variable to 1, and the MIrror command mirrors the text. When MIRRTEXT is 0 (the default setting), text isn't mirrored, and the text still reads the right way. MIRRTEXT is stored in the drawing and carries forward from the template.

OSNAPZ

Sometimes, drawing in 3D can be a little puzzling, especially if you're looking at 3D objects in orthographic views. Say that you're looking at a wireframe cube in Plan view and you want to draw a line between two diagonal corners on top of the cube. By using object snaps to pick the corners, you'll find, as often as not, that the line isn't on the plane of the top of the box, but that it starts at a top corner and ends at a bottom corner. One way around this problem is to set the OSNAPZ variable appropriately. Because OSNAPZ is set to 0 by default, AutoCAD finds the X,Y,Z coordinates of the snapped-to point — but often, you don't want that to happen. Set OSNAPZ to 1, and AutoCAD replaces the Z value of the picked point with the *current* elevation or the Z value of the first point picked.

Because OSNAPZ affects drawing in three dimensions, it's unavailable in AutoCAD LT.

PICKBOX

When AutoCAD prompts you to select an item, the cursor changes into a little square box called the *pickbox*. You can control the pickbox size by setting a value in pixels for the PICKBOX variable. The initial setting of PICKBOX is 3, which is fine for a lower screen resolution, such as 1024 x 768. But when you start running at higher resolutions or when your hair is grayer, your face more wrinkled, and your vision not *quite* as acute as it once was, a higher setting might be more useful. The range is from 0 (turns off the pickbox) to 50, measured in screen pixels.

REMEMBERFOLDERS

Sometimes, you don't want AutoCAD to start in the same Documents folder where every other program stores its files. When you work on a number of projects, you'll want to store drawings in particular folders. If so, REMEMBERFOLDERS is your ticket to Nirvana. By default, it's set to 1, so it defaults to storing files in Documents in Windows 7 and 10/11.

Set this variable to 0, and it opens from, and saves to, the last folder you visit as you work, as well as the folder path you specify in the Properties dialog box of the desktop icon from which you start AutoCAD. REMEMBERFOLDERS is stored in the Windows Registry.

TIP

If you regularly work on several different projects or for specific clients, simply create appropriate copies of the Windows desktop icon that launches AutoCAD. Now right-click each icon in turn, click Properties, and change the Start In window to display your desired default folder for each case. Now when you launch AutoCAD from a specific icon, it will default to finding and storing your drawings in the specified folder.

There are three signs of old age. The first is loss of memory. I forget the other two.

ROLLOVERTIPS and TOOLTIPS

You were just waiting for this one, weren't you? ROLLOVERTIPS controls the tooltip-like message boxes you see when you hover the mouse pointer over objects in the drawing area. When ROLLOVERTIPS is set to 1, pausing the cursor on an object opens a temporary panel showing the object type with its layer, color, and linetype. If (like me) you find the panel intrusive, set its value to 0.

This command can serve double duty. You can not only enter it into AutoCAD but also (when you have a dog named Tips) use it to teach your pet a neat trick. ROLLOVERTIPS is stored in the Windows Registry. And believe it or not (and I would bet that you don't), another system variable is named NOMUTT. Maybe it's AutoCAD-ese for BAD DOG TIPS.

TOOLTIPS does the same for tooltips that appear (by default, anyway) when you hover the mouse pointer over a toolbar or Ribbon button, or just about anywhere in a dialog box. Tooltips are truly useful when you're getting to know your way around a program, but when you have a sense of where things are, they can get in the way. By default, TOOLTIPS is set to 1, to appear whenever you hover the mouse pointer over a program element.

Set TOOLTIPS to 0 if you don't want to see tooltips again. The TOOLTIPS setting is stored in the Windows Registry.

TASKBAR

When you open a second drawing, AutoCAD displays a second button on the Windows taskbar. With each additional drawing you open, AutoCAD adds another button on the taskbar. This makes it easier to switch between drawings. But if you prefer to switch drawings using the document tabs and don't like to see the taskbar cluttered with identical-looking buttons, you can modify this behavior with the TASKBAR variable. This is the first system variable I change with a new installation of AutoCAD, switching TASKBAR from its default value of 1 (displays a button for every open drawing) to 0 (displays a single button for AutoCAD).

VISRETAIN

If you work with xrefs (see Chapter 18), you know the potential for competing layer properties. Sometimes, you want an xref to look different from the current drawing so that you can tell which layers come from which source. You can change the layer colors of xrefs in Layer Property Manager, but whether you have to do that every time you open a drawing with an attached xref depends on the setting of the VISRETAIN system variable. When VISRETAIN (visibility retain) is set to 0, the layer properties in the xref take precedence, so you *would* have to make those changes every time you open the host drawing.

When VISRETAIN is set to 1 (the default value), changes you make to layer properties in the host drawing remain intact every time you open that host drawing; however, those changes have no effect on the external file itself. The VISRETAIN value is stored in the current drawing.

And the Bonus Round

This section describes two system variables that are fun to use, even if they don't boost your productivity.

Using MTJIGSTRING, you can substitute as many as ten characters of your choice to replace the usual *abc* prompt when you start the MText command. For example, `I'm great!` *just* fits.

The last variable I want to tell you about is invoked slightly differently from MTJIGSTRING. Enter the following line to turn on the deprecated DOS screen menu (in use years ago):

```
(setvar "screenmenu" 1)
```

The DOS screen menu was present in the first release of AutoCAD in 1983 and is still present today, albeit turned off. It is fast, has no icons or tooltips, and is fully context-sensitive. No matter how you start a command, its options appear automatically. You don't need to right-click to see command options because it was created before three-button mice even existed.

Clicking AutoCAD always returns you to the root menu, and clicking **** opens a list of the temporary object snap overrides. The bad news is that the DOS screen menu hasn't been updated in years because it's *deprecated* (officially dead but still lying around), so it's missing recent commands. The really bad news is that it cannot be customized by users.

» Custom tool palettes

» Ribbon customization

» Toolsets

» Programming languages

» Vertical versions

» Language packs

» Multiple projects or clients

» Data extraction and linking

» Untying the Ribbon

Chapter 27
Ten AutoCAD Secrets

You may have already figured out that there are a few necessary principles for living your life that your mother never told you about. Well, AutoCAD has a few principles that I don't tell you about elsewhere in this book.

In this chapter, I list ten additional subjects that you may want to explore on your own, using the resources listed in Chapter 25.

Sheet Sets

Large projects involve a great many drawings. Legend has it that in pre-computer days, the British Navy wouldn't launch a new battleship until the weight of paper equalled the weight of the ship. At times, you may want to look at or plot the same set of drawings from a collection. You can use Sheet Set Manager to link individual drawings, and even specific paper-space layouts in drawings, into named sets. See the SHEETSET command. The PUBLISH command performs similar functions.

Custom Tool Palettes

Do you regularly use three or four different hatch patterns? Or the same five or six standard blocks or details? No problem. If you can drag and drop, then you can create custom tool palettes, from which you can drag and drop elements into your drawings and even share them with your office buddies. See the ToolPalettes command.

Ribbon Customization

The Ribbon, and indeed the full specifications for the menu bar and toolbars, are stored in a single CUI (Customizable User Interface) file. It can be edited to customize AutoCAD's interface. As well, you can have AutoCAD access a secondary CUI file that suits your specific needs. See the CUI command.

Toolsets

AutoCAD has a series of specialized toolsets that include special block libraries and functions for the unique needs of mechanical, electrical, civil, architectural, and other users. You can install them through your AutoCAD Desktop app, which is installed alongside AutoCAD.

Programming Languages

AutoCAD supports a variety of different programming languages:

- **Macro recorder:** If you repeatedly perform the same drawing or editing sequence, you can turn on the Macro Recorder to record the sequence as you perform it once, and then replay it endlessly. ("Holy macro, CadMan!") See the ACTRECORD command.

New in 2023: If AutoCAD notices that you tend to use a specific set of instructions fairly regularly, it will automatically create a new macro for you. You don't need to turn on the macro recorder.

- **DIESEL** is generally used in menu macros and to display program data on the status bar.
- **AutoLISP** is best suited to custom applications that primarily stay inside AutoCAD.
- **.NET, Object ARX, and VBA** generally work best in custom applications that interact with other types of programs, such as Excel or Word.

See the Developer area of the online Help system.

Vertical Versions

AutoCAD isn't the only product made by Autodesk. In fact, the company has about 140 products and combinations of products called collections, covering a wide range of markets in the design and entertainment fields. Virtually all Hollywood special effects are produced on Autodesk software (*Avatar* is the best-known example) along with the vast majority of video games and virtually all car ads on TV. That's right; car ads are computer-generated because they don't want to risk the chance of a competitor's car turning up in the background.

The design products include versions of AutoCAD meant for applications such as electrical control design, civil engineering, and architecture, as well as specialty products such as Fusion 360, which is parametric 3D design and analysis software. Do you need to do a stress analysis? Autodesk. How about a virtual wind tunnel? Autodesk. Doodling on an iPad? Autodesk. Managing a construction site? Autodesk. Photographing an object and then 3D printing a copy of it? Autodesk, but be careful. A friend of mine photographed a worn, cracked part for his 1927 Rolls-Royce Phantom II and then had a metal part 3D printed. What came back was a perfect replica of his worn, cracked part.

Language Packs

AutoCAD has been available in a number of languages for many years. After installing AutoCAD, sign into your Autodesk account at `https://manage.autodesk.com/` and then browse to find the Downloads section for AutoCAD. There you can find a number of different language packs — download and install the ones you need. You can then switch between languages simply by launching AutoCAD from the appropriate desktop icon.

Multiple Projects or Clients

If you regularly switch between different projects or different clients or both, you can make your life easier. For example, you can modify the properties of the AutoCAD shortcut icon on your Windows desktop to have AutoCAD start from a specific template file folder, use a specific default folder, use a specific default template file, and start from a specific profile in the Options command. Even better, you can create several desktop icons, one for each set of circumstances. Then, when you are working on a particular project, simply start AutoCAD from the appropriate shortcut, and it comes up ready to rumble.

Data Extraction and Linking

AutoCAD (but not AutoCAD LT) can extract properties of objects, including elements such as the areas surrounded by polylines, and then write them to a table in the drawing or extract them to a spreadsheet (or both). In addition, spreadsheets can be imported into AutoCAD drawings as tables. See the DATAEXTRACTION command.

Untying the Ribbon and Drawings

I grumbled earlier in the book about the space taken up by the Ribbon as screens get wider, not taller. If you want the graphic screen to be closer to the same proportions as a plotted drawing, reduce the screen so that it displays only the panel names. The panels fly out when you hover your cursor over their names.

The Ribbon can also be docked along one side or the other instead of across the top, and individual tabs can be dragged to a second monitor.

Similarly, starting with AutoCAD 2022, you can drag document tabs from the AutoCAD window to another monitor. You can then easily copy and paste between drawings.

Index

B

C

E

F

G

H

I

J

K

L

M

N

O

P

Q

R

S

T

U

V

W

X

Y

Z

About the Author

Ralph Grabowski has been reporting on the computer-aided design software industry since 1985, and has written or updated 240 241 books and e-books about CAD.

In his early teens, he discovered model railroading, along with a fondness for writing. When as a teenager he declared to his parents that he wanted to become a writer, they suggested instead that he take his love of trains to the University of British Columbia, where he received his B.A.Sc. degree in Civil Engineering, specializing in transportation engineering.

While working as a traffic planer at an engineering firm, he bought one of the early personal computers and learned how to take it apart (and put it back together again) and how to program it with a variety of programming languages. In 1985, he came across a demo version of AutoCAD v1.4, and a month later landed the job of technical editor at *CADalyst*, the first independent magazine in the CAD industry. There, he developed standardized benchmark tests for hardware like plotters and graphics boards. During this time, he met Bill Fane, this book's technical editor, at a Vancouver AutoCAD Users' Society meeting. At that meeting, Bill explained object snaps so clearly that Ralph hired him to write regularly for the magazine.

After five years, Grabowski learned all there was to know about running a publication, and so he branched off as a freelance writer, landing book contracts with quite a few publishers, writing regular columns with several magazines, and helping launch three CAD magazines. In 1995, he started *upFront.eZine*, the first independent email newsletter for CAD users — and he is still publishing it, 1,100 issues later. He taught civil engineering at the British Columbia Institute of Technology and has produced a couple of hundred training videos.

In his spare time, he hikes, travels, renovates his home, reads non-fiction, volunteers at local help agencies, and entertains his grandchildren.

Dedication

This book is dedicated to my wife of nearly forty years, Heather, who agreed with me as I made the leap from regular paycheck to the uncertainty of self-employment and has cheered me along the entire path.

Author's Acknowledgments

Sincere thanks to Bill Fane, technical editor of this book, who over the years taught me a lot about AutoCAD.

Publisher's Acknowledgments

Executive Editor: Steve Hayes

Project and Copy Editor: Susan Pink

Technical Editor: Bill Fane

Production Editor: Tamilmani Varadharaj

Proofreader: Debbye Butler

Cover Image: © ngaga/Shutterstock